Construction Superintendent's Operations Manual

Sidney M. Levy

McGraw-Hill
New York Chicago San Francisco Lisbon London Madrid
Mexico City Milan New Delhi San Juan Seoul
Singapore Sydney Toronto

The McGraw-Hill Companies

Library of Congress Cataloging-in-Publication Data

Levy, Sidney M.
Construction superintendent's operation manual / Sidney M. Levy.
p. cm.
Includes index.
ISBN 0-07-141205-0
1. Building—Superintendence. I. Title.

TH438.L3797 2003
690'.068—dc22 2003061571

2 3 4 5 6 7 8 9 0 KGP/KGP 0 9 8 7 6 5

P/N 143003-2
PART OF
ISBN 0-07-141205-0

The sponsoring editor for this book was Larry S. Hager, the editing supervisor was Stephen M. Smith, and the production supervisor was Pamela A. Pelton. It was set in Century Schoolbook by Wayne Palmer of McGraw-Hill Professional's Hightstown, N.J., composition unit. The art director for the cover was Anthony Landi.

Printed and bound by Quebecor World Bogota

This book is printed on acid-free paper.

Contents

Preface

The boss brings in the work, the project manager assembles the team—but it is the project superintendent who's going to get the project *built.*

The project superintendent's role in the construction process is a critical one. He or she is the field general, directing tens or hundreds of skilled tradespeople to ensure the timely completion of the project, provide assurance that the contract requirements will be met, and, hopefully, allow the company to turn a profit. The project superintendent needs people skills, technical know-how, and mastery of the carrot-and-stick management approach. These frontline supervisors are always willing to lend a helpful hand or a sympathetic ear, but they are also tough when required, always mindful of the C's required for a successful project: control over time, control over costs, control over quality.

The *Construction Superintendent's Operations Manual* contains the basic information every project superintendent needs, from interpreting and administering the complex construction contracts of today, to methods of improving organization in the field, to dealing with the intricacies of urban rehabilitation projects. This book is jam-packed with helpful tips and reporting forms required for today's fast-paced projects. The CD inside the back cover includes a whole host of handy forms along with a series of form letters directed to owners, architects, and subcontractors covering most of the documentation concerns arising in the field today—just type in the recipient's name, address, and the project-specific data and send the form on its way.

Today more than ever, a project superintendent's management and technical skills are put to the test. He or she must deal with shortages of trained, experienced workers and learn about new materials and new construction techniques and concepts. As desktop, laptop, and, now, handheld computers become everyday management tools, and as construction companies become more sophisticated in their collection and processing of field information, an entirely new set of skills is required. The successful superintendent must be a master of time management and resist the tugs and pulls of the many attempts by subcontractors and miscellaneous visitors to the construction site to distract him or her from the primary goals of being everywhere on the site and being aware of everything that is taking place within the construction cycle. At day's end when most of the trades have left the site and quiet prevails, the superintendent's work is not quite done. He or she is responsible for processing documents of all kinds—daily logs, drawing reviews, RFIs, updates to the schedule—planning activities for the next day, and making phone calls to track down tradespeople, materials, and equipment. The *Construction Superintendent's Operations Manual* may be of some help in the completion of these tasks.

This book has been assembled to provide insight into many of the facets of the construction process, offer some helpful tips for new entrants into the position of project superintendent, and, possibly, provide the experienced superintendent with a few more tools to add to his or her toolbox.

The superintendent's work day is long, and there are periods that test one's physical, mental, and emotional limits, but there is great satisfaction in knowing that one's efforts have not been wasted, and a project is one day closer to completion. And, when you are riding in a car on a sunny Sunday afternoon and you pass one of those bright, gleaming buildings you've recently completed, you can point with pride and tell the family, "I built that building."

Sidney M. Levy

List of Form Letters

Contents of the CD

Quality control checklist—Miscellaneous Metal
Quality control checklist—Metal Deck
Quality control checklist—Metal Stairs
Quality control checklist—Rough Carpentry
Quality control checklist—Finish Carpentry
Quality control checklist—Roof Insulation
Quality control checklist—Membrane Roofing
Quality control checklist—Flashing and Sheetmetal
Quality control checklist—Sealants and Caulking
Quality control checklist—Metal Doors and Frames
Quality control checklist—Interior Glass and Glazing
Quality control checklist—Painting
Quality control checklist—Gypsum Drywall
Quality control checklist—Resilient Flooring
Quality control checklist—Seamless Elastomeric Flooring
Quality control checklist—Toilet Room Accessories
Quality control checklist—Demountable Partitions
Quality control checklist—Lockers and Benches
Quality control checklist—Metal Toilet Partitions
Quality control checklist—Hydraulic Elevators
Plus, the 66 form letters listed previously

Chapter

1

Introduction to the Construction Industry

The construction industry is a unique business. The size of a construction company can range from a one-person skilled mechanic-entrepreneur to a multibillion-dollar international giant operating throughout the world. Locally operated family businesses compete with New York Stock Exchange–rated companies, but there are opportunities out there for anyone who is willing to work hard and catch a little bit of good luck.

The construction industry is a vital part of the U.S. economy. In 2002, the value of construction put in place amounted to $834 billion. This industry provided jobs for more than 6.6 million people in the year 2000, and this figures rises to nearly 10 million when design work and equipment and the materials and supply businesses are included. New construction represents about 7 percent of this nation's gross domestic product (GDP), and when remodeling and repair work is added, the industry's contribution to GDP climbs to 10 percent. Although real wages in the construction industry have fallen at a more rapid pace than wages for other U.S. workers, the average hourly wage for construction workers in 2001 was $18.34 compared to an average of $14.32 for all workers.

The decline in real wages was partially attributed to the increase in merit shop operations.

Union versus Nonunion Shops

It has been estimated that union labor in the construction industry declined from approximately 70 percent of the workforce in the 1970s to 20 percent in the 1990s. Union membership reached an all-time high of 20 million workers in 1983. In the year 2001, union membership stood at 16.4 million, but in 2002, according to the Bureau of Labor Statistics, unions lost 300,000 workers, even though union workers in 2002 earned on average $740 per week, compared to $587 for nonunion workers. At an annual meeting of the AFL-CIO in February 2003, union president John J. Sweeney vowed to step up organizing efforts.

Employment for the entire construction industry is projected to grow at a rate of 1.2 million between 2002 and 2010, according to government sources. This growth rate is slightly less than the anticipated growth rate of 1.4 million for the economy as a whole. Certain other labor trends were evident in the industry; older craftsworkers have retired and younger entrants to the overall labor pool have chosen career paths other than construction. There is a tendency for workers in the construction business to retire earlier—it is a physically demanding job—and not only is the industry attracting a lower proportion of potential workers, but also there are fewer 18- to 24-year-olds entering the U.S. workforce.

The industry is highly competitive and profit margins are slim. According to statistics compiled by the Construction Financial Management Association (CFMA) for the year 2001, the average percentage of net profit for general contractors, before taxes, is about 2 percent.

Few new technologies have been developed by contractors or equipment manufacturers, and old ways seem to die hard. Although we accept computer technology in the office, the field is still oriented to trowel and mortar, and screw guns. Construction has been compared to an outdoor factory producing a one-off product while being exposed to, and dealing with, all the vagaries imposed by nature.

Other production-oriented businesses such as the automobile industry have long since switched from muscle power to robots to assemble their products. But we still butter up bricks one at a time before placing them side by side in a wall. We still cut and fit individual wood or metal studs every 16 or 24 inches to building interior walls.

We are, however, an industry in transition. As for so many businesses and institutions, the 21st century holds untold opportunities and challenges for construction.

Developing Trends in the Construction Industry

Some changes in the construction industry are apparent while others are more subtle. These changes will affect the way in which projects are designed and built, and the people who are responsible to deliver them will have to recognize shifts in the following areas:

- The organization
- Human resources—the workforce
- Project delivery systems
- Technology
- Productivity
- Quality control
- Safety
- Dispute resolution

The Organization

Many of the trends that developed in the 1980s and 1990s have continued, such as industry consolidation and the infusion of foreign-based contractors seeking the relative stability of U.S. markets.

The move to globalization that permeated much of U.S. industry left its mark on the building business. Contractors from Great Britain, Sweden, Germany, France, Japan, and the Netherlands are firmly established in this country either by buying up U.S. companies or by opening branches that echo the corporate philosophy of "Act globally, think locally." Even such stalwarts as Turner Construction have yielded to the trend, having been bought out by Germany's Hochtief A.G. organization. A look at bidders on many private and public lists reveals such names such as Skanska, Bovis Lend Lease, Philip Holzman, Bouygues, Kvaerner, Balfour-Beatty, Kajima, and Bilfinger+Brau right up there with the more recognizable local construction companies.

Changes in the Subcontracting Industry

Of all specialty contractors, more commonly referred to as subcontractors, 97 percent attained annual revenues of less than $2.5 million, 27 percent reported yearly sales of less than $250,000, but the top 3 percent are getting bigger, expanding by consolidation. This trend began in the mid-1990s and accelerated in 1998 as a few investor groups began to purchase competitors, assuming that consolidation could bring increased profits due to economies of scale and increased geographic exposure. Even the painting industry, previously characterized as being made up of many small, local companies, has not been immune to consolidation. The nation's largest painting contractor, Kenny Industrial Services Company, reported sales of $216.6 million in 2001.

Consolidation and rapid growth do not always bring with them increased profits. The largest subcontractor in the country is Encompass Service Corporation of Houston, Texas, a mechanical and electrical contractor with 2001 revenue of $3.8 billion; but Encompass is saddled with $931 million in debt and in danger of bankruptcy if additional equity capital is not found. Bigger is not always better, and the size of the bottom line is more important than the size of the top line—sales.

Human Resources—The Workforce

The changing workforce is characterized by an impending shortage of workers that was first predicted in the 1980s but has accelerated during the past two decades. In today's marketplace, a shortage of skilled craftspeople and mechanics along with experienced administrators and managers remains one of the major problems facing the industry.

Several years ago, who would have thought that staid construction industry companies would be offering "signing bonuses," just as the NFL or NBA does, to attract experienced and productive employees. Employment projections issued by the Department of Labor in 2002 indicated that employment in the construction industry over the next 8 years will grow at a rate of 1.2 percent, somewhat more slowly than the projected 1.4 percent rate for the economy as a whole.

One of the major challenges of this 21st century will be to recruit and train new entrants to the construction industry, a task that is vital to the interests of the United States at large.

An aging workforce, the lack of technological advances, and the lure of more attractive vocations have all contributed to construction's stodgy image and the difficulty in recruiting the best and the brightest.

The demographics of the U.S. population sounded the warning bell several decades ago, but apparently the construction industry failed to recognize it. By the year 2010 it is estimated that the number of 55- to 64-year-old males will outnumber the 18- to 24-year-old group by at least 1.5 million. The number of males aged 55 to 64 will increase from 10 million to 17 million in the next 8 years. How will this gap be filled?

Immigrant Labor Flocks to the Construction Industry

The influx of immigrant workers, primarily Latinos and secondarily Asians, began in the 1970s and has increased dramatically. Between 1987 and 1997, according to the U.S. Small Business Administration, the number of Hispanic-owned businesses skyrocketed 230 percent to 1.4 million enterprises, a significant number devoted to construction. The United States Hispanic Contractors Association (USHCA), as of 2002, boasted 130,000 member firms in 15 states. An English-Spanish dictionary of construction terminology can be seen on the desk in many field offices these days as superintendents struggle with directing a crew to caulk the *junta de aisiamiento* (isolation joint).

The Growing Female Workforce

Female workers present yet another challenge in the construction workforce. In 2002, nearly 60 percent of women aged 16 and over participated in the workforce. With women entering the workforce, there has been a notable increase in related problems, including sexual harassment. More and more women are entering the field of construction. At first they could only be observed directing traffic on highway projects, but now they are operating excavating equipment on those same types of road construction projects. When the Occupational Safety and Health Act (OSHA) was enacted in 1970, women made up less than 1 percent of workers in the construction trades. Twenty-five years later, that percentage had more than doubled to 2.3 percent and has continued to grow. Ergonomics—fitting jobs to the needs and abilities of workers—has not kept pace with this influx of female workers who complain about personal protective equipment that doesn't fit properly, since it was designed and produced for the male body. Studies by the National Institute of Occupational Safety and Health (NIOSH) revealed that 46 percent of the female construction workers interviewed said that they could not find work shoes that fit properly, and 41 percent had trouble finding appropriately sized work gloves. And, of course, sexual harassment is a constant problem and ranges from overt sexual remarks by male counterparts to offhanded references to women not being capable of doing a "man's" work.

Project superintendents need to become more aware of the specific problems associated with female construction workers and make the workplace a safer and less hostile environment for them.

Project Delivery Systems

Fast-track construction projects have given way to "flash track" as time-compressed projects roll off the drawing board or, more accurately, out of the computer. The design-build process gained greater credibility not only in the private sector but also among government agencies, partially due to its more rapid progression through the design and construction process. Adherents to the design-build process claim that its more efficient design and construction capability combined with lower overall costs has been responsible for its increased popularity as a project delivery system. This project delivery system allows an owner to contract with only one entity to provide both design and construction, thereby creating a single source for responsibility and accountability. This ability to proceed from design to construction quickly, achieving reduced total project costs along with the potential for reducing or eliminating disputes and claims, is why the Design-Build Institute of America predicts that 45 percent of all projects will utilize the design-build concept by the year 2005.

Changes in the design field include new software programs that ultimately benefit the contractor. The architect and engineer, by utilizing some of these software programs, can prepare a list of materials required simultaneously with their design of each component of construction, such as structural steel, framing and drywall systems, flooring and ceiling materials. Incorporating a list of all required materials in the bid document package may accelerate the contractor's estimating process and ensure that all contractors will be including the same scope of work in their bids.

Many architectural and engineering firms in the United States are subcontracting portions of their work to overseas designers operating in different time zones. Taking advantage of Asian time zone differences, a U.S. architect or engineer can punch out in New York at 5:00 p.m. after emailing design criteria to affirm in

Hong Kong or Thailand, where the day is just beginning. Upon return to the U.S. office the next morning, the architect or engineer will find detailed drawings resting in the computer, awaiting printout to full-scale plans.

Technology

Although lasers have more or less replaced optical leveling devices in the field and digital computerized estimating in the office is commonplace, the construction industry has lagged far behind other segments of the economy in advancing technology. Fifteen years ago, when the author was a member of the technology committee of a national contractor organization, a poll of its members revealed that the top priority was a better scheduling system. Third on the list was a request for better *hand tools*. Well, things have changed somewhat, and technology has gained wide acceptance in the administration, scheduling, and management side of the construction industry. Field office computers are on just about every job site these days, digital cameras record day-to-day job progress, and email transmits any questionable details to the office. CDs are replacing those corrugated cardboard boxes as repositories for reams of paper and archival documentation, and Nextel is synonymous with job site communication. Project management software loaded in office personal computers (PCs) is as common as weeds on the dirt pile. Wireless communication is just in its infancy, and handheld computer devices known as PDAs (Personal Digital Assistants), costing just slightly more than those pocket calculators of the 1970s, are coming into their own. Wi-Fi, as wireless technology is known as these days, is looked upon as the next burgeoning telecommunications event.

PDA software including punch list programs, Request for Information (RFI) formats, and daily diary forms is being offered by several companies in anticipation of a growing wireless market. Previously users were hampered by the lack or spacing of special antennas. Now a slew of new developments and new companies are addressing both wireless technology issues, making it easier to transmit longer distances and increasing the potential for PDA use.

But the use of automated building systems and robotic development lag behind other countries. In the mid-1990s the author visited several construction sites in Japan to observe a new lift-slab process for high-rise construction. Not only was the concrete slab for elevated floors poured and finished at ground level, but also the exterior wall and interior walls with mechanical and electrical rough-ins and finishes were completed prior to jacking the floor into place on some upper floor. Imagine the impact that working at ground level as opposed to working on elevated floors could have on productivity, even forgetting about the ease and cost savings accruing from providing "winter protection" and temporary heat for a single ground-level floor instead of an entire building. Robots developed by Japanese contractors were being employed to trail behind conventional compaction equipment to provide real-time compaction tests. Other robots were being used to finish concrete slabs or paint exterior or interior surfaces. Although the "bubble" in Japan's economy burst, putting such costly systems on hold, the Japanese did demonstrate that factorylike technology can effectively be introduced to the construction site.

Productivity

The scarcity of experienced, skilled workers and managers has had an effect on productivity on the construction site. Some contractors are complaining that tasks previously requiring 8 hours to complete now take 12 hours because of inadequately trained tradespeople. Experienced managers are being asked to assume greater responsibility on more projects, and ultimately more things slip through the cracks.

With 85 to 95 percent of all construction dollars being consumed in field operations, the companies with the "leanest and meanest" operations will excel. But for that to occur, both workers and supervisors will have to acquire the necessary skill levels and tools to increase productivity while maintaining acceptable quality levels. More research and development dollars will have to be spent by equipment manufacturers and construction trade organizations to increase the levels of productivity in both field and office operations. This means more highly trained workers and managers equipped with better tools, of all kinds, and the ability and desire to extract that extra effort from the team.

Quality Control

"Do it right, and do it right the first time" will take on greater importance in this decade and future decades. The shortage of skilled workers and experienced supervisors makes it even more important to increase productivity by reducing or eliminating "rework" and callbacks to address poor-quality issues.

Not only is price important to owners, but also they demand a quality product. If your company can't satisfy this demand, there are others waiting in the wings to take your place.

Quality of product means complying with *all* the demands of the contract documents, from submitting shop drawings promptly and ensuring that they meet the plans and specifications, to being responsive to all closeout documents and getting them in on time, to reducing punch list work to zero-tolerance levels.

Safety

There were 503,500 nonfatal injuries and illnesses in construction along with 1154 fatal accidents in the industry in the year 2001. The incident rate for nonfatal accidents was 8.3 per 100 full-time equivalent workers in construction, as compared to 6.1 per 100 hours (h) in all private industry. Both of these records could stand substantial improvement.

The shortage of qualified tradespeople adds another dimension to the need to maintain a safe working environment—not only to polish the industry's image and reduce human pain, but also to retain the integrity of productive work teams. A long-term absence of a skilled carpenter due to an on-the-job injury can affect costs, quality of work, and productivity for the entire crew.

More project owners aware of job site safety from a moral and economic standpoint are requiring contractors to provide them with a history of safe working conditions as part of the bid requirements. Owners do not want media attention focusing on their new construction project where a serious accident or fatality has just occurred. Although construction accounted for approximately 6.6 percent of the total workforce in the United States in 2000, it had the dubious distinction of accounting for 19.5 percent of all workplace fatalities.

As worker compensation insurance rates continue to remain a significant factor in the calculation of a company's overhead costs and thereby affect its competitive position, builders are becoming more aware of the penalties accruing from a poor safety record.

Dispute Resolution

Hopefully the peak of construction litigation has passed, and all parties to the construction process will begin to see the advantages of resolving disputes by other means. The partnering concept initiated in the 1970s revealed an awareness of the factors that create disputes and claims and established a series of procedures to follow to resolve disputes at an early stage, rather than allowing them to fester into claims.

It was encouraging to find that the American Institute of Architects (AIA) in its 1997 revision of Document A201, General Conditions of the Contract for Construction, required mediation as the first step in a dispute resolution process. Lawyers have now added mediation and arbitration services to their practice and recognize their value in an overloaded legal system with clogged court dockets.

The increased popularity of design-build in recent years is partially due to its ability to reduce finger pointing and disputes between architect/engineer and contractor by placing the responsibility for both activities in the hands of a single entity.

The Project Superintendent's Role

Managing the work process with its emphasis on maintaining schedule, controlling costs, and dealing with quality issues in today's complex construction projects can be an overwhelming task at times. But that is what the construction industry is and what makes it so interesting and challenging. *Management* and *control* are the operative words today and can be divided into four basic components in the construction cycle:

- *Construction engineering.* This involves the process of assembling materials, components, equipment, and systems and the selection and utilization of the optimum technology to do so.
- *Management of the construction process.* This process entails establishing the best way to implement the construction process, which includes precise scheduling and the coordination and control of the flow of labor, materials, and equipment to the job site.
- *Human resource management.* Labor productivity and creation of a harmonious working environment are essential elements of a successful project. Control over human resources—the workforce—becomes more important in these days of labor scarcities.
- *Financial management.* Construction is a high-risk business with historically low profit margins. Control over costs, cash flow, and adequate project funding is critical to the success of any business endeavor, but even more so in the building business.

All these key management functions, to some degree or other, will fall to the project superintendent, who forms the first line of defense at the construction site and is one of the most visible and important members of the construction team.

The successful project superintendent will need to manage and control the following seven basic elements of a successful project.

1. The project is completed on time.
2. The completed project meets the company's profit goals.
3. The quality levels expected were achieved.
4. The project was completed with no unresolved disputes and no outstanding claims.
5. The contractor has maintained a professional relationship with the architect and engineers.
6. The contractor has maintained a mutually beneficial relationship with all subcontractors, suppliers, and vendors.
7. The contractor-client relationship was a good one.

Although the responsibilities of the project superintendent may vary from company to company depending, primarily, upon the size and sophistication of the contractor and the availability of support staff, one thing remains constant—the orchestration and management of the construction project.

This is the role that the project superintendent will play in this complex process called construction.

Chapter

2

Construction Contracts and How They Are Administered

The project superintendent should be aware of the ways in which differing forms of construction contracts are administered in the field and how they affect day-to-day operations.

There are five basic types of construction contracts that the superintendent will encounter, with numerous modifications such as exhibits and addendas that customize the otherwise boilerplate provisions.

- The lump-sum or stipulated sum contract
- The cost plus fee contract
- The cost plus fee contract with a guaranteed maximum price (GMP)
- The construction manager type of contract
- The design-build contract

The American Institute of Architects Contract Updates

In the latter part of 1997, the American Institute of Architects (AIA) issued its newly revised contracts, copyrighted 1997, which superseded the 1987 editions. Although some earlier 1987 versions may still be in use, we will address the current versions of these AIA contracts. Another very important contract document is AIA Document A201, General Conditions of the Contract for Construction, which requires its own chapter (Chap. 3) to fully explore its provisions.

Other, less frequently used contracts—the turnkey and joint venture contracts—will also be discussed in this chapter.

A Letter of Intent is often executed as a "contract," generally preceding the issuance of the more all-encompassing standard contract forms. So let's begin our discussion with this document.

The Letter of Intent

A Letter of Intent is generally a temporary document authorizing or initiating the commencement of construction in a limited fashion, anticipating that a more complete and encompassing form of contract will be forthcoming in the near future.

Limits are placed upon the scope of work to be performed, such as the dollar value of work to be performed, often expressed as a "not to exceed" amount, restrictions on what subcontracts or purchase orders can be awarded, and provisions relating to termination of the Letter of Intent. In addition, the Letter of Intent will incorporate provisions to cover settlement costs in the event that a formal contract for additional work is not awarded. However, if a contract is awarded for additional work, that which was spelled out in the Letter of Intent will be incorporated and any payments made under that document will be credited to the new contract sum.

There are a number of reasons for using a Letter of Intent:

- An owner may wish to start demolition of a recently vacated office space while negotiating with a new tenant desirous of a quick move-in.
- Having received verbal loan approval from a lender, an owner may wish to commence a limited amount of construction work on a new project while awaiting full written approval.
- When a project is "fast-tracked," an owner may wish to commit to a certain portion of work while the final budget is being prepared and, via the use of a Letter of Intent, can authorize a contractor to proceed to purchase long-lead-time items necessary to jump-start the project, for example, reinforcing steel or structural steel shop drawings.

The Letter of Intent places limits on the scope of work to be performed and on the dollar value of the work to be performed (often expressed as a not-to-exceed amount) and puts restrictions on what subcontracts or purchase orders can be awarded.

This document is usually the precursor to the issuance of a contract with expanded scope of work and incorporates provisions to include settlement costs if that expanded scope of work fails to materialize.

A Letter of Intent must be specific:

1. It should be specific in defining the scope of work to be performed. If plans and specifications define that scope of work, these documents ought to be referenced. If plans and/or specifications are not available, an all-inclusive narrative should define the exact nature of work to be performed.
2. It should include either a lump sum to complete the limited scope of work or a "cost not to exceed" amount, including the contractor's fee. It is essential to define what is included in *cost* as well as what reimbursable expenses are included.
3. It should include payment terms.
4. It should include a date when the work can commence and, in some cases, when the Letter of Intent expires.
5. If applicable, a statement may stipulate that the scope of the work and its associated costs will be credited to the scope/cost of work included in the expanded construction contract.
6. A termination clause should be included, setting a time limit on the work, or an event that triggers termination, such as the issuance and acceptance of another contract. A termination "for convenience" clause is often added, allowing either owner or contractor to terminate the work upon written notification.
7. The Letter of Intent must be signed and dated by all concerned parties.

The superintendent must therefore be aware of restrictions contained in the Letter of Intent to ensure that no work beyond the scope of work contained in this document is performed.

A typical Letter of Intent might be worded as follows:

> Pursuant to the issuance of a formal contract for construction, the undersigned (Owner) hereby authorizes (Contractor) to proceed with the tree removal in the

> areas designated on Drawing L-5, prepared by Wilton Engineers, dated July 8, 2003. All debris including tree stumps will be removed from the job site. Prior to commencement of the work, all erosion control measures will be installed according to Drawing L-8, by Wilton Engineers, dated July 1, 2003. Maintenance of soil erosion measures will be required from the date of installation until this Letter of Intent is terminated on/about September 15, 2003. All the above work is to performed at cost plus the 15 percent contractor's overhead and profit fee. Daily work tickets will be presented by (Contractor) to (Owner's representative) for signature to provide substantiation for all costs.
>
> Signed: Contractor Signed: Owner

Scope, tasks, and reimbursables included in letters of intent can include shop drawing preparation and cancellation charges for any materials/equipment ordered if a further construction contract is not forthcoming. Reimbursable expenses may include in-house costs incurred by the general contractor for estimating, accounting, and even interim project management and superintendent salaries. The owner should be presented with a list of reimbursable costs appended to the Letter of Intent to avoid any future misunderstandings.

When a formal construction contract is issued, the segregated costs associated with the work performed under the Letter of Intent must be applied against the costs for the total project.

It is important to accumulate and segregate all reimbursable costs as they are incurred. Assigning a separate cost code to all labor, material, equipment, and subcontract commitments will permit easy retrieval of all related costs and the preparation of an accurate exhibit to the owner's requisition.

While operating under the terms and conditions of the Letter of Intent, the general contractor may have to make certain commitments to subcontractors and vendors, and any purchase orders or subcontract agreements issued should contain the same restrictive provisions as are in the agreement between general contractor and owner.

For example, if the owner's letter of intent contains provisions for the preparation of reinforcing steel drawings, placing of an order for some nonstock sizes, and even partial fabrication, the same restriction(s) placed upon the general contractor should be transferred to the reinforcing bar contractor.

This is particularly important if there is a termination clause in the Letter of Intent. A typical termination provision would be as follows:

> Upon receipt of a written directive to cease the work covered under the terms of this Letter of Intent, the Contractor will immediately stop all work. All costs for work-in-place as of that date will cease. Cancellation costs for work-in-progress will be honored upon a detailed explanation for all such costs documented by purchase orders or other commitments and related stop-work orders.

The superintendent must be familiar with the nature of costs to be included in the agreement so that any field-related costs can be segregated, documented, and presented to the owner for payment.

The Five Most Prevalent Types of Construction Contract

1. Cost of the work plus a fee
2. Stipulated or lump-sum
3. Cost of the work plus a fee with a guaranteed maximum price (GMP)

4. Construction management
5. Design-build

Other less frequently used contracts between an owner and general contractor are

6. Turnkey
7. Joint venture

Cost of the Work Plus a Fee

What could be simpler—a contract whereby the contractor invoices the owner for all costs related to the work plus the contractor's fee for overhead and profit? Well, the cost plus fee contract requires a great deal of thought and effort to work successfully.

First, a definition of what constitutes *cost* is often a point of contention between owner and contractor, and this needs to be clearly spelled out. A cost plus fee contract is, as the name implies, one in which the contractor will perform a certain scope of work identified by contract documents or a narrative description. The associated costs will be reimbursed by the owner inclusive of the contractor's fee, generally calculated as a percentage of the work. Cost-plus contracts are used infrequently, but when employed, they require a high level of communication between contractor and owner to avoid misunderstandings and potential disputes.

This form of contract is often used when severe time restraints are imposed on the owner and it becomes necessary to begin construction as quickly as possible—often without the benefit of well-defined plans and specifications.

For instance, the owner of an office building may have an opportunity to lease space in the building to a tenant requiring occupancy in short order, and therefore demolition of previously occupied space must begin immediately in order to facilitate the start of the new construction. The owner may also wish to commence some limited amount of new work as the tenant fit-up drawings are being developed. The scope of demolition work can be easily defined, and partition layout and even partition work may be authorized on a cost-plus basis.

It is to the benefit of all parties to convert this cost-plus agreement into a lump sum or guaranteed maximum price contract when the drawings have reached a stage of completion that would allow for a more defined contract sum. This will eliminate many misunderstandings and potential disputes, and the transition from a cost-plus contract to, say, a lump-sum contract is not difficult.

The owner can either pay for the cost-plus work on a separate invoice or incorporate these costs into the lump-sum or GMP contract.

There are two items that require a detailed explanation and mutual understanding by all parties, when using a cost-plus contract—what costs are to be reimbursed and what costs are *not* reimbursable. Even the cost of hourly labor may become a point of disagreement unless the owner is advised of base labor rates and the upcharge or "burden" applied to the base rate, incorporating such costs as unemployment compensation, social security taxes, workers' compensation costs, as well as employee fringe benefits or union dues. The addition of as much as 40 percent to labor base-pay rates may come as a shock unless explained beforehand.

Reviewing the American Institute of Architects Document A111 (1997 edition), Cost of the Work Plus a Fee with a Negotiated Guaranteed Maximum Price, will establish guidelines for reimbursable and nonreimbursable costs.

Cost to Be Reimbursed as Defined by A111

1. Labor costs
2. Wages, salaries of contractor's supervisory and administrative personnel when stationed at the site *with the owner's approval*
3. Taxes, insurance, contributions, assessments, benefits required by unions
4. Subcontract costs
5. Costs of materials and equipment incorporated in the completed project
6. Cost of other materials and equipment, temporary facilities, and related items fully consumed in the performance of the work
7. Rental costs for temporary facilities, machinery, equipment, and hand tools *not customarily owned by construction workers,* whether rented from the contractor or others
8. Cost of removal of debris from site
9. Cost of document reproduction, fax and telephone calls, postage, parcel post, and reasonable petty cash disbursements
10. Travel expenses by contractor while discharging duties connected with the work
11. Cost of materials and equipment suitably stored off-site, if *approved in advance* by the owner
12. Portion of insurance and bond premiums
13. Sales and use taxes
14. Fees and assessments for building permits and other related permits
15. Fees for laboratory tests
16. Royalties and license fees for use of a particular design, process, or product
17. Data processing costs related to the work
18. Deposits lost for causes other than the contractor's negligence
19. Legal, mediation, and arbitration costs, including attorney's fees arising out of disputes with owner *with the owner's prior written approval*
20. Expenses incurred by contractor for temporary living allowances
21. Cost to correct or repair damaged work provided that such work was not damaged due to negligence or was not nonconforming

Costs Not to Be Reimbursed per A111

1. Salaries and other compensation of contractor's personnel stationed at contractor's principal office or offices, except as specifically provided for in the contract
2. Expenses of the contractor's principal office
3. Overhead and general expenses, except as provided in the contract
4. Contractor's capital expenses
5. Rental cost of machinery and equipment, except as specifically spelled out
6. Costs due to negligence of the contractor
7. Any costs *not specifically included in costs to be reimbursed* (This transfers the responsibility onto the contractor to include a comprehensive list. The contractor cannot claim later that he or she "forgot" to include some miscellaneous costs.)
8. Costs, other than approved change orders, that would cause the GMP price to be exceeded

Changes to these standard costs, both additions and subtractions, need to be clearly delineated in the agreement. It becomes important for the project superintendent to clearly identify all applicable costs generated in the field, by project number and by the proper cost code. Segregation of these costs in the field and in the home office is essential when requisitions are prepared and documentation of all reimbursable costs is to be attached to that request for payment.

Pitfalls for the Superintendent to Avoid When Administering a Cost-Plus Contract without a GMP

1. The scope of the work included in the agreement should have been clearly defined. If the owner, during a site visit, requests the superintendent to perform work that appears to exceed the original scope of work, the project manager should be alerted as soon as possible to determine if these instructions to the field represent increased or possibly decreased costs.
2. A statement of costs to be reimbursed and those not to be reimbursed must accompany any cost-plus agreement and be given to the project superintendent.
3. The project superintendent's time is generally a reimbursable cost; and not only should daily and weekly hours be charged to the appropriate cost code, but also a brief explanation of each day's activities ought to be noted, primarily in the daily log, so that if it is questioned by the owner, all supervisory time can be accounted for.

The superintendent should be meticulous in documenting all reimbursable costs for both labor and materials. Labor costs can be segregated by task by assigning appropriate code costs. Material and equipment costs will also be segregated by applying the correct cost code and will include a brief description of where the material or equipment was used or installed. Sign each ticket to validate that materials were received as indicated.

The author had one experience with an open-ended cost-plus contract that illustrates these pitfalls very clearly.

The Saga of a Cost-Plus Fit-Up Project

The author worked with an owner on a cost-plus contract many years ago, and the problems he encountered and the lessons learned are worth repeating.

The client owned a vacant 80,000-ft^2 one-story building and had been attempting to lease all or a portion of the space for some time, to no avail. Because it was located in a strip shopping mall that had seen better days, he hoped that by attracting a national account tenant, the character of the entire complex could be upgraded.

The building had been a supermarket and consisted largely of open space. Heating and cooling had been accomplished via several large rooftop units with little or no duct distribution system. The floors were covered with vinyl composition tile, and because the heating/cooling systems had been shut down, most of the floor tiles were loose and badly cupped.

The ceilings were made of suspended acoustic tile, and the 2 × 4 ceiling tiles were badly warped due to the lack of conditioned air. The metal grid system, originally white enamel, now exhibited a moldy yellow hue. Recessed electrical fixtures in the ceiling grid were damaged or, in some cases, missing acrylic lenses.

The building's structural system consisted of bar joists bearing on interior steel columns and exterior masonry walls. All in all, the building was in a pretty rundown condition.

A national home builder supply company expressed an interest in leasing at least 40,000 ft^2. This would be one of its first east coast store openings, and the building owner, anxious to close the deal, agreed to major interior and exterior renovations which included a new, large aluminum storefront entrance and a new fascia and canopy on the exterior of that portion of the building to be occupied by the home improvement company.

The new tenant's architectural department was going to submit drawings for all interior renovations, and the building owner was to submit drawings for all exterior work to the tenant.

Oh, yes, and the lease stipulated that all improvements must be in place within 60 days after contract signing, or the tenant could cancel the lease and not be held responsible for any renovation costs incurred if that date was not met.

The owner contacted the author's company, and a cost-plus contract was quickly put together, listing reimbursable and nonreimbursable costs and contractor's fees. On the basis of the scope of work discussed and agreed to by the owner, budget estimates were prepared and delivered to the owner the following morning. He requested that work begin as soon as possible.

The local architect hired by the owner produced drawings for the enlarged storefront entrance within a few days, as laborers and masons had already begun to needle through the existing exterior block walls and create new structural openings in the masonry wall to receive the aluminum storefront work.

Exterior canopy sketches were given to the author, and when this work started, the owner was so pleased with the look that he requested that it be extended along the full 400-ft elevation of the building instead of the 120 ft originally budgeted.

In the meantime minor demolition was taking place inside the building as the local architect awaited the tenant's interior design drawings.

To ensure that a sufficient number of carpenters would be available for the interior work, it was decided to extend the workday for the canopy construction crews to 2 hours of premium time per day per worker. The owner authorized this overtime which was quickly changed to 3 hours per day per worker. With a 15-worker carpentry crew, the cost for the premium time would be significant. At job meetings held at the site every 2 to 3 days to review progress drawings, the owner was kept apprised of the costs to date. Meantime, attention was turned to interior work, and although the existing lighting fixtures were initially budgeted to be cleaned, reballasted, and relocated, when work began, the electrical subcontractor said that most of the fixtures were beyond refurbishing and would have to be replaced. The owner agreed, new fixtures were immediately ordered, and the owner was presented with the cost differential for that portion of the work.

Time was ticking away. The tenant's interior design drawings were late, and therefore the new ductwork distribution system could not be fabricated. When drawings finally appeared, both the sheet metal contractor and the electrician had to work overtime to make up for lost time.

The floor was a mess, but the owner insisted that it be cleaned and loose tiles recemented, using tiles from other sections of the building to replace the broken ones. He was advised by the flooring contractor that an entirely new floor would be only slightly more expensive, and was even quoted a lump sum for that work, but the owner insisted on reusing the existing tiles. The flooring subcontractor proceeded to clean, patch, and repair on a "time and materials" basis.

When the tenant's representative visited the site at 4:00 p.m. the day before the grand opening was scheduled, he rejected the resilient floor installation and insisted that an entirely new floor be installed throughout the building in time for the opening day ceremonies, scheduled for 8:30 a.m. the following morning. If this was not accomplished, he hinted that the lease would be voided—and the owner would be unable to recover any costs to date. A quick trip to the supply house secured enough materials to replace the existing flooring, and frantic telephone calls to workers at their homes brought them swiftly back to the job.

With several teams of laborers and mechanics working around the clock, the last coat of wax was drying as the red, white, and blue buntings were being hung over the front entrance at 8:00 a.m.

The deadline was met, the tenant had started moving in the night before, and opening day was a smashing success.

During all the hectic activity during the 2 weeks prior to the tenant move-in, ongoing costs were reviewed with the owner, although many costs had not been fully developed. All these discussions were verbal, and just a few notes were written.

When the final costs and corresponding scope changes were assembled and presented to the owner, there was a funereal silence. The owner's face turned beet-red, and he said, "These costs are 30 percent higher than the initial budget!" The owner ignored the fact that the initial "budget" was prepared from two rather sketchy drawings prepared by his architect and the fact that he had authorized some rather significant changes during construction. Many of the increased scope items that had been previously discussed, agreed upon by all parties, and implemented during this fast-paced project suddenly were vague in the owner's memory—a condition sometimes known as *selective memory.*

It took 2 to 3 weeks to assemble all the detailed labor, material, and subcontractor cost sheets together with a complete narrative of the progression of scope changes, dates, parties present, etc., for presentation to the owner. This was followed by another 4 weeks of meetings with the owner which mostly ended with a glazed look of bewilderment in his eyes.

There was very little question that he fully understood the extent of all the instructions he gave, instructions that he surely knew had substantial cost impact. But now his tenant was in place, paying rent, and the need to quickly resolve all costs with the general contractor became a less urgent matter, in his eyes.

It took another 3 months to receive final payment, and the author came away with the distinct impression that the owner felt he had been abused, when in fact he had authorized all the work, the project was completed on time, the new tenant greatly enhanced the image of the strip mall, and under the circumstances the costs were reasonable.

The experience certainly made painfully evident the importance of proper and timely *written* documentation when a cost-plus-fee project is undertaken.

There are five critical elements in a cost-plus-fee contract that should be addressed by a project superintendent:

1. Understand completely the scope of work included in the contract.
2. Identify all scope changes as soon as they occur and notify your office.
3. If the owner's representative requests a change, note the time and date that this request was made and pass all the information onto the office.

4. Cost-code and identify all field-related costs for labor, material, and equipment as these costs are incurred.
5. Do not authorize any subcontractors or vendors to proceed with the changes until advised to do so by your office.

The Stipulated or Lump-Sum Contract

A stipulated or lump-sum contract is most frequently used in competitively bid work, in either the public or private sector, where a complete set of plans and specifications has been prepared by the owner's design consultants.

Contractors are expected to estimate the cost of the work contained in a specific set of bid documents—no more, no less.

Any deviation from the scope of work contained in these bid documents, except if amended later by other contract provisions, will result in a change of scope, and the associated costs will be dealt with by change order.

Although this may appear to be a rather straightforward approach to the administration of a construction contract, it is not as simplistic as it seems. The *intent* of the plans and specifications can often be interpreted in many ways by each participant to the construction process—the owner, the architect/engineer team, other design consultants, the general contractor, and subcontractors.

The perfect set of plans and specifications, based upon the author's experience, has yet to be produced, and changes are almost always inevitable, to include not only items inadvertently omitted from the scope of work but also items added by the owner to include additional amenities or upgrades.

Since the architect is, by contract, generally designated as the "interpreter" of the plans and specifications, the final decision on what constitutes the obligation of all parties to the contract rests with that authority—unless challenged and resolved by negotiation, arbitration, or litigation.

In the case of renovation or rehabilitation work, the stipulated sum contract may include a contingency allowance to cover the costs of unanticipated problems that generally occur in these types of projects. A contingency in the amount of 5 percent of the total contract sum is not unreasonable, and a contingency in the amount of 10 percent of the total contract sum for some types of renovation work is proper. The purpose of any contract contingency amount needs to be clarified. Is the contingency for the sole use of the owner or sole use of the contractor, or shared by both parties in some definable way?

The contractor's fixed fee in the stipulated or lump-sum contract is dependent upon the tabulation of final project costs. If everything goes well and costs track favorably with the estimate, the contractor will achieve the anticipated fee, and possibly more. If, inadvertently, costly items have been omitted from the estimate or grossly undervalued, or adverse job conditions occur for which no owner reimbursement is received, the contractor's fee may diminish or disappear.

The project superintendent must be alert to discovering any and all deficiencies in the plans and specifications and their associated costs. These comments should be passed on to the project manager to determine which unanticipated "extra" costs should have been reasonably expected and absorbed by the contractor and which costs should be presented to the owner as a change order.

Pitfalls to Avoid When Supervising a Lump-Sum or Stipulated Sum Contract

A thorough review of the contract documents—the plans and the specifications—is essential to uncover any ambiguities, errors, and omissions before they surface during construction. This review should be made as early as possible, either during the period when the project's mobilization is taking place or in the initial stages of construction. Discovery of problem areas can be presented to the architect or engineer in the form of a *Request for Information (RFI)* or a *Request for Clarification (RFC)*. Upon receipt of the response to the RFI or RFC, the project superintendent and project manager can determine whether the architect/engineer (A/E) interpretation or ruling warrants a scope change and a request to increase or decrease the contract sum via change orders.

Some of the more common errors and omissions encountered during a plan and specification review are as follows:

Common Drawing Shortcomings

1. Architectural, structural, mechanical, electrical, and plumbing drawings are not coordinated dimensionally.
2. Openings for mechanical and electrical work, louvers, and other exterior wall penetrations may be shown on the architectural and mechanical, electrical, and plumbing (MEP) drawings but not on the structural drawings—or else the size of an opening, if indicated, may vary from one drawing to the next.
3. Dimensions on structural foundation drawings may be at variance with dimensions shown on the architectural drawings—or individual dimensions do not add up to overall dimensions.
4. Elevator shaft openings on the architectural drawings may vary in size and location from those on the structural drawings.
5. Housekeeping pads for electrical and mechanical equipment as indicated on the MEP drawings may not be represented on the architectural or site drawings.
6. Partition types as shown on the architectural floor plans may be at variance with larger-scale details or partition schedules.
7. Finish schedules may be at variance with finishes indicated on the architectural drawings.
8. Reflected ceiling plans may not accurately locate electrical fixtures, HVAC diffusers, or sprinkler head locations or may not be coordinated with these drawings.
9. Ductwork and other above-ceiling work will not fit into the space assigned.
10. Note on one drawing referring to a detail on another that does not appear, or the detail does not apply to the condition to which it is supposed to apply.
11. Invert elevations on utilities leaving the building are not compatible with inverts indicated on the mechanical/electrical drawings.

Common Specification Review Problems

1. Reference is made to information contained in another section, but is not provided in that other section.
2. Specification sections do not apply to the project at hand.
3. Specifications do apply, but contain sections that are inappropriate to the project at hand, e.g., a requirement for TV brackets for "patients" that is included in a school project.
4. References to painting are made in the mechanical and electrical specification section but are not included in the painting section, or vice versa.

5. The requirement for equipment starters is not included in the appropriate section, or is not included in any section.
6. Equipment for the same item is listed in two different specification sections, i.e., hardware for aluminum entrance doors is indicated both in the door specification section and in the hardware section.

A thorough review of the plans and specifications by the project superintendent early on may avoid some of the panic situations that occur during construction. The superintendent should also urge all subcontractors to review their scope of work prior to coming on site, so that any errors, omissions, or inconsistencies in their respective trades can also be brought to the architect/engineer's attention in the initial stages of the project.

The Cost Plus Fee Contract with a Guaranteed Maximum Price

The cost plus fee contract with a guaranteed maximum price (GMP or G_{max}) is frequently used because it allows the owner to gain the protection of the maximum cost of the construction while retaining the potential for cost savings. It is basically a cost plus fee contract with a cap on it.

The GMP contract is often used for fast-tracked work, and for the faster flash-tracked projects when incomplete or sketchy construction documents are all that is available at the time of contract preparation. The GMP contract is quite often used when design-build work is being considered.

The contractor's fee is usually prenegotiated, based upon a percentage of cost, and generally there is a provision in the contract allowing additional fees for work above and beyond the initial scope of work, i.e., change order work.

A GMP contract will most likely include a *savings clause* specifying that any savings are to be shared by the owner and the contractor. Some owners prefer to have the contractor receive a greater portion of the savings, theoretically creating greater incentive for the builder to search for potential savings. Many contracts stipulate that any savings will be shared on a 50-50 basis, giving the owner an incentive to review any scope changes or accept value engineering suggestions proposed by the contractor.

A standard feature of the GMP contract is a requirement for a cost certification audit when the project has been completed. The owner will have the authority to audit the contractor's books to verify costs and to determine the extent and nature of any potential savings. The audit may also be used to ensure that all costs charged to the project are proper ones and were actually expended for that project.

The project superintendent should carefully review, identify, and isolate all costs as they are incurred in the field. This will save countless hours and reduce any owner frustration at the end of the project, when various job-related costs may be called into question. Identification of costs for labor, materials, and equipment will be similar to the procedures described in the administration of cost-plus contracts.

Scope changes, whether initiated by the owner, subcontractor, or general contractor, should be analyzed to determine if the guaranteed maximum price should be increased or decreased. Since the GMP contract is based upon a specific scope of work, any change in that scope should theoretically increase or decrease the guaranteed maximum price. Even if costs to date reveal that substantial savings have accrued which will allow any added costs due to scope increases to be absorbed by these savings, the contract sum must be revised. Remember, your company will share in any eventual savings, and absorbing these costs will only decrease the amount of savings to be shared.

One of the many functions of a project superintendent and project manager administering a GMP contract will be to look for savings throughout the life of the project. All subcontractors should be requested to review their work with an eye toward developing possible cost savings. As these suggestions are received, they must be reviewed and analyzed to determine whether a savings by one trade may increase costs in another trade, thereby eliminating any net savings and, in some cases, resulting in higher costs. This process of *value engineering* can be costly, not only in dollars, but also in testing the project superintendent and project manager's professionalism.

For example, the substitution of two small rooftop exhaust fans for one larger unit may result in a savings in the mechanical portion of the work but may add costs for two electric circuits, two breakers, associated labor, one additional framed rooftop opening, roof curb, and flashing.

All such value engineering suggestions should be routed through the architect, engineer, and any related subcontractors for their comments before being formally submitted to the owner.

The more adept a contractor becomes at developing meaningful cost savings or value engineering suggestions, the easier it will be to build a solid reputation as an effective administrator of GMP contracts.

Pitfalls to Avoid When Administering a GMP Contract

1. All material and equipment receiving tickets should be identified with the project name, number, cost code, and an indication of where the item was used. This will make it much easier to document costs, if and when a detailed audit is required.
2. If the GMP was based upon less than complete plans and/or specifications, the project superintendent and project manager should carefully review all design development drawings as they are produced to ascertain that the scope of work is neither increased nor decreased. When scope changes occur, alert the owner immediately and attempt to quantify those changes. If additional costs are involved, request confirmation of acceptance. If the scope of work can be modified or reduced in other areas to offset these added costs and create a tradeoff, notify the owner in writing.
3. When you receive value engineering cost savings, review them carefully with all related suppliers and subcontractors to verify that the actual savings being suggested are, in fact, true savings and that no hidden costs lurk somewhere down the line.
4. Do not allow scope increases to occur without increasing the guaranteed maximum price, even though it appears that some costs could be absorbed by the savings that exist at the time. (These savings may disappear rapidly as unanticipated costs suddenly appear.)
5. Remember that if the architect or owner deletes items of significant value from the contract, she or he will expect a credit that lowers the GMP.

In GMP-type projects, identify all labor, material, and equipment costs so that an audit by the owner at the end of the project will clearly indicate that all costs assigned to the project are properly documented. When scope increases or decreases occur, report them to the project manager so that the owner can be notified of the extent of the change and its associate cost impact.

Construction Manager Contracts

Once associated with megaprojects, the construction manager (CM) form of contract has been utilized in significantly smaller projects, and most general contractors today also market themselves as construction managers.

Although there was once much confusion over what constitutes construction manager work, the *Construction Management Association of America (CMAA)* defines it as follows:

> Construction management incorporates the following elements:
>
> - A project delivery system consisting of a program of management services;
> - Defined in scope by the specific needs of the project and the owner
> - Optimally applied to a construction project from conception to completion in order to control time, cost and quality
> - Performed as a professional *service* under contract to the owner by a construction manager as the owner's *agent.*

Note: The word *agent* is a key word in discussing CM work. Unlike a general contractor (GC) who issues subcontracts and purchase orders in the name of the GC, a CM is an *agent* for the owner, and so all subcontracts, purchase orders, and the like are issued in the owner's name when approved and recommended by the CM.

The construction manager is

- Selected on the basis of experience and qualifications of the firm operating as a construction manager, as opposed to experience as a general contractor
- Compensated on the basis of a negotiated fee for the scope of services rendered, generally expressed as a percentage of the total cost of construction but also as a fixed fee in some cases

The construction manager concept can be subdivided into

- *CM for fee.* All required construction management services are performed by the CM, and the CM receives a fee plus a previously agreed upon list of reimbursable expenses. The CM does not guarantee the final cost of the project.
- *CM at risk.* The CM in this case also provides all required construction management services for which he or she receives a fee plus listed reimbursable expenses. However, the CM guarantees that the final cost of the project will not exceed a stipulated sum, excepting approved changes in scope.

CM Contracts—A Two-Part Arrangement

Quite often an owner will select a CM and award one contract for preconstruction services with a termination clause prior to the award of a second contract for the actual administration of the project. If the owner is unhappy with the performance of the CM during the preconstruction phase, the owner can terminate the second contract and seek a more qualified firm for the administration of the work and possibly to review the preconstruction portion as well.

The division of services provided by a construction manager is rather straightforward:

- *Preconstruction services.* The CM provides professional staff to the owner prior to or during the design phase. In some cases staff is provided during various stages of project development, including estimating, scheduling, purchasing, and project management. The purpose is to assist in development of the owner's construction program by working closely with the design team to ensure that it meets the client's schedule and budget restraints. The CM may provide these services as a lump-sum proposal or cost plus fee, with or without a GPM.

- *Construction services.* The CM provides the staff and related field office facilities to administer and manage the construction project, acting as the owner's agent. The CM during construction will be awarded either a for-fee contract or an at-risk contract.

And, of course, many owners engage CMs to perform both functions—preconstruction and construction services—in one contract incorporating both phases.

CM work has come in for some criticism, directed not so much at the system, but at those firms that profess to be qualified construction managers but lack the experience and staff to perform the required services at the anticipated professional level.

The Construction Manager and the Preconstruction Phase

One major advantage of the CM process is that an owner can obtain the services of a team of construction professionals to act on the owner's behalf during the preparation of the project's design. The CM's staff of experienced professionals, having day-to-day contact with subcontractors, local labor pools, equipment manufacturers, and material suppliers as well as a detailed database of construction component costs, can provide invaluable assistance in determining the most cost-effective design commensurate with the owner's program, budget, and project delivery dates.

The CM entering into a preconstruction services contract must have a complete understanding of the services the CM is expected to provide, so that the CM can establish the fee accordingly and include a list of corresponding reimbursable expenses for staff and services.

Some owners prefer to enter into a lump-sum contract once the scope of these preconstruction CM services has been defined.

Preconstruction services generally include

1. Consultation with the client and the architect/engineering team during project development with respect to building systems and components
2. Preparation of a schedule to include dates for completion of bid documents; time required to solicit, receive, and analyze bids; contract award and commencement of construction; and a detailed outline of construction activity from start to completion of the project
3. Preparation of an initial budget which is to be updated as design development proceeds.
4. Coordination and assistance in the preparation of bid documents
5. Selection of qualified bidders and solicitation of bids
6. Review and analysis of bids and award recommendations

Note: Items 5 and 6 are sometimes included in the construction phase of the CM contract.

Construction-phase services include

1. Selection of a qualified bidders list, review, analysis of bids, and recommendation for contract award (if not included in preconstruction phase)
2. Project supervision during construction to include project superintendents, project managers, and project engineers, depending upon the size and complexity of the project
3. Field office and related expenses and administrative staff
4. Cost control and scheduling services

5. Assistance in obtaining all required permits
6. Establishing procedures for change order preparation, review, and approval/rejection recommendations
7. Consultation with the owner and design consultants involved in the project
8. Inspection of the work to ensure compliance with the contract documents
9. Acting as a conduit between subcontractors and vendors and owner and design consultants over matters of contract interpretation and compliance
10. In collaboration with the architect and engineer, establishing a shop drawing processing and monitoring procedure to ensure its smooth operation
11. Reviewing, recording, and processing all reports and site documents
12. Determination of substantial completion in conjunction with the architect and preparation of a list of incomplete and unacceptable items
13. Monitoring the start-up and testing of equipment with the architect and engineer and supervising the turnover of equipment in accordance with the contract provisions
14. Developing and monitoring the punch list process in collaboration with the design team
15. Collecting, reviewing, and approving all closeout documents in conjunction with the design consultants
16. Assisting in ensuring that all warranty and guarantee work is provided and performed in full accordance with the contract requirements
17. Assisting the design consultants in the completion of all project closeout procedures and documentation

Some CM contracts do not prohibit a general contractor from performing certain work tasks with his or her own forces if the GC is experienced in these tasks and can demonstrate that the GC's involvement will be cost-effective. If such an award is made, the CM will be allowed to include a certain percentage for administrative costs and profit just as though that portion of the work had been subcontracted to another firm.

CM Fees

The fee charged for construction manager services varies depending upon whether preconstruction *and* construction services are required and whether the CM will be a for-fee or at-risk contract.

Fees in either case are significantly lower than those charged by general contractors performing lump-sum or GMP work because the CM is reimbursed for most field and many office-related expenses. Therefore most of the fee will go to the contractor's bottom line.

The *reimbursable expenses* are specifically listed in the CM's proposal, and their *cost* includes a percentage allocated for overhead and profit. These reimbursables including overhead and profit are referred to as *reimbursables with a multiple*; in other words, the owner will pay the CM for specified costs multiplied by a factor of 1.5 or 2.0 or whatever is agreed upon. Therefore the cost of a superintendent's weekly salary including fringe benefits may be, say, $1500.00. At a multiple of 1.5, the owner will agree to pay $2250.00 to the CM for the "cost" of this superintendent's services; at a multiple of 2, the CM would be reimbursed $3000.00.

Exclusive of preconstruction services and reimbursables, compensation for a for-fee CM will be in 1.5 to 3 percent whereas the fee for a CM at risk could be between 4 and 7 percent.

As stated previously, preconstruction services are often quoted on a lump-sum basis and include the CM's specific duties and responsibilities associated with administering these duties and responsibilities.

Pitfalls to Avoid When Administering a Construction Manager Contract

As the owner's agent, a great deal of responsibility shifts from the contractor to the owner since the CM will *recommend* rather than *decide* certain issues involving costs. One of the pitfalls to avoid is failure to include a complete and inclusive list of reimbursables for field-related expenses. (And that is why a CM will administer a project from the field, incorporating all office-related project functions and thereby transferring those costs from their central office overhead to reimbursable field-related costs.

A typical CM list of standard reimbursables might include the following:

Project office field setup

1. Office complex, trailers, security fencing
2. Office equipment, duplicating machine, computers, miscellaneous supplies (staplers, hole punches, etc.)
3. Utility connections for telephone, data, water/sewer (if applicable), and electricity
4. Signage
5. Rental of reproduction equipment (even if owned by the CM)
6. Office furniture—desks, chairs filing cabinets, conference table

Project field office expenses

1. Supplies for office equipment, periodic maintenance
2. Copy machine supplies
3. Fax machine supplies
4. Utilities—power, telephone, sanitary, water
5. Postage, package deliveries, overnight delivery service
6. Field radios, beepers
7. Blueprints
8. Automobile and trucking expenses and repairs
9. Travel expense
10. Reproducibles
11. Office maintenance and cleaning services
12. Travel expense
13. Security
14. Office maintenance and cleaning services

Site-related expenses

1. Engineering (if requested)—initial survey, interim layouts, final survey (if not supplied by owner)
2. Testing, geotechnical services (if not contracted for directly by owner)
3. Erosion control—installation and maintenance (if not awarded to another contractor)
4. Shop drawing receipt, review, and transmission to A/E and return to the appropriate subcontractor/vendor
5. As-built drawings (either preparation or review of drawings prepared by others)
6. Safety/first aid
7. Site security
8. Photographs—progress photographs and others required for documentation

Project maintenance

1. Access roads—supervision of installation and maintenance
2. Fire extinguishers and maintenance of same
3. Personal safety equipment
4. Portable toilets—delivery and periodic maintenance

5. Cleanup and dust control
6. Dumpster services
7. Trash chutes
8. Pest control

The project superintendent's role in administering a construction manager (owner's agent) type of construction contract will be not much different from that supervising any other project. However, the low CM fee is dependent upon the owner reimbursing the construction manager for all costs as set forth in the reimbursable portion of the contract. Therefore the project superintendent must diligently track, isolate, and document all such reimbursable costs, being careful not to include any nonreimbursable costs.

The Joint Venture Agreement

In today's marketplace, local builders may discover that large national and international contractors are venturing into their geographic area, seeking work to increase or maintain their already huge annual volume. Sometimes these incursions are the result of one of the national account clients deciding to build in your area, and at other times these construction giants are just looking to increase their market share.

Expansions into local markets by these companies may present opportunities for established local contractors and may not necessarily be viewed negatively. Without any long-term relationships with local subcontractors and suppliers and lacking knowledge of local job conditions, these large contractors may seek out local partners to work with in what is known as a *joint venture.* Advantages can accrue to both firms if the joint venture agreement is properly prepared, and shared responsibilities and shared profits can enrich all parties.

A joint venture contract is often used when a minority contractor and a nonminority contractor, having a higher financial capacity or higher bonding ability, team up to bid on a public project requiring an affirmative action program.

A typical joint venture (JV) agreement (contract) is set forth in Fig. 2-1. The joint venture agreement defines the *entity* that will actually be engaged in construction and is independent of the type of construction contract executed between owner and JV builder. Any one of the basic five types of construction contracts can be employed by the owner—cost-plus, GMP, lump-sum, CM, or design-build.

Turnkey Contracts

Although there are several variations on the *turnkey contract* concept, the most universally accepted definition is a project whose costs will not be reimbursed until the contractor completes the project and "turns the keys" over to the owner. In a turnkey project, the cost of financing the project until turnover is borne by the turnkey contractor, and all such costs are included in the contract sum. Monthly requisitions are not submitted to the owner for payment; but once the project has been completed and accepted, the contractor receives payment in the amount of the contract or adjusted contract sum. A turnkey contract can be used for a variety of types of construction projects. The author prepared a joint venture proposal for a private university desirous of building a new dormitory building complete with furniture and related equipment. Turnkey contracts have been used successfully in certain types of public housing projects where, in competition with others, developers select a site and submit a proposal to construct the required type and number of housing units on that site. Some turnkey contracts require the builder to not only construct the project but also provide interior equipment and furnishings.

JOINT VENTURE AGREEMENT

THIS AGREEMENT made and entered into this ________ day of ________, 19___ by and between

WITNESSETH:

WHEREAS, ____ (hereinafter Owner) has advertised for bids for the construction of ____

____

____

which bids are to be submitted on or about ____

and

WHEREAS, the parties hereto have agreed to form a Joint Venture to submit a joint bid for and, if possible, to obtain a contract with Owner for such construction.

WHEREAS, the parties desire to enter into a joint Venture agreement in order to fix and define between themselves their respective interest and liabilities in connection with the submission of such bids and the performance of such construction contract in the event that it is awarded to them.

NOW, THEREFORE, in consideration of the mutual promises and agreements herein set forth, the parties hereby agree to constitute themselves as joint venturers for the purpose of submitting a joint bid to Owner for the performance of the construction contract hereinbefore described, and for the further purpose of performing and completing such construction contract in the event that it is awarded to them on such joint bid, and the parties hereby agree that such bid shall be filed and such construction contract, if awarded to them, shall be performed and completed by them as a joint venture subject to the following terms and conditions:

1. NAME AND SCOPE OF JOINT VENTURE

1.1 The bid shall be filed and the construction contract, if awarded to the parties hereto, shall be entered into in the names of the parties as joint venturers, and the obligations of the parties under such bid and construction contract shall be joint and several. Such construction contract, if awarded to the parties hereto based on their joint bid, shall be carried out and performed by them in the name of "____, a Joint Venture" (hereinafter Joint Venture) all money, equipment, materials, supplies and other property acquired by the Joint Venture shall be held jointly in that name.

1.2 No payment shall be made by the Joint Venture to any of the parties hereto in reimbursement of expenses incurred by such parties in connection with the preparation of the bid for and securing the award of the construction contract, unless otherwise agreed in writing.

1.3 It is the intent of the parties hereto that the joint bid contemplated and provided for herein shall be satisfactory and acceptable to both parties. If the parties are unable to agree upon a joint bid this Joint Venture shall terminate upon written notice by the dissenting party to the other party or parties, which written notice shall be delivered to the other party or parties prior to the time of the bid, and, in such event no party shall have any liability to any other party or parties.

1.4 This Agreement shall not be interpreted or construed so as to extend beyond the submission of such joint bid and the performance of such construction contract, nor to create any permanent partnership or permanent Joint Venture between the parties and shall not limit any of the parties in their right to carry on their individual businesses for their own benefit, including other work for the Owner.

2. PROPORTIONATE SHARES

2.1 Except as otherwise provided in Paragraphs 4.2 and 4.3 hereof, the interest of the parties in any profits and assets and their respective shares in any losses and liabilities that may result from the filing of such joint bid and/or the performance of such construction contract, shall be as follows:

______________________________, ________%

______________________________, ________%

______________________________, ________%

______________________________, ________%

with such percentages being referred to hereinafter as the respective party's Proportionate Share.

2.2 It is the intention of this Agreement, and the parties hereby agree that in the event of any losses arising out of, or resulting from the performance of said construction contract, each party hereto shall assume and pay its Proportionate Share of such losses. If for any reason any one of the parties hereto sustains any liabilities or is required to pay any losses arising out of or directly connected with the performance of such construction contract, or the execution of any surety bonds or indemnity agreements in connection therewith, which are in excess of its Proportionate Share in the losses of the Joint Venture, the other party or parties shall reimburse such party in such amount or amounts as the losses paid and liabilities assumed by such party exceed its Proportionate Share in the total losses of the Joint Venture, so that each member of the Joint Venture will then have paid its Proportionate Share of such losses; and to that end each of the parties hereto agrees to indemnify the other party or parties against and to hold it harmless from any and all losses of said Joint Venture that are in excess of such party's Proportionate Share. Provided, however, that the provisions of this sub-paragraph shall be limited to losses that are directly connected with or arise out of the performance of said construction contract and the execution of any bonds or indemnity agreements in connection therewith, and shall not relate to or include any speculative, prospective, incidental or indirect consequential losses that may be sustained or suffered by any of the parties hereto.

2.3 The parties shall, from time to time, execute such applications for bonds and indemnity agreements, and other documents and papers as may be necessary in connection with the submission of said joint bid for and the performance of such construction contract. Provided, however, that the liability of each of the parties hereto under any agreements to indemnify a surety company or surety companies shall be equal to the Proportionate Share of each of said parties in the Joint Venture.

Figure 2-1 Joint venture agreement.

2.4 If any of the parties hereto is a subsidiary of another corporation, the performance by such subsidiary of the obligations assumed hereunder shall be guaranteed by the parent corporation of any such subsidiary.

3. MANAGEMENT OF JOINT VENTURE

3.1 Authority to act for and bind the parties to this Joint Venture in connection with all or any part of the performance of said construction contract shall be vested in the Management Committee, which may, from time to time delegate all or any part of such authority to one of the parties and/or to any individual or individuals upon unanimous consent of the parties. Neither the Management Committee nor any party hereto shall have the authority to act for or bind any other party except in connection with the performance of said construction contract. Except as provided in Paragraphs 4.2 and 4.3 each party shall have a voice in the Management Committee equal to its Proportionate Share. Except as otherwise noted herein the Management Committee may act upon consent of the party or parties having a Proportionate Share or Shares totalling more than fifty percent (50%). The parties hereby designate the following indviduals to represent them on the Management Committee:

PARTY	REPRESENTATIVE
____________________	____________________ Representative Alternate
____________________	____________________ Representative Alternate
____________________	____________________ Representative Alternate

Such designations may be changed by any party at any time upon written notice to the other parties by the Chief Executive of such party.

3.2 ____

is hereby designated as the Managing Party, subject, however, to the superior authority and control of the Management Committee. The Managing Party, shall appoint the Project Manager through whom the Managing Party shall have direct charge over and supervision of all operational matters necessary to and connected with the performance of said contract, except as otherwise provided herein.

3.3 Any delegation of authority to any individual or party may be revoked by majority vote of all the parties; provided, however, that if the authority of the individual serving as Project Manager is revoked, the Managing Party shall have the right and obligation to appoint another individual to serve in that capacity who is acceptable to the parties hereto.

3.4 Decisions (or, at their option, the establishment of Guidelines to be followed by the Managing Party) regarding the following matters shall be made upon the unanimous vote of the Management Committee:

a.) Significant financial matters such as borrowing, debt guarantees, lease commitments and the investment policy with respect to Joint Venture funds.

b.) The sale of Joint Venture assets; including the terms of such sale and the agent therefor, if any.

c.) Settlements, in excess of a threshold amount set by the Management Committee, of claims or litigation by or against the Joint Venture.

d.) Transactions between any joint venturer and the Joint Venture, including agreements, if any, concerning rates of payment or reimbursement for employees, equipment, temporary or permanent materials, or management, data processing and/or other services.

3.5 The Managing Party shall be designated as the "tax matters partner" (as said term is used in sections 6221 through 6232 of the Internal Revenue Code) for the Joint Venture.

3.6 Management Committee meetings shall be held as needed, but in no event less frequently than once every three months. Job progress reports, a recently updated construction schedule, and the most recent copies of the financial reports described in Paragraph 5.2 shall be presented and reviewed by the parties at such meetings. Any other matters of interest to the Joint Venture may be investigated at, or as a result of, such meetings. Any party may, upon written notice fifteen (15) days in advance of same, call a Special Meeting of the Management Committee.

3.7 Notwithstanding the provisions of Paragraph 3.8, if any dispute between the parties affects or threatens the orderly or timely progress of the Work, the Joint Venture shall proceed with the Work as directed by the Managing Party in writing, whose decision with respect to matters affecting the prosecution of the Work shall be final and binding unless an objecting party provides written notice of its objections within twenty (20) days after receipt of the Managing Party's written directive. In no event shall any dispute be permitted to delay the progress of the Work.

3.8 In the event of any dispute, including those which have been the subject of a formal objection pursuant to Paragraph 3.7, the parties shall exhaust every effort to settle or dispose of same. If, after the Chief Executive Officers of all of the parties have met on no less than two separate occasions in an attempt to settle or dispose of such dispute, then such dispute shall be settled by arbitration under the American Arbitration Association Construction Industry Rules, and judgement upon the award rendered by the arbitrator (s) may be entered in any court having jurisdiction, and the arbitration decision shall be final and binding on the Joint Venture and on all parties. The venue of such an arbitration shall be ____________________.

4. CAPITAL CONTRIBUTIONS AND DEFAULT

4.1 The Management Committee shall from time to time determine the amount of working capital required to carry out and perform said construction contract, and each party shall contribute its Proportionate Share of such working capital whenever requested to do so. Such contributions shall be made within ten (10) days after request therefore.

4.2 If any party fails or is unable to provide its proportionate share of the funds required by the Joint Venture, the interest of said party in the return of investment and profits of this Joint Venture shall be decreased to the proportion that the amount actually provided by it bears to the total amount of the funds provided by all parties, and the interest of any party which may have contributed more than its proportionate share of such funds shall be increased in the same proportion. Nothing contained herein shall increase or decrease the proportionate liability of the parties hereto for losses suffered or sustained by the Joint Venture. The amount unpaid, plus simple interest which shall be charged at the rate per year of 3% above the prime rate of interest charged by the Morgan Guaranty & Trust Company of New York (but not exceeding the maximum allowed by law), shall continue to be a charge against the defaulting party until repayment. It is understood that the subsequent payment of working capital in arrears by any party hereto, which has failed or refused or was unable to contribute its appropriate share of the working capital and funds, shall not cure the default of such party, except by the express written consent of the other parties hereto not so in default.

Reduction in a defaulting party's share of the profits and increases in the share of the other parties shall be calculated as of the time of each default in contributions and as of the time of excess contributions by other parties. The profit shares as so adjusted may be further adjusted to reflect any subsequent default or excess contributions.

Figure 2-1 *(Continued)*

The defaulting party shall have no representative on the Management Committee and shall have no right to participate in the affairs of the Joint Venture until either (1) all of the defaulted contributions and default interest have been paid to the Joint Venture or (2) distributions to the non-defaulting parties have included repayment of all of the excess contributions and payment of all default interest.

4.3 If any party hereto shall dissolve; become bankrupt, or shall file a voluntary petition in bankruptcy, the remaining party or parties shall do all things necessary to wind up the affairs of this Joint Venture, including the completion of said construction contract, the collection of all monies and property due to the Joint Venture, the payment of all debts and liabilities of the Joint Venture, and the distribution of its assets. Such dissolved or bankrupt party shall have no further voice in the performance of said construction contract or in the management of the Joint Venture, nor shall any trustee, legal representative, or successor of any type. The participation of such dissolved or bankrupt party, or its representative, in the profits of the Joint Venture shall be limited to that Net Proportion (Contributions of defaulting party minus default interest charges) which the contributions of such party to the working capital of the Joint Venture bears to the total contributions to such working capital made by all of the parties, but such dissolved or bankrupt party and its representatives shall be charged with and shall be liable for its full share, as fixed in Section 2 hereof, of any and all losses that may be suffered by the Joint Venture under said construction contract, or any additions or supplements thereto or modifications thereof.

5. BANKING AND ACCOUNTING

5.1 All contributions of working capital made by the parties hereto, and all other funds received by the Joint Venture in connection with the performance of said construction contract, shall be deposited in such bank or banks as the parties may designate in separate bank account(s) bearing the name of this Joint Venture. Withdrawals of such funds may be made in such form and by such persons as the parties may from time to time direct. All persons authorized to draw against the funds of the Joint Venture shall be bonded in such company or companies and in such amounts as the parties shall determine.

5.2 Separate books of account of the transactions of the Joint Venture shall be kept in accordance with generally accepted accounting principles. Such books, and all records of the Joint Venture, shall be available for inspection by any party at any reasonable time. Periodic audits shall be made of such books at such times and by such persons as the parties may direct, with a certified audit performed annually (unless otherwise agreed, in writing, by the parties hereto) and copies of the audit reports shall be furnished to each party. Monthly financial statements and cost reports shall be prepared, with contract profit reported on a Percentage-of-Completion method. No less frequently than every three months forecasts of cash flow, final contract revenue, cost and profit and reports setting forth the status of change requests, shall be prepared and copies furnished to each party. Upon completion of the construction contract, a final audit shall be made and copies of such audit report shall be furnished to each of the parties.

6. MISCELLANEOUS

6.1 The parties may determine from time to time during the course of this Agreement that some of the assets held and acquired by the Joint Venture may be divided among or paid to the parties in accordance with their Proportionate Shares. Upon the completion of the construction contract, the assets held and acquired by the Joint Venture shall be divided between the parties and the profits or losses accrued in the performance of said contract shall be divided between or paid by the parties, as the case may be, in accordance with the terms of this Agreement, and this Agreement shall then terminate; provided, however, that if claims of any nature or legal action of any type are brought against the Joint Venture or any of the parties at any time after such distribution by any third party or parties not signatory to this Agreement and such claims and/or legal action relate to or arise out of this Agreement, the performance of the Construction Contract and/or the work product thereof, this Agreement shall be considered to have remained in full force and effect and the rights and obligations of the parties hereto with respect to such matters shall be determined by this Agreement, the passage of time notwithstanding.

6.2 The interests and rights of each party in this Joint Venture shall not be transferable or assignable, except that any party may assign its share in any money to be received by it from the Joint Venture for the purpose of obtaining a loan or loans from any bank or other lending agency.

6.3 The scope and limits of insurance which shall be obtained by the Managing Party on behalf of the Joint Venture and, as appropriate, the joint venturers, individually, shall be as mutually agreed by the parties. Said insurance program shall not necessarily be limited by the minimum requirements set forth in the contract with the Owner and shall clearly define what liabilities, if any, are to be insured against by each joint venturer.

6.4 This Agreement shall be construed and governed by the laws of ____________.

6.5 The following Attachments are attached hereto and made a part hereof:

A. Policy Statement on Business Conduct

IN WITNESS WHEREOF, the parties have caused this Joint Venture Agreement to be executed by their duly authorized officers or agents on the date first written above.

ATTEST:

____________________ ____________________

ATTEST:

____________________ ____________________

ATTEST:

____________________ ____________________

Figure 2-1 *(Continued)*

Contracts with Government Agencies

Many local, state, and federal agencies have their own contract forms. Although these forms may borrow heavily from many standards of the American Institute of Architects or Associated General Contractors of America (AGC), they also include pages of various local, state, and federal laws and ordnances and executive orders. Under the canopy of *equal employment opportunity* (*EEO*), provisions for fulfilling requirements for *disadvantaged business enterprises* (*DBEs*), *minority business enterprises* (*MBEs*), *women-owned business enterprises* (*WBEs*), war veterans, and the handicapped may also be included in supplemental or special conditions in print so small as to defy readability, but nevertheless needs to be read, understood, and complied with.

Minimum hourly rates for skilled and unskilled labor are most likely established and included in the contract documents and certification of compliance required (Fig. 2-2), with the submission of weekly certified payroll costs on special forms provided in the bid documents (Fig. 2-3). These certifications, following the provisions of the Davis-Bacon Act, are to be strictly followed since falsification is a violation of federal law.

For the project superintendent embarking on the first public works project, it is critical to read all the bid documents, contract, and specification boilerplate. Noncompliance with some of those provisions contained in the documents may result in nonpayment of monthly requisitions, penalties of various sorts, and the potential for violation of public laws, resulting in fines and even jail.

In the special, general, and supplementary conditions, there are stipulations that will affect the way in which negotiations are to be concluded with subcontractors and methods by which EEO requirements are to be met. Some requirements may dictate how a site logistics plan is to be established, and other requirements may govern the various closeout procedures and requirements that must be addressed even before the job begins.

The Notice to Proceed in a Public Works Project

This document, generally in the form of a letter, sent to the contractor by the government agency is the official notification of the starting date of the contract—the date from which the contract time will be charged.

In some cases there are two notices to proceed. The first one is issued for mobilization of the contractor's field office (the contract time clock generally does not start with this notification), and the second one stipulates when the contractor is to commence construction—and this one *does* start the contract clock.

Public Works Provisions That Can Affect Subcontractor Negotiations

Prompt-payment provisions are being included in an increasing number of public works projects. The pay-when-paid clause in most general contractor-subcontractor agreements is therefore nonenforceable when these prompt-payment clauses exist. These prompt-payment provisions typically state that the general contractor is obligated to pay the subcontractors within 30 days of receipt of payment, and these subcontractors, in turn, must pay their subcontractors within 30 days of receipt of their payment. An exception can be made if payment is being withheld pending resolution of a dispute or claim documented by the general contractor.

Wage and Hour Division

Budget Bureau No. 44-R1093

STATEMENT OF COMPLIANCE

Date ____________

I, ______________________, ______________________ do hereby state:
(Name of signatory party) (Title)

(1) That I pay or supervise the payment of the persons employed by __________________ on the __________________;
(Contractor or Subcontractor)

that during the payroll period commencing on the _____ day of ______, 19__ and ending the ___ day of _________, 19__, all persons employed on said project have been paid the full weekly wages earned that no rebates have been or will be made either directly or indirectly to or on behalf of said ______________________ from the full weekly wages earned by any person and that no deductions have been made either
(Contractor or Subcontractor)

directly or indirectly from the full wages earned by any person, other than permissible deductions as defined in Regulations, Part 3 (29 CFR Subtitle A), issued by the Secretary of Labor under the Copeland Act, as Amended (48 Stat. 948, 63 Stat. 108, 72 Stat. 967; 76 Stat. 357; 40 U.S.C. 276c), and described below:

(2) That any payrolls otherwise under this contract required to be submitted for the above period are correct and complete; that the wage rates for laborers or mechanics contained therein are not less than the applicable wage rates contained in any wage determination incorporated into the contract; that the classifications set forth therin for each laborer or mechanic conform with the work he performed.

(3) That any apprentices employed in the above period are duly registered in a bona fide apprenticeship program registered with a State apprenticeship agency recognized by the Bureau of Apprenticeship and Training, United States Department of Labor, or if no such recognized agency exists in a State, are registered with the Bureau of Apprenticeship and Training, United State Department of Labor.

(4) That:

(a) WHERE FRINGE BENEFITS ARE PAID TO APPROVED PLANS, FUNDS, OR PROGRAMS

☐ In addition to the basic hourly wage rates paid to each laborer or mechanic listed in the above referenced payroll payments of fringe benefits as listed in the contract have been or will be made to appropriated programs for the benefit of such employees, except as noted in Section 4(c) below.

(b) WHERE FRINGE BENEFITS ARE PAID IN CASH

☐ Each Laborer or mechanic listed in the above referenced payroll has been paid as indicated on the payroll, an amount not less than the sum of the applicable basic hourly wage rate plus the amount of the required fringe benefits as listed in the contract, except as noted in section 4(c) below.

(c) EXCEPTIONS

EXCEPTIONS (CRAFT)	

Remarks

Name and Title	Signature

The wilful falsification of any of the above statments may subject the contractor or subcontractor to civil or criminal prosecution. See section 1001 of title 18 and section 231 of title 31 of the United States code.

Form WH-348 (1/68) Purchase this form directly from the Supt. of Documents

Figure 2-2 Statement of compliance.

U S DEPARTMENT OF LABOR
WAGE AND HOUR DIVISION

Form Approved
Budget Bureau No. 44-R1093

PAYROLL
(For Contractor's Optional Use; See Instruction, Form WH-347 Inst.)

NAME OF CONTRACTOR ___ OR SUBCONTRACTOR ___	ADDRESS

PAYROLL NO.	FOR WEEK ENDING	PROJECT AND LOCATION	PROJECT OR CONTRACT NO.

(1) NAME, ADDRESS AND SOCIAL SECURITY NUMBER OF EMPLOYEE	(2) # OF EXEMP TIONS	(3) WORK CLASSIFICATION	OT OR ST	(4) DAY AND DATE / HOURS WORKED EACH DAY							(5) TOTAL HOURS	(6) RATE OF PAY	(7) GROSS AMOUNT EARNED	(8) DEDUCTIONS FICA	WITH HOLDING TAX			OTHER	TOTAL DEDUCTIONS	(9) NET WAGES PAID FOR WEEK
			O																	
			S																	
			O																	
			S																	
			O																	
			S																	
			O																	
			S																	
			O																	
			S																	
			O																	
			S																	
			O																	
			S																	
			O																	
			S																	
			O																	
			S																	
			O																	
			S																	
			O																	
			S																	

FORM WH-347 (1/68)

Figure 2-3 Certified payroll form.

It is always good practice to document any disputes or claims involving subcontractors where the withholding of funds is being considered. Where these prompt-payment clauses are enacted into law, the project superintendent must notify the office of any reasons for withholding payment and provide the necessary documentation to substantiate the action.

The special conditions portion of the contract specifications may include limitations on fees that can be applied to change order work, and these restrictions need to be incorporated into any subcontractor negotiations. A typical requirement may be as follows:

> On work performed by a general contractor with his or her own forces, their allowance for overhead and profit will be as follows:
>
> | | |
> |---|---|
> | For amounts up to and including $5,000 | 15 percent overhead and profit |
> | For amounts between $5001 and $25,000 | 10 percent overhead and profit |
> | For amounts exceeding $25,000 | 5 percent overhead and profit |
>
> For work performed by subcontractors, the general contractor is allowed to include 5 percent overhead and profit.
>
> Subcontractors are allowed the same overhead and profit as outlined above for the general contractor performing work with the GC's own forces.
>
> Total subcontractor fees, including those of their subcontractors, cannot exceed the amounts stipulated above.

Requirements for payment for off-site storage of materials and equipment are frequently spelled out in these government contracts, and the methods by which payment is allowed are another important consideration when you are negotiating subcontract agreements. If materials or equipment is stored in a location any distance from the job site and off-site payment has been approved, there may be provisions in the general, supplemental, or special conditions requiring the contractor to reimburse the government-appointed inspector for any costs incurred to travel to the area and inspect the item being requisitioned.

Subcontractors need to be made aware of these provisions at the time of contract negotiations to avoid misunderstandings at a later date. The project superintendent should never assume that the subcontractor has read and understood all these miscellaneous provisions. It has been the author's experience that very few subcontractors thoroughly read all these special, general, and supplementary conditions; and when this is brought to their attention during the progress of the project, they seem upset that they are expected to abide by such provisions.

Change Order Clauses in Government Contracts—Enrichment and Betterments

Some contracts may contain clauses stating that no changes other than those for project *enrichment* or extra work ordered by the owner's representative or architect will be approved. The term *enrichment* can have one meaning for the owner but a different one for the contractor.

For example, prior to the issuance of a building permit, plans and specifications are presented to, and reviewed by, local building officials and the fire marshal. If they are approved on the basis of this initial plan and specification review, a building permit is issued; it signals that all local and state requirements have been met and construction can begin. As the building nears completion, however, local building officials and the fire marshals visit the site to begin a series of final inspections. And often these officials may point out the need to install more exit lights or relocate ones already installed or to add more heat or smoke detectors or to install a few more emergency lights or possibly another floor drain in the kitchen, and so forth.

Typically this means that the subcontractor will request extra money to perform this work, and the general contractor will feel justified in preparing a change order for the owner—who will be reluctant to recognize these costs as *extra* to contract. Their argument will be based on the fact that these added items, and their costs, do not represent *enrichments,* but merely compliance with local and state requirements.

Payment of enrichments can be defended on the basis that if they had not been installed, no *certificate of occupancy* (*C of O*) will be issued, and without the C of O the building cannot be occupied and is therefore worthless. This argument can also be used when you meet resistance from owners in the private sector as well, even though there is no enrichment clause in their contract. (*Note:* The provisions of AIA Document A201, 1997 edition, specifically mention that the contractor is *not* required to ascertain that the contract documents are in accordance with building codes. There is more on this subject in Chap. 3 under general conditions review.)

Administering contracts with public agencies can be a demanding task. The project superintendent should be thoroughly familiar with all phases of the contract requirements because they may be enforced to the letter.

Looking for Those Onerous Provisions in the Contract with the Owner

Although the plans and specifications can contain some rather strict requirements or procedures, the contract for construction is where most of the onerous provisions find a home.

Rare is the owner who merely fills in the blanks in a contract form and attaches a standard AIA A201—General Conditions document to it. Owners will invariably amend both documents, adding pages of small print in the form of exhibits, special conditions, or supplementary conditions—and that is where may traps lie for those who don't take the time to read and understand *all* the terms of the agreement. Here is a sample list of some of the onerous provisions frequently added by owners and the sections of the general conditions contract, as amended, where they may be found:

Article 3—Contractor

1. With respect to shop drawings: "The contractor shall submit complete and accurate submittal data at the first submission. If the submittal is returned requiring re-submittal, only one (1) additional submittal will be reviewed at the Owner's cost. Any additional submittals will be reviewed at the Contractor's cost."
2. Schedules: "If any of the work is not on schedule, the Contractor shall immediately advise the owner, in writing, of proposed action to bring the Work back on schedule. In such event the owner will require the Contractor to work such additional time over regular hours, including Saturdays, Sundays and holidays, at *no* additional cost to the Owner, to bring the Work back on schedule."
3. Schedules: "If the contractor fails to take prompt and adequate corrective action (to get the project back on schedule) to the Owner's satisfaction, the owner reserves the right to perform such work as it deems necessary and to back charge the cost thereof against payments due the Contractor."
4. Contractor responsible for details not shown on the drawings: "The contractor has constructed several projects of this type and has knowledge of the construction and finished product." (If some minor portions of work are omitted from the contract documents, and as the contractor, you are purported to have constructed several similar structures previously, you may be required to perform any such extra work at no cost to the owner.)

Article 4—Administration of the Contract

1. Extension of time and related costs: "Any extension of time in which to complete the Work granted by owner for items beyond the Contractor's control shall be the sole remedy for any delay, hindrance in performance of the Work, loss in productivity, impact damages or similar claims." This is known as a no damages for delay clause.
2. Deletion of the arbitration clause: "No arbitration clause is enforceable by any party under the contract documents. It is the intention of the parties under the contract documents that all disputes that cannot be resolved by negotiations shall be resolved by appropriate proceedings at law or in equity. "This puts a clamp, somewhat, on resolving small contractor claims quickly and economically. Are you willing to sue an owner for $15,000 when the lawyer's fee may meet or exceed that number?

Article 7—Changes in the Work

1. "The owner at all times shall have the right to participate directly in the negotiations of Change Order requests with subcontractors and Material Suppliers."
2. "If the Owner and Contractor are unable to agree on the amount of any cost or credit to the Owner resulting from a change in the work, the Contractor shall promptly proceed with, and diligently prosecute, such change in the work and the cost or credit to the owner shall be determined on the basis of reasonable expenditure and savings." This means that you cannot refuse to perform extra work even if the cost of this work is in dispute prior to starting the work.

Article 9—Payments and Completion

1. Definition of *substantial completion.* This phase adds yet another requirement to meet: "Notwithstanding anything contained in the contract documents, the Work shall not be deemed substantially complete unless and until a Certificate of Occupancy is issued."
2. "Unless otherwise agreed to in writing by the Owner, the project shall not be considered substantially complete if the items on the Punch List would reasonably require more than two calendar weeks to complete."

Article 13—Miscellaneous Provisions

1. Applicable primarily to work in buildings being renovated or rehabilitated: "The contract shall review the structural capability of the structure prior to allowing installation of temporary lifting devices or staging equipment or the temporary off-loading and storage of materials. Costs associated with the architect's review or re-design of structure to accept the temporary construction loading shall be borne by contractor." In other words, you break it and you own it.

Bonds

Although the project superintendent's duties and responsibilities may not extend to issues relating to bonds, a general understanding of the types of bonds and their functions may be of value.

The high-risk nature of the construction industry, coupled with the ever-changing financial stability of individual contractors and subcontractors, creates the need to provide a layer of assurance to project owners against the possibility of default by the builder prior to completion of the project. Bonds provide this assurance.

Three Basic Types of Bonds

The Bid Bond

Requirements for bid bonds will be clearly indicated in the bid documents and generally require a bid security in an amount equal to not less than 10 percent of the bid amount. The purpose of the *bid bond* is to assure the owner that when the low bidder has been determined, the low bidder will be able and willing to proceed to enter into contract. The bid bond gives the owner protection against a contractor's declining to accept a contract after being declared low bidder. If such a situation occurs, then the contractor will forfeit his or her bid bond and all proceeds, up to the dollar value of the bond, will be used to cover any losses if another bidder is awarded a contract.

Payment Bonds

The *payment bond* is sometimes referred to as a *labor and material bond* because its purpose is to ensure that the contractor will pay the subcontractors and material/equipment suppliers and all labor costs for work incorporated into the structure. The penal sum of the payment bond is usually the amount of the contract and requires the principal (contractor) to pay for all labor and materials promptly. If subcontractors or vendors are not paid within 90 days after they last worked on the project or delivered materials or equipment to the project, these claimants may *sue on the bond,* in other words, call the bond.

Performance Bonds

The *performance bond* assures the owner that the contractor (principal) will promptly and faithfully perform in accordance with the terms and conditions of the contract. If the contractor does not perform, the owner (obligee) will notify the bonding company that will investigate the nonperformance. If the builder refuses or is unable to meet the performance standards incorporated in the contract, the bonding company may elect to engage another contractor to complete the work and use the proceeds from the defaulting contractor's performance bond to pay all costs to complete the work, up to the penal sum of that contractor.

Some Bond Terminology

Calling the bond—notification to the bonding company that the contractor or subcontractor has failed to live up to the commitment as stated in the contract and the bonding company (surety) is being requested to provide sufficient funds to cover these unsatisfied commitments.

Consent of surety—When a construction project has been successfully completed, all bills have been paid, and the provisions of the bond have been met, the contractor will request that the project owner sign off the bond so that the bonding company can be notified that all conditions have been met and the bond can be effectively terminated.

Dual obligee—When two parties have a financial interest in the project, such as the owner and the owner's lending institution, the bonding company will have a financial obligation to these two entities.

Guarantor—the underwriter or surety company.

Obligee—the project owner and others if there are dual obligees.

Penal sum—the amount of the bond (generally the amount of the contract).

Premium—the cost of the bond.

Principal—the contractor.

Surety—the bonding company (not the insurance agency transmitting the bond).

Other Types of Bonds

There are *maintenance bonds* which provide an owner with assurance that the contractor will provide the required maintenance stipulated in the contract for the period of time so stated.

There is a *supply bond* which is usually required when special materials or equipment crucial to the completion of the project is needed. Custom-made equipment, possibly manufactured overseas, often requires supply bonds which include the cost to the project if the equipment is lost or damaged in transit, say, if the boat on which it was stored sank, or the train on which it was being transported was derailed.

A *lien bond* indemnifies the owner against the cost to remove liens filed against the property.

There are also license or permit bonds required by various government agencies. Many states require contractors to post bonds before they are given licenses to operate as contractors. Certain contractors such a road builders or excavating contractors are often required to post bonds to ensure that disturbance of an existing public roadway will result in its being repaired/replaced in the exact manner prescribed by that government agency.

Federal Agencies and Bonds

On all federal public works projects, where taxpayer money is involved, the Miller Act mandates that surety bonds on all such contracts in excess of $100,000 be required. Most states have passed similar laws where projects involving state funds are concerned, and these laws are referred to as *little Miller Acts.*

The Letter of Credit

This is another type of safeguard against contractor or subcontractor default. A *letter of credit* is issued by a bank, usually for 10 percent of a contract sum and callable by an owner if a general contractor defaults or by a general contractor if the subcontractor defaults. In case of default, those contractor or subcontractor funds secured by the bank would be dispersed to the owner or general contractor, as the case may be, if and when the default conditions spelled out in the letter of credit occurred. A letter of credit does not guarantee sufficient funds to complete the work, but limits the payment to the sum specified in that instrument. Unlike the bond, a letter of credit does not ensure that subcontractors or vendors will be paid or, in fact, that the project will be completed in accordance with the terms of the contract—it is an instrument that merely passes on a specific sum of money to the party so designated. Letters of credit are often used when an owner or contractor requires some form of financial surety and the requested party does not have bondability or has reached the limit of his or her bonding capacity.

Subcontractor Bonds Required by General Contractors

Surety bonds are effective tools for shifting risks for possible subcontractor failure to a surety company. These surety bonds are three-way instruments: The surety guarantees to the general contractor that the third party, the subcontractor, will perform its obligation under the terms of their subcontract agreement. There are other instruments, such as subcontractor default insurance, that appear to offer the same protection of a bond, but there are distinct differences between the two. The bond covers costs up to the amount of the contract, but default insurance can have deductibles or copayments and defines *cost* and *expenses* above a certain minimum that the contractor must pay as a result of subcontractor default.

The project superintendent will generally not deal with bond issues, except in those cases where subcontractors have been required to submit a bond as part of the terms and conditions of their subcontractor agreement. The superintendent should be alert to any signs that the subcontractor may be failing in the contract responsibilities and alert the project manager to those events that may portend calling the subcontractor's bond.

Chapter

3

General Conditions of the Construction Contract

There is often a great deal of what appears to be *boilerplate* material in the contract for construction and the specifications manual. The tendency is to skip over this material, because it is tiresome and boring; but it should be thoroughly read because it contains valuable information. One such document that the project superintendent will frequently encounter is the *general conditions* to the contract. This document generally takes the form of American Institute of Architects Document A201, General Conditions of the Contract for Construction.

Most construction contracts are supplemented by documents that include general, supplementary, and special conditions. These supplementary contract obligations are frequently incorporated into the project specifications manuals or in stand-alone documents, and the most widely used contract supplement is AIA Document A201, General Conditions of the Contract for Construction. The project superintendent may wonder what kind of document prepared by the project architect will contain contractor-friendly material to help manage the construction project. The answer is that a number of provisions in AIA A201 will prove very helpful to the project superintendent who has read and understood these provisions.

The American Institute of Architects periodically updates its contract forms, the latest of which are the 1997 editions released to the profession in December 1997. The previous issue of AIA A201, General Conditions of the Contract for Construction, was dated 1987 and may still be referenced in some construction contracts.

To the project superintendent who has not taken the time to completely read and understand the provisions to AIA A201, it may come as a shock that this contract document, prepared by an organization of architects, contains quite a few provisions that protect the contractor. Every project superintendent should read this document from cover to cover, at least once. She or he will undoubtedly be rereading selected sections from time to time to support a position or defend against one.

Let's review the 1997 edition of AIA A201 and point out some sections of importance to the general contractor.

Article 1—General Provisions: The Contract Documents

Along with defining the components of the contract documents, section 1.5.2 requires the contractor to stipulate that he or she has visited the site and become somewhat familiar with local conditions under which the work is to be performed and concluded personal observations with respect to the contract requirements. This is not a statement to be taken lightly. Quite often a contractor's failure to visit the site and observe conditions that are apparent, even though not specifically shown on the drawings, may hinder her or his ability to claim these conditions as extra to contract and will result in denial of certain types of claims.

If construction has just gotten underway and a site condition not indicated on the drawings is encountered—one that would have been apparent after a visit to the site—it may be very difficult to initiate a claim for extra work on the basis that its condition was not noted in, or on, the contract documents.

For example, if an abandoned well were found in the area of a proposed footing and the well cap or well cover were clearly visible but not shown on the drawings, the architect could invoke the provisions of Article 1 as the reason for disallowing a contractor's claim for additional costs for structural fill required to raise the subgrade under the footing or lower the elevation of the footing to achieve proper bearing. Not only would the abandoned well have been visible, but also a prudent contractor should have noted its location and even removed the well cap/cover to determine its depth and its impact on costs.

Article 2—Owner

One provision in this article directs the owner to designate, in writing, a representative who shall have the authority to bind the owner with respect to matters requiring an owner's approval or authorization. This should speed up the communication process between contractor and owner when field conditions arise that require a prompt owner/architect decision in order to maintain job progress.

This section of the general conditions also stipulates that the owner is obliged to present reasonable evidence that financial arrangements have been made to satisfy the requirements of the construction contract. The contractor may obtain a copy of this financial commitment by writing to the owner and requesting same.

Another provision deals with reproducibles. Unless the contract stipulates to the contrary, the owner shall furnish the contractor, free of charge, sufficient copies of plans and specifications that are *reasonably necessary* for the execution of the work. (Does this include sufficient copies for each subcontractor as well? A strong case could be made for several sets for major subcontractors and one set for minor subcontractors.)

The owner has the right to subcontract portions of the work by giving written notice to the general contractor. But suppose the general contractor is a union contractor and the owner hires, say, a nonunion electrical contractor to install the data communications work? Although the owner has the right to do this, a prudent general contractor should respond by requesting that she or he be held harmless from any jurisdictional labor disputes that may arise out of these arrangements.

Article 3—Contractor

This is a very important article for the general contractor to read and comprehend fully. First, it deals with shop drawings and requires the contractor to take field dimensions of any existing conditions related to the work; any errors, omissions, or inconsistencies discovered shall be reported promptly to the architect. In subparagraph 3.12.4 and 3.12.5, it is stated that shop drawings are not *contract* drawings. (This voids a contractor's argument that approval of a shop drawing is proof that the item has been accepted by the architect. If it has been accepted, but is deemed of lesser quality than the specified item, the architect may request, and in fact is entitled to, a credit.)

Contractors would be wise to read completely subparagraphs 3.12.4 through 3.12.10 relating to shop drawing submissions, review, and disposition.

Second, with respect to errors or omissions noted by the contractor, it is recognized that the drawing review is being performed by a contractor who is not a licensed design professional, and the contractor is not required to ascertain that the plans and specifications are in conformance with laws, statutes, ordinances, building codes, and rules and regulations.

Paragraphs 3.2.2 and 3.2.3 of this article absolve the contractor for damages resulting from errors, inconsistencies, or omissions in the contract documents or for differences between field measurements and conditions and the contract documents, unless the contractor recognized such error, inconsistency, omission, or difference and failed to report it to the architect.

Article 3 provides the contractor with responsibility and control over construction means, methods, techniques, and sequences unless the contract documents dictate otherwise. In another section, it is restated that the contractor has no responsibility to ascertain that the contract documents are in accordance with applicable laws, statutes, ordinances, building codes, and rules and regulations. *Superintendents take note:* This may diffuse arguments about responsibility for costs for those last minute additions to the scope of work when inspecting building officials require more exit lights, fire alarm devices, etc. The contractor is obligated, in this section of the general conditions, to submit a construction schedule for the architect's information. (Previous issues of this A201 document required submission of a schedule for the architect's approval.)

The question of costs to be included in an allowance is often raised by the contractor. Does an allowance include the contractor's overhead and profit?

Article 3.8 answers this question and others concerning the reconciliation of an allowance item. Unless otherwise stated in the contract, an allowance includes

1. Cost to the contractor of all materials, labor, and equipment for the item, delivered to the site and including taxes and trade discounts.
2. Contractor's cost to unload and handle and all related labor and installation costs
3. Contractor's overhead and profit for the allowance are to be included in the contract sum and not in the allowance category.

If the allowance is more or less than stated in the contract, the contract sum is to be adjusted by change order and shall reflect the difference between actual cost and the allowance.

And lastly, Article 3 states that the contractor shall not be required to provide professional services which constitute the practice of architecture or engineering unless specifically called for in the contract.

Article 4—Administration of the Contract

This article repeats, once more, the contractor's right to control construction means, methods, techniques, sequences, or procedures since these are "solely the Contractor's rights and responsibilities."

The architect is charged with the duty to review shop drawings with reasonable promptness. (The exact time frame for shop drawing review is frequently spelled out in other parts of the contract documents, i.e., special, supplementary conditions, but the word *reasonable* does restrict the review time frame to some arguable degree.)

The architect is also charged with the authority to interpret and decide matters concerning performance or requirements of the contract documents or interpretations and decisions consistent with the intent of or reasonably inferable from the contract documents. (*Note:* There appears to be a lack of checks and balances in this arrangement. The person who prepared the plans and specifications is given authority to interpret these documents and rule on their intent? This is like asking the fox to watch the chicken coop!)

Article 4.2.7 stipulates that the architect approval of a specific item does not indicate approval of an assembly to which the item is a component. This provision could apply to a value engineering proposal presented by the contractor in which one component of an assembly is substituted and approved, but its substitution invalidates the assembly. For example, the substitution of the type or gauge of roof coping or flashing may be approved by the architect, but will not meet the manufacturer's recommendations; hence a roofing bond will not be issued.

Article 4 includes procedures for filing a claim and sets a time limit of 21 days after the occurrence of the event as the time frame within which the claim must be filed. Claims for concealed conditions or unknown conditions are set forth in subparagraph 4.3.4, which is a very important section to read and understand. The sentence about *differing conditions* may be helpful for those contractors filing a claim for unsuitable soils or excessive rock when actual conditions encountered differ *materially* from those in the geotechnical report accompanying the bid documents. When the contract includes all site work as *unclassified,* this, in effect, means that whatever deleterious or unsuitable material the contractor uncovers is to be removed and replaced with acceptable fill, at no cost to the owner. However, if such a claim is based upon conditions uncovered that vary *significantly* from those anticipated after a reasonable review of the geotechnical report, invoking the differing-conditions provisions of subparagraph 4.3.4 may allow the contractor to recover some of these extra costs.

Another important part of this article is subparagraph 4.3.10, which disallows any claims for consequential damages. This theoretically denies the contractor the right to use the Eichleay formula (fully discussed in Chap. 10), which is a method to arrive at the contractor's attempt to recoup unabsorbed or underabsorbed corporate overhead costs primarily in delay claims.

Dispute resolution procedures are included in Article 4, and mediation is listed as the first step in pursuing claims resolution. If mediation fails to resolve the claim, arbitration is the next step in the process, and several paragraphs in this section outline the steps to be taken to commence mediation and arbitration.

Subparagraph 4.5.4 is entitled "Limitation on Consolidation or Joinder," which may be of interest to general contractors if they have reached the arbitration stage. The term *joinder* means to join with. If, for example, both the general contractor and the subcontractor have a claim, ostensibly, against the owner, they cannot join in one arbitration proceeding but must pursue their claims individually. The subcontractor would file for arbitration against the general contractor, and the general contractor would file for arbitration against the owner, as the case may be.

Article 5—Subcontractors

The general contractor is directed, in this section of A201, to submit in writing a list of proposed subcontractors to the architect, who is to promptly review and reply whether the owner has any objections to any of the subcontractors so presented.

The contractor is cautioned not to proceed to contract with a subcontractor to whom the owner/architect has objected; however, if the rejected subcontractor could have been deemed reasonably capable of performing the work, the contract sum will be increased or decreased by the substitution of an owner-acceptable subcontractor. If the general contractor is considering invoking this provision, the GC should request a written statement from the owner listing the owner's reasons for the rejection of any subcontractor. If the general contractor does not agree with the owner's criteria for selection, the GC should respond to the owner, by letter, stating his or her objections to the choice, properly and completely documented. If problems arise when the owner's selected subcontractor is working on the job site, this letter of concern sent by the general contractor may prove helpful.

The author had such an experience when requested to obtain bids on low-voltage alarm systems for a senior living community. The owner rejected all subcontractors' bids and requested that a bid be solicited by a low-voltage systems supplier whom the owner had employed on several recent projects in other parts of the country. This *preferred* supplier failed to submit the final estimate for the work in a timely fashion, failed to submit a proper schedule of values for comparison with the other bidders, and failed to deliver materials and equipment in a timely fashion. It was apparent that this vendor felt that his or her preferred position with the owner would cover a multitude of sins, until the author sent a letter to the owner, advising of serious delays in the offing unless the *owner's vendor* met the current project's demands. Within a day or two the much needed equipment arrived at the job site.

Other provisions in this article require the contractor to bind all subcontractors to the contractor by the terms of the contract documents and to assume all obligations and responsibilities that the contractor maintains toward the owner. This is a standard *pass-through* provision which is actually beneficial to the general contractor.

According to Article 5.4, Contingent Assignment of Subcontracts, the general contractor must include in the subcontract agreement a provision allowing the assignment of the subcontract to the owner under specific conditions are spelled out in detail in this subparagraph. This affords the owner some degree of protection if the general contractor defaults on the contract. Rather than renegotiate with existing subcontractors to complete the work, possibly at much higher costs, the owner merely "takes over" all active subcontract agreements.

Article 6—Construction by Owner or by Separate Contractors

This article gives the owner the right to subcontract certain portions of the work. However, as stated previously, if the general contractor anticipates that a conflict could arise due to collective bargaining agreements, the owner must be advised of these potential problems in writing. Will a nonunion millwork installer working on a union project create a jurisdictional dispute that may lead to a job action and delays that could reverberate through several other trades?

Article 7—Changes in the Work

This provision of A201 deals with change orders and the construction change directive. A *change order,* according to Article 7, is based upon an agreement between owner, contractor, and architect; but a *construction change directive* (*CCD*) requires only agreement between owner and architect inasmuch as the costs associated with these changes may not have been agreed to by the contractor.

If the CCD issued by the architect affects the contract sum, an adjustment in the contract sum will be based upon one of the following methods:

1. A mutual acceptance of a lump sum, properly itemized and documented
2. Unit prices contained in the contract (if, in fact, there are any in the contract)
3. Cost to be determined in a manner agreed upon by all parties as well as a mutually acceptable fixed amount or percentage for contractor's fee and overhead
4. Time and materials cost approach, which is to include
   a. Cost of labor and fringe benefits
   b. Cost of materials, supplies, and equipment, including transportation costs
   c. Rental costs of equipment whether rented or company-owned
   d. Costs of bond, insurance premiums, and any related fees
   e. Costs of supervision and field office personnel directly attributable to the work

Subparagraph 7.3.6 of this article contains a full explanation of the CCD process.

Subparagraph 7.3.8 is also of importance to contractors and states that all amounts in the CCD not in dispute can be included in the current application for payment. This means that an interim billing can be included in a current application for payment for change order work completed and accepted but for which no formal, fully executed change order has been received. In a complicated change order involving participation by multiple trades, the architect or owner may agree with most of the costs submitted but may take exception to, say, the plumber's or electrician's costs. Before this, issuance of this provision payment for an entire change order would be delayed until a few isolated costs were resolved. Subparagraph 7.3.8 allows the general contractor to invoice for all approved costs, receive payment, and pay the respective subcontractors and vendors while providing the owner with additional documentation to support the plumbing or electrical portion of the change order.

Don't forget to consider potential changes in contract time when you are formulating the CCD. Will the contract time remain unchanged, decrease, or increase?

Article 8—Time

The definition of *contract time* is included in this article. Unless stated otherwise, the contract time is the period of time, including authorized adjustments (change orders), required to attain substantial completion of the project. The term *day,* unless otherwise stated, is meant to be a calendar day.

Methods by which delays and extensions of time are to be treated are included in this section of the general conditions.

Article 9—Payments and Completion

The contractor is directed to submit a *schedule of values* for approval by the architect prior to submission of the first application for payment. If there are no objections from the architect, this schedule of values will be used for review of the contractor's monthly requisition requests and serves as the basis for determining percentage of completion for each line item in that requisition.

This article deals with payment for on-site and off-site storage of materials and equipment. Any request for off-site storage payments is to be made to the architect in writing and is conditional upon meeting the following terms and conditions:

1. Presentation of a procedure by the general contractor to ensure that title to the materials and equipment will pass to the owner. This can be accomplished by submitting a bill of sale which will automatically transfer title to the owner once payment is received.
2. Presentation of evidence that the cost of insurance, storage, and subsequent transportation to the site will be paid by the contractor.
3. Furnishing of an insurance certificate documenting that coverage will remain in effect during storage and transportation to the site.

Article 9 includes conditions that allow the architect to withhold certification for payment on the monthly requisition:

1. Defective work has not been repaired/replaced.
2. Third-party claims have been filed.
3. The contractor has failed to make payments to subcontractors or pay for labor, materials, and equipment incorporated into the building.
4. There is reasonable evidence that the work cannot be completed for the unpaid balance of the contract sum.
5. There is damage to the owner or another contractor.
6. There is reasonable evidence that the project will not be completed within the contract time.
7. The contractor has consistently failed to perform the work in accordance with the contract documents.

This article in the general conditions document also directs the contractor to pay each subcontractor, upon receipt of the owner's payment, the amount which each subcontractor is entitled to receive. This, in effect, is confirmation of the standard "pay when paid" clause.

Article 9 also stipulates that the subcontractors may, upon written request to the architect, obtain information regarding percentage of completion paid by the owner to the contractor for their portion of the work. A subcontractor can call the architect and find out if the general contractor has been paid for work that was included in the subcontractor's current requisition and, if so, can invoke the pay-when-paid provision to demand payment.

Article 9 includes the subject of substantial completion, defining both this stage of completion and the payments due to the contractor when this phase of construction has been achieved. Partial occupancy and/or use of the project is described in Article 9.8, and the owner and contractor responsibilities that are to be concluded at that time are spelled out in this section. Last, but not least, final payment procedures are described, stating that no payment will be made until the contractor completes the following procedures:

1. Provision of an affidavit stating that all labor, materials, and equipment incorporated into the building have been paid
2. A certificate evidencing that insurance will remain in effect for 30 days after final payment and will not be canceled prior to that date

3. A written statement that the contractor is aware of no reason that current insurance coverage will not be renewable to cover the period set forth in the contract documents
4. Consent of surety to final payment
5. Release from the general contractor to the owner of any claims, liens, or other encumbrances arising out of the contract

Article 10—Protection of Persons and Property

Safety precautions and contractor procedures to protect persons and property during construction are delineated in this article. Article 10 requires the owner to advise the contractor of the absence or presence of any hazardous materials likely to be encountered on the site. If hazardous materials are discovered on the site, the contractor is to cease work and notify the owner, requesting further direction on how to proceed.

Article 11—Insurance and Bonds

In conjunction with specific limits of insurance usually contained in the bid documents, this article provides more details about insurance coverage. Unless otherwise stated in the contract, the owner will purchase and maintain builder's "all-risk" insurance. When partial occupancy occurs, it shall not commence until the insurance company or companies have consented to partial occupancy, or use, by endorsement or otherwise.

Article 12—Uncovering and Correction of Work

If a portion of the work to be inspected by the architect/engineer has been covered or enclosed contrary to the architect/engineer instructions, Article 12 requires the contractor to uncover the work if requested by the architect/engineer and also recover, both at the contractor's expense.

If the architect/engineer did not previously request to inspect the work before being covered but then decided to do so, then the costs to uncover and replace would be cause for a change order. If, when uncovered, the work is found to comply with the contract documents, all related costs will be borne by the owner; but if defective work is exposed, all costs to remove, repair, and replace will become the contractor's responsibility. Of importance to the contractor is that section of Article 12, paragraph 12.2.2, that deals with corrective work after substantial completion has been attained.

According to the provisions of this paragraph, if during the 1-year warranty period the owner fails to notify the contractor of work to be corrected, thereby not affording the contractor an opportunity to do so, then the owner waives all future rights to require the contractor to perform that warranty or guarantee work.

Article 13—Tests and Inspections

Procedures for inspections and testing are provided in this section. Also included in Article 13, seemingly misplaced, is a statement concerning the payment of interest on late remittances to the contractor. Article 13.6.1 stipulates that, in the absence of an agreement to the contrary, interest will accrue on payments to the contractor that remain unpaid from the date when payment is due. The interest rate will be the prevailing one at the place where the project is located.

Article 14—Termination or Suspension of the Contract

The contractor may terminate the contract for the reasons set forth in this article and conditions under which the owner may terminate the contract *for cause* are also included.

Article 14 allows the owner to suspend or terminate the contract *for convenience,* and the circumstances surrounding termination for convenience are set forth in subparagraphs 14.3 and 14.4.

The 1987 Edition of AIA A201

The current, 1997, issue of the A201 document contains 154 substantive changes from the 1987 edition. If any contracts currently administered include the 1987 general conditions provisions, the project superintendent should carefully review this older document and take note of its predecessor provisions.

AIA Document A201CMa—General Conditions of the Contract for Construction, Construction Manager-Adviser Edition

Although many of the provisions contained in AIA A201 are similar to those included in the construction manager version, there are enough fundamental differences to warrant a project manager administering a construction manager contract to read this document from cover to cover—at least once.

Article 3 of the CM version of the general conditions defines the *contractor* as the person or entities who perform the construction which is *administered* by the construction manager. The *contractor* is to "carefully study and compare the Contract Documents with each other," and is not liable to the owner, construction manager, or architect for "damage resulting from errors, inconsistencies or omissions" unless the contractor recognized these deficiencies and "knowingly failed to report it." It would appear that it might be difficult to prove that the contractor *knowingly* failed to report problems concerning the quality of the drawings.

This article requires the contractor to take field dimensions, as required, and compare them with those indicated on the contract drawings. The contractor is directed to report any errors, omissions, and inconsistencies to the construction manager and architect promptly.

A provision in paragraph 3.7.3 states that the contractor's responsibility does not extend to verification that the contract documents comply with applicable building codes, laws, ordinances, and other appropriate rules and regulations.

The topic of *allowances* is discussed in this article and defines the elements of cost allowed for incorporation in the allowance item.

Similar to the provisions in the non-CM A201 document, the contractor's cost to unload and distribute the materials and equipment in the allowance item and the cost to install that work are to be included in the contract sum, and not the allowance item. The same is true of the contractor's overhead and profit—it is to be included in the contract sum.

Provisions for the submission of the contractor's construction schedule are included in this article. Although it is to be submitted for the owner's and architect's information, it is submitted to the construction manager for approval.

This is quite different from the schedule provisions contained in the standard AIA. A201 document which requires submission of the construction schedule for information. This subtle difference between *for information* and *for approval* can mean quite a bit when a contractor is assembling a claim for delays. An *approved* schedule becomes a baseline schedule, and deviations from that baseline schedule are generally used as the basis for requests for compensable costs associated with the delays.

When the schedule is submitted for informational purposes, the contractor should include enough details and documentation backup that it could also qualify as a baseline schedule.

Article 4 of A201CMa establishes the basic responsibilities of the construction manager:

1. The CM will determine, in general, if the work is being installed in accordance with the contract documents and will inspect for defects and deficiencies in the work.
2. The CM will provide coordination of activities required for the work of the contractors under their supervision.
3. The CM will review and certify all requests for payment.
4. Although the architect will have the right to reject work that, in her or his opinion, does not conform to the contract documents, no such action will take place until the construction manager has been notified. Subject to the review by the architect, the construction manager also has the responsibility to reject nonconforming work.
5. The CM will receive all shop drawings, review and approve them as consistent with the contract requirements, coordinate them with information received from other contractors, and pass them on to the architect.
6. The CM will prepare change order and construction change directives for presentation to the architect and owner.

Article 4.7.4 is entitled "Co.ntinuing Contract Performance" and states that pending final resolution of a claim, the contractor shall proceed with the performance of the contract. All too often, a contractor, out of extreme frustration in the attempt to resolve a dispute, will tell the owner, architect, or CM, "I've had it; I am going to stop the job until this claim (or change order, or dispute) is settled." Unfortunately, if that threat is carried out, the contractor will be declared in default of contract and his or her bargaining power will be significantly decreased.

Article 4 also establishes arbitration as the first step in the dispute resolution process. Article 7 changes in the work follow, more or less, the same procedures as those in the non CM version of AIA A201 and include similar provisions for the preparation of a construction change directive in the event that the contractor and construction manager cannot reach an agreement on a lump sum for the work.

Article 8—Time includes discussions regarding delays in the work and contains a specific provision in paragraph 8.3.3 that allows the contractor to recover damages under other provisions of the contract documents.

Associated General Contractors' Version of General Conditions between Owner and Contractor—AGC Document No. 200

The Associated General Contractors of America, Inc. (AGC) developed a series of construction contracts. AGC Document No. 200 combines a standard form of agreement and general conditions between owner and contractor in one document.

Contractors using the AGC contract document will find some provisions in their general conditions portion similar to those in the AIA A201 document, while other provisions may have the same effect but are worded in a slightly different manner.

With respect to uncovering errors, omissions, and inconsistencies in the contract documents, AGC requires the contractor to promptly advise the owner of any defects in the plans and specifications, but recognizes the fact that the contractor is not acting in the capacity of a licensed design professional. In other words, the contractor should not be held accountable for building code or building regulation violations.

The matter of warranties is treated quite differently from that in the architect's version of the general conditions. Although the contractor is bound by a 1-year warranty, any extended warranties required by the contract will be assigned to the owner after the standard 1-year warranty period has expired. The owner then assumes responsibility to notify the vendor or subcontractor of warranty issues, and the contractor is bound only to provide *reasonable* assistance in enforcing the provisions of these extended warranties.

Paragraph 3.10.3 requires the contractor to designate a safety representative whose duty will be to enforce safety rules and regulations on the site.

Article 10.2, Mutual Waiver of Consequential Damages, may be looked upon as a two-edged sword. Recovery of consequential damages, most often created by delays, cannot be pursued by the contractor; but by the same token, the owner cannot threaten the contractor with consequential damages if the contractor is seeking recovery of costs caused by contractor-generated construction delays.

Exhibit 1 is a standard attachment to AGC Document No. 200 and is a dispute resolution menu that allows the owner and contractor to resolve any potential disputes in one of five ways, as indicated on this check-off type of form, which becomes an integral part of the construction contract.

1. *Dispute Resolution Board* (*DRB*) is comprised of one member selected by the owner, one member selected by the contractor, and a third member chosen by two owner- and contractor-selected members. This board will meet periodically to track the construction process and make advisory recommendations along the way to avoid or settle any potential disputes or claims as they arise, instead of letting them drag on through the construction process.
2. Advisory arbitration will be conducted in accordance with the Construction Industry Rules of the American Arbitration Association.
3. In a minitrial, top managers from the owner's and contractor's organization will submit their individual positions to a mutually selected individual, who will make a nonbinding recommendation to the parties. (This process is similar to a mediation proceeding.)
4. Arbitration is binding pursuant to the Construction Industry Rules of the American Arbitration Association.
5. Litigation.

The General Conditions of the Engineers Joint Contract Documents Committee

In 1990 a joint committee composed of the National Society of Professional Engineers, the American Consulting Engineers Council, the American Society of Civil Engineers, and the Construction Specifications Institute prepared several contract documents for use by engineers. In cases where roadwork, infrastructure, or other civil engineering projects are concerned, engineers are the designers, and it was felt that these designers should have their own contract and general conditions documents.

The Standard General Conditions of the Construction Contract, prepared by the joint committee, contains a rather definitive index that requires a full 12 pages of text. This document contains several unique features.

Article 2, Preliminary Matters, requires that a preconstruction conference be arranged within 20 days after contract signing. The purpose of the conference is to establish a working understanding of the project and to discuss schedules and procedures for handling shop drawings and other submittals.

Article 3, Contract Documents: Intent, Amending, Reuse, states that although it is the intent of the contract documents to describe a "functionally complete project," the "intended result will be furnished and performed whether or not specifically called for." This phrase puts the contractor on notice that the *intent* of the documents is of equal importance to the scope defined by the plans and specifications. Although several articles prohibit the engineer from dictating means, methods, techniques, sequences, or procedures of construction, the contractor is prohibited from working overtime on Saturdays, Sundays, or legal holidays without the written consent of the engineer.

Throughout this document the term *engineer* is substituted where the word *architect* would normally be used. All communications between the owner and contractor must pass through the engineer. Procedures for change orders, rejection of work, and approval of progress payments are similar to those in the AIA General Conditions document, but Article 16, Dispute Resolution, is rather unique. The Engineers Joint Committee requires that a dispute resolution method and procedure be spelled out in a separate document designated *Exhibit GC—A Dispute Resolution Agreement.* If no such agreement has been reached, the owner and contractor may use whatever remedy is at their disposal as long as it does not conflict with any contract language.

A project manager must take the time to read all the pertinent contract documents including the general, supplementary, and special conditions. These important contract documents may contain conditions and elaboration of each party's duties, rights, and obligations so necessary for the intelligent and professional administration of the construction project. And as the warning printed on the front page of AIA A201 states, *"This document has important legal consequences."*

Chapter

4

Organizing the Project—Before and During Construction

Organization and planning are two of the most important tasks facing the project superintendent. In today's fast-paced projects, rapid communication is a fact of life, and without organization, the nearly instantaneous response that all parties to the construction contract seem to require, will only add to an already hectic day on the job.

Before construction actually begins, the superintendent will need to review all the contract documents to begin to understand what is required by the owner, the architect and engineers, the general contractor, and the subcontractors.

The plans and specifications need to be scrutinized for completeness, and the contract reviewed with the owner to determine if project-specific requirements have been added. Are there a number of allowance items that will require further information from the owner in order to start the work? Are there several alternates which require a response from the owner as to whether they will be accepted, and whether these decisions are to be made in a timely fashion?

Does the contract contain *unit prices* to be applied to certain items of *contract* or *extra* work, and do the appropriate subcontract agreements contain these same unit price requirements (hopefully, less the general contractor's overhead and profit)? Proper administration of the contract with the owner depends upon answers to these and other such questions, and these issues should be brought to light and resolved prior to the start of construction.

The Preproject Handoff Meeting

Between the time that the estimate has been assembled and revised and the contract award, any number of changes to the bid documents may have taken place. Quite often in a competitive bid situation, the owner will select a contractor and begin to negotiate both scope of work and contract sum. These sessions often end up with changes in scope and price, both minor and major, taking place, and hopefully these modifications will be documented when the contract for construction is executed.

There are other occasions when concessions are made by the estimating department in the heat of assembling a bid and these "concessions" are not documented. Promises may have been made to select a certain subcontractor or vendor in return for a favorable quote. Assurance may be given that a substitute product or piece of equipment will be strongly considered in return for a vendor or subcontractor offering a value engineering option at reduced cost in the quote. The estimating department may have inadvertently omitted a cost item or included a suspiciously low bid in the estimate in order to be competitive. The boss may have relented on some terms and conditions in order to obtain a contract with the owner. Unless all these conditions are brought to light early in the project, disagreements will occur as subcontract awards are made or purchase orders for materials are issued.

Some contractors hold a *preconstruction handoff meeting* attended by all parties having a hand in the preparation of the project's bid or contract. The vice president who negotiated the job will attend along with the estimator who prepared the initial bid as well as the future key players in the construction process—the project manager and the project superintendent. This handoff meeting will fill in some of the blanks not evident in the construction contract or the plans and specifications. Each attendee will contribute his or her knowledge of modifications to the contract documents, inadequacies in the estimate, any tips picked up during the bidding process, and any deals made with the owner, subcontractor, or vendors that would not be apparent by reviewing the project documentation. This handoff meeting is usually documented either by notes prepared by one of the attendees, usually the project manager, or by the use of a preprinted form to ensure that all pertinent topics are covered. Figures 4-1 through 4-3 are examples of a preproject handoff meeting form. Figure 4-1 reflects information gleaned by discussing the project with the estimating department and the boss (in this case the division manager or partner). Figure 4-2, the PM/superintendent review, will cover issues to be addressed at the architect's or owner's meeting. Figure 4-3 contains a checklist of responsibilities to be shared between the project manager and superintendent.

These forms, when completed, will become part of the project's basic files, to be referred to at some later date if appropriate questions arise.

Organizing the Job in the Office

The project manager has a responsibility to organize the project files in the office and to note any particularly important documents that should be sent to the field. Any, and all, significant or unusual modifications to the standard contract form must also be passed on to the project superintendent.

If you recall, Article 3 of the 1997 edition of AIA Document A201, General Conditions, requires the contractor to "carefully study and compare the various drawings and other contract documents relative to the work...[and] any errors, inconsistencies or omissions discovered by the contractor shall be reported promptly to the Architect as a request for information."

A great many potential disputes can be avoided if this procedure is followed by both project manager and project superintendent as soon as the contract with the owner is signed.

Coping with Addenda

If during the bidding process the architect or engineer issued a number of addenda to the plans and/or specifications, it is now time to assemble, categorize, and identify them for future use in the administration of the project.

There may be confusion, at times, between the difference between an *addendum* and a *bulletin,* since both signify changes to the plans and/or specifications. The term *addendum* applies to any changes made to the plans and specifications *before* the award and issuance of a contract to the general contractor. Changes made to the plans and/or specifications after contract signing are referred to as *bulletins.* Usually, changes effected by addenda are included in the contract sum while

PREPROJECT HANDOFF MEETING

Date of meeting ______________

- ☐ PM has reviewed with Estimating, other PM, Division Manager or Partner as applicable their knowledge of the terms of the deal.

People at the handoff meeting

______________ ______________
______________ ______________
______________ ______________

1) Were there any problems, or critical issues, reviewed as part of this meeting?

____

2) What was the anticipated fee that was supplied during this meeting? ______________

3) What is the anticipated start, finish, and/or duration time? ______________

4) MBE/WBE requirements: ______________

5) Wage scale: ______________

6) Special requirements, Division 1 critical issues: ______________

____

7) Were there any subcontractors that you were requested to consider for this project?

______________ ______________
______________ ______________
______________ ______________

Date of follow-up meeting if required to resolve all outstanding issues: ______________

cc: Attendees, Division Manager

Figure 4-1 Preproject handoff meeting form.

changes made and issued as bulletins require investigation by the general contractor to determine if these changes result in increases or decreases to the contracted scope of work.

It is important that all parties to the contract recognize this difference and assign the proper designation to such changes. And it is interesting that not all architects and engineers abide by these standards.

The author was involved with a $42 million project several years ago, and the contract with the owner included Addenda 1 through 4. However, the architect, after reviewing the "contract" drawings, made several *changes* to the drawings and issued the "For Construction" drawings as Addenda 1 through 5 in the title block. The author immediately contacted the architect and requested that the drawings be reissued, reflecting Addenda 1 through 4 *and* Bulletin 1.

Although the architect recognized the difference between the two terms and agreed that Addendum 5 should have been designated Bulletin 1, the firm had already printed 25 sets of drawings for distribution (and each set cost $195!), and was reluctant to throw away nearly $5000 and reprint all new sets. But the architect assured the author that any additional costs associated with Addendum 5 would be recognized as extra and reimbursable costs.

PM/Superintendent Review

Job name ______________________ Date of meeting __________

PART I – *To be used for Architect, Owner's Representative meeting*

People in attendance

________________ ________________

________________ ________________

- ☐ Complete current contract set of drawings reviewed.
- ☐ Specifications book reviewed with major sections highlighted.
- ☐ What pre-installation conferences have been agreed to as required for this project?
  1) ______________________
  2) ______________________
  3) ______________________
  4) ______________________
  5) ______________________
- ☐ Completed list of required project submittals reviewed (see attached).
- ☐ Review of project Change Order procedures.
- ☐ Schedule review.
  1) Was a full schedule completed and reviewed? ______________________
  2) Major issues related to schedule that need to be addressed or considered. ______________________
- ☐ General open/deviation items review.
  1) What major issues need to be addressed or resolved at this point?
- ☐ Review contact list, Architect, Owner's Rep., etc.

rev 10/00

Figure 4-2 Project manager and project superintendent review.

In the spirit of cooperation, no reprinting was demanded, but this incorrect designation plagued the project for months to come. The purchasing agent issued subcontract agreements that referenced the addenda and bulletins. This resulted in lots of phone calls from subcontractors requesting to see Bulletin 1 and taking issue with the fact that this document had been included in the scope of their work. Eventually this disparity was straightened out, but not before lengthy discussions took place between the purchasing agent, subcontractors, and vendors receiving copies of the "contract documents."

In situations like this, verbal notification is not enough. A follow-up written acknowledgment should be prepared by the project manager and sent to the architect to memorialize this event. It would be helpful to send "information" copies to all concerned subcontractors.

When addenda have been issued as changes to the specifications, they can be incorporated into the contract specifications books in one of several ways. One method involves cutting out the line item(s) changes and pasting them directly over the lines they supercede. To identify the addenda from which they were extracted, the addendum number can be written in the margin. If these cutout portions are taped on one end only, they can be lifted for comparison between the original document and the revised one.

PART II - *To be used between PM and Superintendent*

People in attendance

____________________ ____________________

____________________ ____________________

- ☐ Who is responsible for RFI log? ____________________
- ☐ Meeting minutes: Who will run subcontractor's meetings and keep minutes? ____________________
- ☐ Who will be responsible for upkeep of full schedule? ____________________
- ☐ Subcontract review.
  1) Reviewed who the potential subcontractors were to be for the project. Was there a directory of subcontractors supplied? (attached) ____________________

  2) What specific contract scopes have not been reviewed with the Superintendent at this time? ____________________
- ☐ General open items review.
  1) What major issues need to be addressed or resolved at this point?
- ☐ Review of direct Superintendent purchase procedures. ____________________
- ☐ Budget overview. ____________________
- ☐ Permit status. ____________________

Date of follow-up meeting if required to resolve all outstanding issues: ____________________

cc: Attendees, Division Manager

rev 10/00

Figure 4-3 Project manager and project superintendent checklist.

Changes can also be handwritten above or below the affected lines in the specifications, if they are not too wordy and don't affect too many lines.

When addenda add full pages or even full sections to the specifications book, it may be difficult to add these pages or section to an already bound volume. If that is the case, remove the binding and put the specifications in a large-capacity three-ring binder. The binder will also allow the project superintendent to place separators or tabs on the various sections, making it easier to find a particular section when required.

Addenda may also take the form of sketches, details, or drawing clarifications printed on 8½ × 11 inch paper, and these types of addenda can be taped directly to the construction drawing where the changes occurred. If this is not done, it is possible that the area changed by the addenda may go unnoticed and work may proceed with a detail that had been subsequently revised. There again, tape only one edge so that this sheet can be lifted to reveal the original detail.

It is helpful to maintain another booklet in the field that contains all the addenda in sequential order—both printed page and 8½ × 11 inch sketches. In spite of the efforts to update contract specification manuals and drawings to reflect the addendum changes, this additional binder will prove invaluable as a cross-check of what changes took place and which addendum directed that change.

Job Files

Most of the written materials coming into and going out of the construction field office will end up in files, and it is important that they are easily retrievable. Although this sounds so simplistic as not to warrant discussion, how many times have you filed an important document and could not remember in which file it had been placed?

Of course letters to and from the architect, engineer, and owner will be filed in folders entitled "Correspondence with Architect," "Correspondence with Engineer," and "Correspondence with Owner," but not all document filing is so easily compartmentalized. By requesting the office to forward two copies of every letter to the job site, a chronological file can be set up.

The Chronological File

At times, this "chrono" file can be a lifesaver. In addition to the central, compartmentalized files, filed by subject matter for a particular project, another dual-tracking filing system can be established to supplement correspondence in the central files. All it takes is to make another copy of each outgoing document.

Company Letterhead

Oriole Construction Company
566 Southway
Baltimore, Maryland 21200

A/E Collaborative
888 Airport Road
Towson, MD 21240

Re: The Academy
Project No. 5732

Attention: Mr. Arch Teck

Dear Mr. Teck:

Over the past several (days, weeks, months) we have received (either include the number of small sketches or simply state *numerous*) $8^1/_2$ x 11 (or whatever size the small sketches are/were) sketches containing changes (or clarifications, or both) to the contract documents.

The number and type of changes documented in this manner creates a problem for our field personnel and our subcontractors, who are concerned that they may not have received or posted all of the changes issued by your office.

We request that revised drawings for (name the drawings to which these multiple sketches apply) be reissued, incorporating all of the changes represented in your previously issued series of sketches. Without the issuance of revised, full-size drawing incorporating these changes, we feel that we cannot be held responsible for their implementation.

With best regards,

Will Spencer
Project Superintendent

Letter 1 When the architect/engineer issues lots of small sketches containing changes, confusion can reign—send this letter.

These duplicate documents will be filed *chronologically* instead of by subject matter. There may come a time when the project superintendent can remember *when* an important letter was sent or received, but because of its subject matter, it may have been filed in any number of subject files.

If a chronological file is created, a quick flip through the time period when it was supposedly sent or received will retrieve the document in a hurry.

And now that the contract has been reviewed, all addenda have been accounted for and properly posted, and project files have been set up, the project superintendent can proceed with the project administration.

Rereading the Specifications

The specifications section related to project start-up and project closeout proceedings is generally Section 0, Bidding and Contract Requirements, and/or Section 1, General Requirements. These sections should be reread thoroughly at the beginning of every project and especially before commencing any subcontractor or vendor negotiations, since some of the provisions in these two specification sections may affect those subcontractor or vendor negotiations. These specifications may contain certain conditions that are required before actual construction can begin and will also include guidelines to be followed during construction to satisfy closeout procedures. If project closeout requirements, for example, include a provision for as-built drawings to be maintained by the mechanical and electrical trades during construction, allowing the owner the right to inspect these drawings and withhold payment if they are not current, then all parties must be made aware of this requirement.

If a special lien waiver form is required from each subcontractor to be submitted with the monthly applications for payment, these forms should have been incorporated into the packet accompanying the subcontract agreements or given to the subcontractor at the first project meeting.

Quick Review of Closeout Requirements

A quick review of the specifications prior to start-up of the new project might reveal some of the following conditions that require immediate attention:

1. Specific insurance requirements and insurance certificates for both the general contractor and the subcontractors are to be submitted to the architect before commencement of any work on the site. And those insurance requirements may be in excess of those normally required. Copies of all insurance certificates should be sent to the field office as soon as they are received. A copy retained in the field will alert the project superintendent to any policy expiration dates.
2. Requirements must be spelled out for field offices and temporary utilities, possibly including new telephone service and computer terminals for the architect and clerk of the works. This may require some time to arrange for temporary cell phones.
3. Some architects require the contractor to submit a site logistics plan for approval prior to the start of any mobilization. So the project superintendent must begin to formulate such a plan as soon as she or he has been assigned the project.
4. Project identification signs may be required not only for deliveries but also to comply with specific contract requirements. Some government projects require elaborate signage to be placed on-site prior to the start of construction, and there might be zoning restrictions that conflict with these size and location requirements.

5. If the project is in a congested urban area, are there restrictions on storage, access for deliveries, or noise abatement ordinances?

The saying "The project begins before the project begins" means that there may be many preconstruction activities that are to be met before actual construction can commence.

A quick read of the special conditions, requirements, and restrictions outlined in the specifications will ensure that the project gets off on the right foot, an image that is important to convey to an owner, architect, and engineer who will begin to feel that they are dealing with a professional who will exhibit the same degree of professionalism throughout the project.

Changes in CSI's MasterFormat

The 40-year-old system of identifying various specifications sections, referred to as *MasterFormat,* is changing. Currently the major division of building construction components and related data begins with Division 1, General Conditions, and ends with Division 16, Electrical. Occasionally a Division 0 is added and includes bidding instructions, and often a Division 17 is included for catch-all items such as control systems. But this will all change sometime in the year 2003.

The *Construction Specifications Institute* (*CSI*) will publish a new version of MasterFormat that will commence with Division 1 and will stop at Division 14, starting up again at Division 20. This will allow for expansion of the numbering system as new technologies develop over time. Divisions 20 through 25 will contain specifications for mechanical and electrical systems, including separate divisions for telecommunications systems, environmental engineering systems, and high-voltage electrical work. In Division 30 site engineering will be included, and in Division 40 power systems and waste treatment construction components will be found.

Submission of a Schedule of Values

After the contract for construction has been fully executed and the job is a go, the project manager will update the original estimate to include any last-minute changes that may have been reflected in the final, often negotiated, contract sum. This updated and revised estimate will also serve as the basis for a *schedule of values,* generally required by the architect as part of the monthly requisition process.

This current estimate will serve as the basis of a job cost code report so that purchase orders for materials, labor costs, and subcontracted work, when issued, can be cost-coded against the proper category for future budget analysis.

Reviewing Allowance and Alternate Items Included in the Owner's Contract

A review should be made of any specification section or contract exhibits to the contract for construction having to do with allowances and/or alternates. Allowances must be reconciled; alternates must be accepted or rejected.

Allowances

When the exact scope of a portion of work is either not known or not fully developed, the owner often includes a specific sum of money in the construction contract to be applied against this *allowance* item. When the scope of work covered by the allowance is finalized, the contractor will prepare an estimate for this work. If the estimate is higher than the allowance, a change order request will be submitted to the owner, requesting an increase in the contract sum for the amount of the overage plus the contractor's fee. Conversely, if the actual cost of the work is less than the allowance, a credit will be due to the owner.

At some point during the construction process, these allowances will have to be reconciled with the amount set aside in the contract, adjusted accordingly, and the work scheduled.

The superintendent should review all such allowance items to determine when this reconciliation needs to take place, in order to have the work fit into the construction schedule. In the early stages of a project, reconciliation of an allowance for vinyl wall fabric for the conference room may not be a priority; but depending on the delivery time for this product, selection and resolution of this product may be required sooner rather than later, if the schedule is to be met. Resolution and reconciliation of a finish hardware allowance will most certainly require expediting, particularly if hollow metal doors and frames or factory-machined wood doors are called for, since these frames and doors cannot be released for production until an architect-approved finished hardware schedule with shop drawings is submitted to the hollow metal frame/door supplier.

Alternates

Alternates present a somewhat different problem. One problem with alternates has to do with the date by which they must be selected. Too often neither the bid documents nor the contract establishes a time limit on the selection of an alternate. Each alternate may require a different time frame selection. For example, if one alternate relates to lobby quarry tile set in a mortar bed instead of carpet, obviously the owner must make a decision well in advance of the placing of a concrete slab in that area, since the slab must be depressed to receive the tile.

If an alternate deals with the application of stain versus paint on certain wood surfaces, the selection process is less urgent.

When alternates require the purchase and installation of equipment, the time limit for selection must consider the requirement to place an order for the equipment, receive shop drawings, obtain the architect's approval, and allow for enough time to receive the equipment so it can be incorporated into the project in its proper sequence.

The project superintendent, after reviewing both allowance and alternates, should discuss any concerns with the project manager.

Shop Drawings and the Shop Drawing Log

A review of the specifications will establish the procedures for shop drawing submissions, such as

- Is there a special stamp required by subcontractor and general contractor that must be used on each shop drawing submission?
- How many copies of shop drawings are required for submission, and how many sepias or other reproducibles are also needed?
- How are samples to be handled, identified, and submitted?

- Are all types of shop drawings to be sent to the architect, or can structural drawings be mailed directly to the structural engineer with only a copy of the transmittal going to the architect?

Shop drawing submittals and the creation and updating of the shop drawing log are the responsibility of the project manager; however, the project superintendent has a prime interest in this entire process.

A shop drawing log should be prepared which will serve many purposes:

1. Establish a submission date for each required shop drawing.
2. Track receipt of shop drawings from vendors and subcontractors.
3. Track transmission of the shop drawing to the architect and engineer.
4. Track receipt of approved or rejected shop drawings.
5. Record the date when these drawings were returned to the vendor(s) or subcontractor(s).
6. Alert the superintendent to any delays in submissions review and return of critical materials.

Some contracts require the general contractor to submit a shop drawing log listing each required shop drawing and its planned submission date, while other contracts require the general contractor to incorporate a shop drawing submission schedule into the critical path method (CPM) schedule for the project.

It is critically important that subcontractors submit their drawings promptly. Often subcontractors will delay critical submissions because they continue to negotiate with the suppliers to obtain the best possible price. Little do they realize that significant delays in key product submissions may create delays that will ultimately prove costly to them.

At the first job meeting, major subcontractors and/or material suppliers should be presented with a time frame for submission of a preliminary shop drawing schedule. This schedule should include the major pieces of equipment for which shop drawings are required and the date when each drawing will be submitted. This schedule should also indicate projected product delivery dates, taking into account the review period allowed by the architect/engineer. (Refer to the specification section dealing with shop drawings.) A sample shop drawing submission schedule is shown in Fig. 4-4. Note that along with the proposed date of the drawings to be submitted, given 2 weeks for review by the architect/engineer, a proposed delivery date for each piece of equipment is inserted. Figure 4-5 contains a typical shop drawing submission log which will be used to track the flow of shop drawings to and from the architect/engineer along with action taken and parties to whom shop drawings are to be distributed after being reviewed.

More and more architects and engineers are demanding that general contractors actually *review* each drawing to ensure compliance with the plans and specifications instead of merely passing them through will little or no scrutiny. Some contract provisions allow the architect/engineer to charge contractors for additional reviews beyond the first or second submission.

If compliance with the plans and specification is questionable, contact the subcontractor or vendor submitting the shop drawing and discuss the problem. If there are deviations from plans and specifications, it might be best to highlight them to alert the reviewer as they are being transmitted to the architect/engineer. A note on the submittal indicating the deviation and the reason for it will go a long way in maintaining or improving relations with the architect and engineer.

PRELIMINARY DWG. SUBMISSION & EQUIPMENT DELIVERY SCHEDULE			
	10/09/92 ********		SUBMITTED BY ABC ELECTRIC CO., INC. PROJECT: WOODBRIDGE OFFICE BUILDING
EQUIPMENT (DESCRIPTION)	DWGS SUBMITTED	DELIVERY AFTER APPROVAL	DELIVERY DATE
1. Wiring devices 16140	11/19/92	04 Weeks	01/06/93
2. Panelboards 16160	10/26/92	04 Weeks	12/14/92
3. Safety switches 16170	10/26/92	04 Weeks	12/14/92
4. Switchboard 16440	10/26/92	10 Weeks	01/25/93
5. Motor control center 16480	10/26/92	10 Weeks	01/25/93
6. Lighting fixtures 16500	11/19/92	10 Weeks	02/18/93
7. Lighting protection system 16601	11/19/92	06 Weeks	01/21/93
8. Emergency lighting system 16610	11/19/92	10 Weeks	02/18/93
9. Fire alarm system 16721	11/19/92	10 Weeks	02/18/93
10. Time switch 16930	11/19/92	06 Weeks	01/21/93

Figure 4-4 Sample shop drawing submission schedule.

SHOP DRAWING LOG

SUBCONTRACTOR__________ PROJECT__________

CONTACT & PHONE NO.__________ TRADE__________

				Date sent to Arch./Eng.						Action				Distribution					
															No. of Copies				
Date Received	Shop Drawing Number & Title	No. of Copies	Sent To	Initial Submission	Resubmission		Final Submission	Date Returned	No. of Copies Returned	No Exceptions Taken	Note Corrections	Resubmit	Rejected	Distribution Date	Subcontractor	Jobsite	File		Remarks

Figure 4-5 Tracking shop drawings to/from architect.

When the shop drawings are submitted to the architect/engineer and an expedited review is requested, so state on the transmittal. As long as this privilege is not abused, an architect will usually process them more rapidly than required by contract.

Many project managers provide the project superintendent with an "information" copy of a shop drawing for review and comment either prior to or concurrent with

Company Letterhead

Oriole Construction Company
566 Southway
Baltimore, Maryland 21200

A/E Collaborative
888 Airport Road
Towson, MD 21240

Re: The Academy
Project No. 5732

Attention: Mr. Arch Teck

Dear Mr. Teck:

We are attaching X copies of (describe the shop drawings) for your review. Although specification section (insert the spec section pertaining to shop drawing review) allows for a 2-week review period, we would appreciate an expedited review of these drawings (or you can be more specific – a 1-week turnaround, 5-day, etc.) to ensure that (product or equipment) delivery can occur on-site when required.

With best regards,

Will Spencer
Project Superintendent

Letter 2 Requesting an expedited review of shop drawings. (Do not abuse the privilege.)

Company Letterhead

Oriole Construction Company
566 Southway
Baltimore, Maryland 21200

A/E Collaborative
888 Airport Road
Towson, MD 21240

Re: The Academy
Project No. 5732

Attention: Mr. Arch Teck

Dear Mr. Teck:

On (date) we sent X copies of (describe the shop drawings) to your office for review. As of this date, we have not received these drawings and assume they will be forthcoming in the next few days so that we can proceed to order the (equipment or product) in time to meet our schedule commitments.

With best regards,

Will Spencer
Project Superintendent

Letter 3 When the architect/engineer is overdue in returning shop drawings.

Company Letterhead

Oriole Construction Company
566 Southway
Baltimore, Maryland 21200

A/E Collaborative
888 Airport Road
Towson, MD 21240

Re: The Academy
Project No. 5732

Attention: Mr. Arch Teck

Dear Mr. Teck:

On (date) we submitted X copies of (describe the shop drawings). As of this date, we have not received those drawings from your office. Due to the delivery time required to receive this (equipment or product), we need to have these shop drawings in hand not later than (date). Your expedited review and return will be appreciated.

With best regards,

Will Spencer
Project Superintendent

Letter 4 When previous letters to the architect/engineer still do not produce the requested shop drawings.

Company Letterhead

Oriole Construction Company
66 Southway
Baltimore, Maryland 21200

A/E Collaborative
888 Airport Road
Towson, MD 21240

Re: The Academy
Project No. 5732

Attention: Mr. Arch Teck

Dear Mr. Teck:

On (date) we requested the return of (describe shop drawings) and indicated that the construction schedule would be affected if these drawings were not returned by (date). To date we have not received the (describe the drawings) and we are assessing the impact on our schedule.

Or substitute this sentence for the last one above:
As of this date, we have not received (describe shop drawings) and therefore anticipate a delay in completing the (describe the work or task that will be impacted by these late deliveries; e.g., drywall, finish carpentry, painting, millwork), which will delay this operation by (number of days or weeks) and will also affect work subsequent to this trade's work.

With best regards,

Will Spencer
Project Superintendent

Letter 5 Final letter to the architect/engineer when previous letters fail to produce shop drawings.

Company Letterhead

Oriole Construction Company
566 Southway
Baltimore, Maryland 21200

A/E Collaborative
888 Airport Road
Towson, MD 21240

Re: The Academy
Project No. 5732

Attention: Mr. Arch Teck

Dear Mr. Teck:

On (date) we submitted (describe the shop drawings or sketches) to your office with a request to review and comment. On (date when second or third request was made) we again requested review of these shop drawings.

As of this date we have had no response from your office and we have been advised by (subcontractor or vendor) that delivery of (equipment or product) will be seriously delayed to the point where it may impact our construction schedule.

Upon receipt of the reviewed shop drawings, we can obtain the current delivery date of the (equipment or material) and determine what impact, if any, this delivery has on the overall project completion date. We will advise your office accordingly.

With best regards,

Will Spencer
Project Superintendent

Letter 6 Notifying the architect/engineer of a delay because of the need for a shop drawing review.

the submission to the architect/engineer. If there are any comments, the superintendent should quickly mark up the drawing and return it to the project manager for review and discussion with the architect/engineer. If there are no comments, the shop drawing should be immediately discarded so that it cannot be confused with the approved shop drawing when issued.

During the subcontractor's meeting, the shop drawing log is to be reviewed with each attendee; and if a submission is late, the subcontractor should be placed on notice, in the meeting minutes, that this late submission must be expedited to avoid any potential backcharges.

The project manager at the owner/architect/engineer meeting will review the status of drawings at the A/E's office to ensure that drawings have not languished there too long. If these procedures are practiced religiously, one major weapon in the battle to assign responsibility in case of late product or equipment deliveries and related delays will have been developed.

Care must be taken to discern which subcontractors or vendors need to receive informational copies of approved shop drawings. For instance, when an approved copy of a boiler shop drawing is being returned to the mechanical subcontractor, the electrical subcontractor should receive an informational copy. Too often a piece of equipment is ordered with electrical characteristics at variance with the voltage requirements indicated on the contract drawings. If an error such as this is caught in the shop drawing stage, there may be little or no additional cost required to make the equipment compatible with the building's electrical system. If it is not caught at this early stage and the equipment is delivered with the wrong voltage, it is easy to envision the problems that will occur.

Remember that even if the architect/engineer mistakenly approves an equipment shop drawing containing the wrong electrical characteristics, it does not relieve the general contractor of the responsibility to provide the correct (*contract*) equipment.

Abbreviations and Acronyms Referred to in the Specifications

As the specifications are reviewed, there will be repeated references to certain alphanumeric abbreviations relating to design and performance criteria and corresponding sources. These are some examples:

Item	Abbreviation/acronym in specification
Precast manholes	Conform to ANSI/ASTM C-478
Concrete quality	To meet American Concrete Institute (ACI) specification 213R
Wood doors	Per Architectural Woodwork Institute (AWI) Section 1300 to ANSI/NWMA I.S., 2d revision

These references to trade and professional publications assume that every general contractor has a complete library including ANSI (American National Standards Institute), the ASTM (American Society for Testing and Materials) handbook, and the five-volume set of *ACI Manuals of Concrete Practice.*

Inspections and Testing

Another section of the specification book that should be reviewed carefully prior to the start of the job concerns inspections, testing, and quality control procedures. The responsibility for these services is usually fixed in the specifications and will include, as far as testing is concerned, the type and number of tests required during excavation and site work, tests required prior to and during placement of concrete, and if the structure is a steel frame, the testing and inspections required as the frame is being erected. Some engineers may require an inspection of a structural steel fabricator's shop if the shop is not certified by the American Institute of Steel Construction (AISC).The contractor may be required to reimburse the engineer for the inspection of a non-AISC-certified shop.

If some particular testing is the responsibility of the owner, the project superintendent must give the owner, or the owner's designated representative (architect or engineer), sufficient notification when said testing will be required.

When testing is to be performed by the general contractor, the specifications may require architect or engineer approval of the testing laboratory before it can be engaged. As for the reports generated by the tests, the specifications will designate how they are to be distributed. Generally a copy of each such test is sent directly to the architect, with copies to the general contractor.

Company Letterhead

Oriole Construction Company
566 Southway
Baltimore, Maryland 21200

The American Steel Company
855 Industrial Circle
Owings Mills, MD 21240

Re: Waterfall Plaza
Project No. 6444

Attention: Mr. Jim Beam

Dear Mr. Beam:

We request that you forward a shop drawing submission schedule not later than (date). This schedule should include not only the anticipated date of submission of the specific shop drawing(s) but, allowing for a 2-week review turnaround, also the proposed delivery date of (either materials, equipment, or product)

.

With best regards,

Will Spencer
Project Superintendent

Letter 7 Letter requesting a shop drawing submission schedule.

Company Letterhead

Oriole Construction Company
566 Southway
Baltimore, Maryland 21200

The American Steel Company
855 Industrial Circle
Owings Mills, MD 21240

Re: Waterfall Plaza
Project No. 6444

Attention: Mr. Jim Beam

Dear Mr. Beam:

On (date) we requested that you submit a shop drawing submission schedule. To date we have not received this document. It is important that you prepare and submit this schedule for our review to ensure that these submissions are acceptable within the context of the project schedule. Failure to comply with this request in a timely manner may result in backcharges to your account.

With best regards,

Will Spencer
Project Superintendent

Letter 8 Shop drawing submission schedule request follow-up letter.

Company Letterhead

Oriole Construction Company
566 Southway
Baltimore, Maryland 21200

A/E Collaborative
888 Airport Road
Towson, MD 21240

Re: The Academy
Project No. 5732

Attention: Mr. Arch Teck

Dear Mr. Teck:

In accordance with the provisions of (specification section dealing with close-in inspections, or drawing reference), we are requesting an inspection that would allow us to close in (the area/wall–describe by drawing/detail, elevation, column location if necessary).

We plan to proceed with the work in this area as quickly as possible and would appreciate a prompt response to our request.

With best regards,

Will Spencer
Project Superintendent

Letter 9 Requesting a close-in inspection.

Company Letterhead

Oriole Construction Company
566 Southway
Baltimore, Maryland 21200

A/E Collaborative
888 Airport Road
Towson, MD 21240

Re: The Academy
Project No. 5732

Attention: Mr. Arch Teck

Dear Mr. Teck:

In accordance with the provisions of (specification section or drawing reference), we have completed the excavation (be specific in describing the location and the type of excavations: storm/sanitary trench, footing excavation, etc. Use dimensions from column lines or property lines to locate) and are requesting an inspection prior to backfilling.

Since we are continuing to work in this area, your prompt response to our request will be appreciated.

With best regards,

Will Spencer
Project Superintendent

Letter 10 Request for an inspection prior to backfilling.

Many local building departments or building officials may also require copies of compaction tests, concrete test breaks, and steel inspection reports.

It will be the project manager's responsibility to establish the distribution list for all test reports at the beginning of the project, so that all interested parties are assured of receiving the reports they require.

The project superintendent should attempt to witness all field tests so that he or she can become aware, first hand, of whether a particular test or series of tests has failed or passed. If the architect or engineer requests immediate action in the case of a failed test, the superintendent should obtain a handwritten notification from the A/E; if that is not forthcoming in the field, the superintendent should prepare a written memo of any instructions and send it to the project manager.

The A/E should be requested to expedite any inspection reports requiring corrective action.

Job Scheduling

A job progress schedule is initially prepared as the job is being estimated. Its purpose at that time was primarily to determine the duration of construction so that an estimate of the general conditions or general requirements containing time-related costs could be prepared and included in the project's estimate.

Some bid documents require the general contractor to furnish a bar chart with the bid proposal, but in both cases these schedules are normally broad-brush or milestone type. Now that the project has become a reality, the job progress schedule needs to be reviewed and refined for other purposes.

The specifications may require that an initial, detailed job progress schedule be submitted to the architect within a specified period of time after contract signing. This schedule will become the *baseline schedule,* and all future schedule updates will refer to this initially, accepted version. The contract may also include provisions for periodic updates to the baseline schedule, usually coinciding with the monthly progress payment schedule. Although the primary purpose of creating a schedule is to provide all participants in the project with an orderly, time-related sequence of events to follow in order to effect a timely completion of the project, it can become the general contractor's friend or foe when delays are encountered and the project completion date is extended.

If the schedule is prepared properly, with subcontractor and supplier input, and is updated judiciously, and delays are identified and accounted for, it can be a critical document in resolving a claim. If not treated with the proper attention, a poorly administered schedule can be used against the general contractor, if and when any delay claims arise during the project.

The Bar or Gantt Chart

The bar chart is often referred to as the *Gantt chart* in deference to its originator, Henry Gantt. This schedule lists the various construction activities, or tasks, in a vertical column on the left side of the schedule, and a calendar, represented by either weeks or months, is placed horizontally along the top of the chart. A bar placed alongside each activity or task listed in the left-hand margin will extend to the duration represented by the time line spread horizontally across the top of the chart.

Some projects are such that a simple bar chart can be implemented effectively to display and track job progress. The bar chart is simple to create and is easily

Company Letterhead

Oriole Construction Company
566 Southway
Baltimore, Maryland 21200

The American Steel Company
855 Industrial Circle
Owings Mills, MD 21240

Re: Waterfall Plaza
Project No. 6444

Attention: Mr. Jim Beam

Dear Mr. Beam:

Please submit a schedule of your work tasks indicating the start and finish of each essential operation.

You can also add, if you wish:
We will incorporate these tasks and time frames into our overall construction schedule which, when completed, will be submitted to you for your final review and comments.

With best regards,

Will Spencer
Project Superintendent

Letter 11 Requesting a schedule of work.

Company Letterhead

Oriole Construction Company
566 Southway
Baltimore, Maryland 21200

The American Steel Company
855 Industrial Circle
Owings Mills, MD 21240

Re: Waterfall Plaza
Project No. 6444

Attention: Mr. Jim Beam

Dear Mr. Beam:

We are enclosing a copy of our preliminary baseline schedule. Please review and either accept or comment on that portion pertaining to your trade. If we do not receive your response by (date), we will assume that to mean acceptance.

With best regards,

Will Spencer
Project Superintendent

Letter 12 Request for a subcontractor to review baseline schedule.

understood by professional and nonprofessional alike. Most field supervisors feel comfortable with a bar chart; and the simplicity of the bar chart makes for an excellent presentation in large groups. But this chart does have limitations:

- It cannot graphically display a great deal of detail in complex projects.
- It cannot adequately display the interdependence of one work task with another.
- Updating the schedule does not permit displaying of the delay impact and the cause and effect of delays on the entire project's completion.
- The bar chart cannot reflect the impact that a delay in one activity will have on a prior or subsequent work activity.

While the bar chart is often used in the bidding process to depict various milestone dates and start-finish dates for broad categories of work tasks, in today's fast-paced construction project where complex, detailed, and interrelated activities must be shown, the critical path method schedule is what is required.

Critical Path Method

We won't deal with the mechanics of creating a CPM schedule but will discuss the important elements and concepts of this scheduling method. The critical path method of construction scheduling involves the preparation of a graphic display of most major and minor operations taking place during the life of the project.

A simplistic explanation of CPM can be described by the use of an arrow diagram. One end of the arrow indicates the start, and the other end indicates the completion, of an activity. The length of the arrow indicates the length of time apportioned to each activity. Some construction operations precede others on a straight-line basis and cannot start until a prior operation has been completed. Other operations can start prior to the completion of the preceding activity, and some operations are performed simultaneously or concurrently with others.

Although this may sound elementary to those project superintendents versed in complex, computer-generated CPM schedule preparation, it is the kind of thinking that is applied whenever such schedules are being prepared.

The CPM schedule provides

- Concise information regarding planned sequences of construction
- A means to predict with reasonable accuracy the time required for overall project completion and time required to reach milestone events
- Proposed calendar dates when activities will start and finish
- Identification of critical activities
- A matrix that can be manipulated to change the project's completion time, if required
- A basis for scheduling subcontractors and material and equipment deliveries
- A basis for balancing scheduling, workforce, equipment, and costs (if the schedule is resource-loaded)
- Rapid evaluation of time requirements of alternative construction methods
- A vehicle for recording and reporting progress
- A basis for evaluating the impact of delays and changes

There are six basic phases in the CPM schedule development process.

1. Understand the project, not only from a technical standpoint but also from the perspective of the contract requirements.

2. Interact with others. How did the estimator put the job together? Discussions with the project superintendent, vendors, and subcontractors are essential to develop sequences and time frames. Bringing the owner, architect, and engineer into schedule development, if practical, makes them a part of the process.
3. Physically create the schedule based upon access to a computer and scheduling software, at which point the project is divided into subnetworks—site work, foundations, structure, building envelope, interior finishes.
4. Develop activity codes.
5. Specify subcontractor networks.
6. Draft the logic diagram.

The project superintendent must participate in the draft review of the project schedule and comment on the sequence of events, the time allotted to various segments or components of construction, and whether there is ample time allowed for *float.*

Are the number of days designated for some activities reasonable or too few or too many? What about weather delays? What about time allocated for various inspections, such as framing and mechanical and electrical rough-ins?

Do any events require some *lag,* or will another activity be created to accommodate this time needed between the start of one activity and the ending of a previous activity? For example, if cast-in-place concrete foundation walls are included as an activity, and the contract requires 3 days' cure time before the forms can be stripped, do you include 3 additional days in the form-pour activity or create a separate activity designated as *curing time?*

There are three abbreviations that should be kept in mind when you prepare a CPM schedule:

PLDF: Predecessor lag duration float

SLDF: Successor lag duration float

FS: Finish–start

The project superintendent's experience is necessary to create a workable, effective end product.

The first schedule released will be designated the baseline schedule, and all other updates will be compared with the activities displayed on that issue.

This is where the project superintendent can play a key role. At subcontractor meetings or upon review of the schedule with the project manager, any changes to the baseline are to be reported to the project manager along with the reasons justifying these changes. When a delay claim or notification is being prepared by the project superintendent or project manager as the project progresses, it is essential that all changes to the baseline schedule be precisely defined by showing not only the affected activity, but also the date on which the CPM schedule was revised to reflect the change.

Although some CPM schedules may be complex, by becoming familiar with basic schedule terminology, some of the mystery will be lifted.

Activity flow—the sequence of work from one task to the other.

Order of activity or order of Precedence—an indication of which work event or task precedes or follows the other.

Duration—the time required to complete a work task.

Nodes—the graphic representation of specific tasks displayed as a rectangle, hexagon, or rounded box and usually containing a number identifying the task it represents. A node may also include start and finish dates. *Note:* The current computer scheduling software programs do not make use of a node, but this may be found on earlier CPM schedules.

Early start—a date earlier than that initially anticipated for the start of an activity, but which can be accomplished if its predecessor has been completed.

Late start—a date indicating a start later than initially proposed but which will still allow for on-time completion of the task.

Early completion—completion of a task prior to its initially scheduled date.

Late completion—completion of a task later than originally scheduled but which has no effect on the overall scheduled completion date.

Float—contingency time allotted to a specific work task or to a series of tasks to compensate for unforeseen delays.

The Importance of Float and Who Owns It

It will be a rare case, indeed, when every planned event in the construction schedule takes place as indicated. Weather delays, workforce shortages, equipment and materials delivery problems, and just plain mistakes all take their toll on job progress. To compensate for the unknown or unanticipated, a contingency needs to be added to each schedule in anticipation of occasional delays.

In the language of CPM, this contingency time is called *float,* and who owns this float can become an important issue, especially when the construction contract includes a liquidated damages provision and the general contractor will be assessed added costs for late delivery of a construction project. It is important for the contract language to indicate whether the contractor "owns" the float that can be used to compensate for delays incurred during the life of the project. Or does the owner "own" the float to be used to respond to construction decisions raised by either design consultants or the general contractor during the course of construction which, if not resolved in a timely manner, can create delays?

Determination or definition of the party owning the float time is important when liquidated damages, penalty, or bonus clauses are included in the construction contract. A builder who completes the project on time, not using any float time, may be adjudged as having actually completed the project *ahead of* schedule and therefore may qualify for an early completion bonus, if such a provision was included in the contract.

Conversely, if by using the float time the completion of the project is extended by the number of float days, the *contract completion* date may arguably be defended, thereby avoiding liquidated damages. However, if the owner owns the float, both situations described above would not be defendable.

Chapter

5

Organizing in the Field

The Site Logistics Plan

Quite often the contract requirements include a provision directing the contractor to prepare a site logistics plan for submission to the architect for review and approval. Depending upon the contract language, this plan may be required prior to moving the field office on-site, while other contracts may give the general contractor a little more time to submit the site logistics plan.

Each site is different from the last one, and that is one of the reasons why the construction business is so interesting—new situations arise on each and every project. Urban sites present a whole list of problems different from those encountered in building in a suburban or campuslike setting. Construction projects even in the heart of a small or mid-sized city with their crowded building sites can sometimes become a logistical nightmare—so planning ahead becomes even more critical.

Dealing with site logistics can be a time-consuming process, but proper planning in the early stages of a project can obviate many problems as construction activity speeds up. A costly site fence hurriedly installed, with little or no consideration for traffic flow, or security considerations may need to be relocated several times often at considerable expense.

The site logistics plan ought to include provisions for the following:

1. Maintain a secure site, making sure that the height of the fence is adequate for the site and that it can be secured at the end of the day with a minimum amount of time and effort.
2. Gates should be placed to allow ease of entry and egress, not only for small trucks but also for ready-mix concrete trucks and tractor-trailers with long turning radii.
3. There should be parking areas for the general contractor's field personnel, visitors, owner's representatives, and subcontractor (if space is available). Consideration of worker parking must be addressed, and if it is not available on-site, an off-site location should be investigated.
4. Access points around the site will be required for trucks to travel and to unload their materials and return to the exit.
5. Placement of the field office should be done so that it does not have to be moved and offers easy access to visitors.
6. Placement of the field office should take into consideration close proximity to site utilities such as existing electric, telephone, and water mains.

7. Placement of the field office should afford visibility to a maximum portion of the area under construction.
8. Storage areas for those materials and/or equipment are to be supplied by the general contractor.
9. Subcontractor field office and storage trailer and material storage areas require discussions and input from the major subcontractors.
10. If the building has a structural steel framework, a *lay-down* area for the structural steel members needs to be designated along with the path for the erection crane. The same will be true if precast plank structural or architectural panels are required on the project.
11. Lay-down areas will also be required if any substantial amount of masonry work is involved in the project.
12. The path where underground utilities will be located must be considered early on so that the relocation of materials, trailers, etc., is minimized when these utilities are being installed.

A fairly simple way to begin to develop a site logistics plan is to take a spare copy of a site or site utilities drawing and begin to lay out the security fencing and entrance and exit gates, keeping in mind the location of gates and the turning radius of tractor-trailers entering from the street.

By referring to the scale on the site and site utilities plan (1 inch equals 40 feet, as an example), prepare a series of templates to the proper scale that represent the size of an office trailer, a storage trailer, and other potential material storage areas. These templates can be cut out of an old manila folder and labeled *office trailer, storage trailer,* etc. Cut out enough of these office/storage trailer and storage area templates to represent the possible requirements for the major subcontractors, and place all these templates on the site plan to see if they will fit. They may have to be moved several times before a final location is determined along with access roadways to these locations. Before the plan is finalized, review the site logistics plan with the subcontractors who will be occupying the site. And it might be best to have them all in the room at the same time. One subcontractor may require more storage space while another may require less. Resolving all space and access issues at one time is preferable to dealing with a separate meeting with each subcontractor.

When all templates have been moved to their final location, merely trace their perimeters; label and identify the trailers, storage yards, and roadways; and send the site logistics plan off to the architect for review. When it is approved, hang it on a wall in the office trailer so that it can be easily accessed and viewed by all interested parties. Changes will inevitably be made during the life of the project, but hopefully they will be minor and will not disrupt the flow of workers and materials during intense construction activity.

Setting Up the Field Office

When the field office has been physically established on the site, it will become home to the superintendent for the duration of the project, and proper attention needs to be devoted to this home away from home. Proper organization, orderliness, neatness, and cleanliness of the field office will all reflect upon the superintendent's professionalism. Easily read signage directing visitors, vendors, and deliveries to the proper location on the site will become an early necessity.

The need to control the flow of workers and visitors on the site will also become an early order of business.

Visitor Control

A superintendent does not want people roaming the site without knowing who they are and what their business is. Visitor control necessitates a highly visible sign requiring all visitors to report to the field office. A simple sign in bold letters is easy to make (Fig. 5-1).

A visitor sign-in sheet should be placed in a prominent position right inside the field office trailer. A sample sign-in register for visitors is shown in Fig. 5-2. Sign-in procedures should be strictly enforced and ought to be reviewed at the first project meeting.

ALL VISITORS

Please Report to (Company's) Field Office

BEFORE

Proceeding Anywhere on the Site or in the Building. Please Register on the Sign-In Sheet.

THANK YOU

Figure 5-1 Sample visitor sign format.

JOB SITE VISITOR LOG

Project ________ Plat # ________ Tax # ________

Location ________ Street Address ________

Job # 8034 Project Superintendent ________ City ________ State ____ ZIP ________

	Date	Name	Representing	Purpose Of Visit	Person Contacted	Time In	Time Out
1							
2							
3							
4							
5							
6							
7							
8							
9							
10							
11							
12							
13							
14							
15							
16							
17							
18							
19							
20							
21							
22							
23							
24							

Figure 5-2 Sample visitor log.

Did You Remember to Bring or Order Everything You Need?

A checklist similar so the one displayed in Fig. 5-3 may be helpful. Along with ample supplies of pads, pencils, pens, and various payroll and field reporting forms, the field office must be organized to receive, store, and retrieve all the paperwork, reports, and drawings coming from the office, subcontractors, and design consultants.

Shop drawings for structural steel, precast, or cast-in-place concrete structures stored in the field office will be voluminous. Ductwork, heating and plumbing piping, and sprinkler shop drawings will be required not only for construction purposes but also to record as-built conditions.

Don't Let Filing Get Away from You

Equipment and material catalog sheets and samples of products to be incorporated into the structure all require controlled storage space. Unless there is organization in the field office, chances are that critical drawings and documents will become lost—always when they are needed the most.

PROJECT START-UP CHECKLIST

Order telephone	_____	Hand tools	_____
Pay phone for subs	_____	Power tools	_____
Temporary power	_____	Generator	_____
Access to temporary water	_____	Laser, transit	_____
Portable toilets	_____	Wooden stakes	_____
Temporary fencing	_____	Mason's line	_____
Hay bales, siltation fence	_____	**Office Supplies**	
Construction sign	_____	Lumber crayon	_____
Emergency numbers	_____	Pencils, pens, pads	_____
Field Office Supplies		Cans of spray paint	_____
Building permit	_____	File folders	_____
OSHA documents	_____	**Telephone Numbers**	
EEO documents	_____	Owners	_____
First-aid kits	_____	Architect	_____
Lock for door	_____	Engineer	_____
Security system tie-in	_____	Testing lab	_____
Temporary stairs	_____	Building official	_____
Equipment		Suppliers	_____
Foul weather gear	_____		

Figure 5-3 Project start-up checklist.

A condensed version of the office job files should be prepared for the field office. Correspondence from the architect and engineer, memos from the office, job meeting minutes, and letters to the owner and architect are to be filed in their respective files—promptly. Unless filing is current, papers pile up on the desk, get misplaced, or are mistakenly discarded. Filing daily is key to prompt and accurate retrieval.

Equipment catalogs for the mechanical and electrical trades can be placed in the folders designated for each trade.

Shop drawings should be installed on a plan rack or in a file drawer *only* when they are approved. Some superintendents prefer to receive an advance copy of a shop drawing to review either before or during its transmission to the architect or engineer. In many cases a review by the superintendent may reveal problems that may have been missed by the project manager, and this procedure should not be discouraged.

But what is important is that all *unapproved shop drawings* must be either discarded or filed in an area that is not accessible to anyone but the superintendent, and they must be clearly marked, "Not Approved—Do Not Use."

It is important to state, once more, that unapproved shop drawings left on the plan table in the field office scrutinized by a subcontractor seeking information can spell trouble for everyone. So put away those unapproved shop drawings and put away those superceded plans before someone inadvertently refers to them for current, approved field information.

The Daily Log and Its Function

One of the many chores that a superintendent must face is the entries into the daily log or daily diary (Fig. 5-4). At the end of a long and tough day, it is sometimes difficult to spend the time required to properly fill out a log describing many of the key events during that long, hectic day. But the daily log or diary can be one of the most important pieces of documentation when a claim or dispute arises, either when construction is underway or long after the project has been completed.

The information required to be posted daily may seem to serve little function at the time it is recorded. But suppose the company is involved in a lawsuit and the case comes to trial 3 years after the project was completed—and you are called to the stand to testify. The lawyers will have accumulated and introduced lots of documents into the trial, one of which will be the daily log.

You are on the stand, being questioned about an event that took place about 3½ years ago and the lawyer asks you whether it did, in fact, rain for 2 days while the foundation was being excavated at the northwest corner of the building. The company's justification for defending a delay claim may hinge on the answer to this question. Can you clearly remember this situation? If you have a fantastic memory, you might; but most of us could not answer with certainty. Your lawyer will ask, "Would it be of value for you to review your daily log and refresh your memory before you answer that question?" You'll respond in the affirmative, and after turning to the pages representing the dates involved, you can unequivocally say, "Yes, here it is in my daily log. It rained all day on both days you asked about, and we had been excavating for footings along the A line from column 2 to 7."

Would you be able to help your company in a case like this? Or, in the absence of a well-documented daily log, would you stumble and hem and haw on the witness stand and get some pretty angry stares from the boss?

DAILY LOG

Day: ______________

December 30, 20___ Weather: 8:00 A.M. ___ Noon: ___ 4:00 P.M. ___

WORK FORCE		SUBCONTRACTORS ON PROJECT & THEIR ACTIVITY
Supt.	___	______________________
Foreman	___	______________________
Carpenters	___	______________________
Laborers	___	______________________
Masons	___	______________________
Oper, Engrs.	___	______________________
Iron Wrkers	___	______________________
Electricians	___	______________________
Plumbers	___	______________________
Steam Fittrs	___	______________________
Sheet Metal	___	______________________
Glazers	___	______________________
Roofers	___	______________________
Sprinkler	___	______________________
Painters	___	______________________
Tile Setters	___	______________________
Carpet Lyrs	___	______________________
Controls	___	______________________
___	___	DAILY ACTIVITY - VISITORS - INSPECTIONS
___	___	______________________
___	___	______________________
___	___	______________________

Supervisor's Signature: ______________________

Figure 5-4 Daily log.

The proper application of a daily log serves many purposes:

1. It provides a chronological, day-to-day account of the number and type of workers on the job and the type of activities taking place each day.
2. It provides a record of all visitors to the site. (It does *not* take the place of a visitors' sign-in sheet.)
3. It provides a log of the weather, not only on a daily basis, but if filled out correctly, the weather at various times of the day (morning, early afternoon, late afternoon).
4. It provides a record of the types of materials, and sometimes the quantities, delivered to the site each day.
5. It provides a record of the types of equipment on the site each day, whether the general contractor's or subcontractor's. The equipment record should record *active time* and inactive or *down time.*
6. Record any inspections by either government officials or consultants.
7. Provide a short description of any problems occurring on the site or unusual conditions uncovered.
8. Include any discussions with the owner or the owner's consultants regarding problems, extra work items, or specific directions received from the owner or the

owner's consultants (even though they indicate that they will be sending a letter to the office confirming their field instructions).

Organizing the Subcontractors' Meetings

Developing and maintaining a good relationship with subcontractors on the job site is one of the project superintendent's most important jobs. This book devotes a full chapter to dealing with subcontractors, but we will now deal with some of the procedures required to organize an effective subcontractors' meeting agenda.

Depending upon company policy (does the superintendent or project manager conduct subcontractor meetings?) and the frequency of these meetings (initially once every 2 weeks, increasing to weekly as project reaches peak activity), these subcontractor meetings require advance preparation to be effective and productive. After the first such meeting, documentation of prior meetings will have been prepared and sent to all participants and other interested parties in advance of the currently scheduled meeting.

It is not enough to review old business. Items of interest or concern should be presented by the superintendent at every meeting.

Management by Walking Around

Several decades ago when management gurus seem to be everywhere, coming up with new and often bizarre methods to train managers so they could increase productivity, one such scheme really did seem to make sense—management by walking around. The basic idea behind this concept was that managers cannot sit behind their desks all day long, cranking out procedures and policies without actually going out to the factory floor to observe real working conditions and real problems. By walking around, not only does the manager become familiar with what the workers are actually doing during their workday, but also he or she is highly visible and is therefore perceived as having an interest in observing what workers do, experiencing some of the problems they often bring before him or her.

In construction this concept of managing by walking around is almost one of the golden rules. The superintendent cannot sit in a warm or air-conditioned office all day, going out on the site only when an emergency arises. Effective management means knowing what is going on in every corner of the site and the structure, and this can only be accomplished by walking the site and project several times each day.

Coordination—A Prime Topic for the Subcontractor Meeting

Coordinating the sequence of trade work is one of the prime responsibilities of a project superintendent, but there is another coordination activity that is also very important—ensuring that everything fits in the space allotted to it, as shown on the plans. Many contract specifications require the general contractor to prepare a *coordination drawing* that includes all the mechanical, electrical, plumbing, and sprinkler work to be installed above the ceiling. When a specific requirement such as this exists, the superintendent will be responsible for distributing related drawings to all subcontractors involved in installing above-ceiling work. When it is determined that everything will fit into the space shown on the drawings with no deviation from locations as designed, this drawing will be transmitted to the architect for review. However, often modifications to ductwork, piping arrangements or lighting fixtures, or ceiling heights must be made to ensure a proper installation.

Appropriate remarks are included on the submittal to the architect for review, acceptance, or comments.

The problem often arises when there is no contract requirement for a coordination drawing but the general contractor is still responsible to ensure that everything fits into its designated space. The project superintendent must be proactive in this regard, and the subcontractors' meeting is the perfect place to discuss and resolve any coordination matters. With all subcontractors sitting around the table, a potential problem can be reviewed and commented on by each subcontractor, in turn, to ascertain whether there are concerns about performing the work per plans and specifications. If there is such an agreement, it will be important to document each subcontractor's acceptance, and this can be achieved by sending each subcontractor a letter or including a statement in the subcontractor meeting minutes. This will avoid any problems that may arise, e.g., as the work is being installed above the ceiling and one of the subcontractors at that meeting now requests extra costs to modify the installation to fit. Their prior agreement, confirmed in writing, by letter or meeting minutes, will promptly end that discussion.

Other Subcontractor Meeting Requirements

The first subcontractor meeting will consist of introducing each subcontractor to the other members of the team and to the architect, engineer, and owner, if they attend that first kickoff meeting. Discussions regarding shop drawing submissions, the date by which subcontractors are to submit their requisitions, general safety rules, and guidelines will also take place. At this meeting, if the subcontractors have not worked together before, each one will size up the other to anticipate how their current working relationship will be. The project superintendent who has not worked with any of these subcontractors previously can also get a sense of which subcontractors will be "team" members and which ones are likely to be irritants throughout the project. The superintendent can then establish the ground rules that will be implemented and enforced to manage this project.

The Kickoff Subcontractor Meeting

Although project closeout procedures may be required at some future date, perhaps a year or two down the road, one of the first items that should be discussed and memorialized as new business (subsequently becoming old business) is the requirement for closing out the project. Why put a closeout statement in the minutes of the first subcontractor meeting? It should stay there throughout the life of the project and be read at every subsequent meeting to remind all subcontractors that these closeout requirements are key to contract completion and release of retainage and final payment. How many projects have you witnessed where the closeout process at the end of a job went smoothly, where all as-builts, operating and maintenance manuals, warranties, and the like were submitted on time and approved by the architect/engineer? Very few, if the experience of most other project superintendents is any indication.

So we stress the closeout requirements in that very first subcontractor meeting, keeping this topic at the top of all subsequent meeting minutes and emphasizing the urgency to comply with these procedures, now that the completion of the project is only months or weeks ahead.

The superintendent should review the specifications for the current project and prepare a list of general closeout requirements obtained from Division 1 and specific closeout procedures culled from each specification section. This list can be distributed as an attachment to the first subcontractor meeting minutes. Although it may take some time to prepare initially, it may save time on the back end of the job. And remember, the general contractor's retainage and final payment will not be released until all the subcontractors' closeout procedures have been submitted and approved.

A Sample List of Closeout Requirements

A fast walk through the project specification book can uncover the items required for project closeout which can then be converted to a list similar to this one:

1. Permits and inspections, including the Certificate of Occupancy (C of O), also referred to as the Use and Occupancy (U and O) Permit.
2. Certificate of substantial completion. Read Article 9.8 of AIA Document A201 to fully understand the term *substantial completion.*
3. Certifications and signoff from architect, mechanical and electrical engineers, and structural and civil engineers (if applicable). This may include a final inspection report from the MEP and structural engineer.
4. Final property survey.
5. Maintenance bond (if applicable).
6. Final lien waivers from each subcontractor and a general release of liens from the general contractor.
7. Warranties and operating and maintenance manuals (O&Ms)
   - *a.* Roofing and flashing warranties
   - *b.* Joint sealant warranty
   - *c.* Doors and hardware warranties + O&M
   - *d.* Flooring—carpet, vinyl composition tile, sheet vinyl, ceramic, epoxy
   - *e.* Windows—aluminum,wood, vinyl, steel, + O&M
   - *f.* Curtain wall and storefront work including antichalking of aluminum, color retention of members, air/water infiltration
   - *g.* Waste compactor and trash chute, + O&M
   - *h.* Window covering—blinds, curtain, shades, + O&M
   - *i.* Toilet and bath accessories, + O&M
   - *j.* Transmittals of trades, generally in separate three-ring binders
     - (1) Plumbing and mechanical + O&M, including air and water balancing reports.
     - (2) Electrical + O&M
     - (3) Fire protection + O&M
     - (4) Elevator + O&M
     - (5) Data/communication systems + O&M
   - *k.* Attic stock
     - (1) Extra flooring materials
     - (2) Extra cans of paint in various colors
     - (3) Hardware
     - (4) Toilet accessories
     - (5) Sealants
     - (6) Masonry materials—brick, concrete masonry unit (CMU)
     - (7) HVAC—spare filters, fusible links
     - (8) Plumbing—filters, trim
     - (9) Fire protection—sprinkler heads, fire extinguishers
     - (10) Electrical parts—wiring devices, fixture lenses, lamps
   - *l.* Start-up and test reports
     - (1) Boilers
     - (2) Chillers
     - (3) Air-handling units (AHUs)
     - (4) Makeup air unit (MUAU)
     - (5) Water treatment
     - (6) Balance reports for air and water
     - (7) Fireman's test report
   - *m.* Valve charts, tags, piping and equipment identification, directories
   - *n.* As-built drawings

Walking the Job prior to the Subcontractor Meeting

Prior to the next subcontractor meeting, usually the day before that next meeting, it is important to walk the site and observe the progress made since the last meeting. This is often best done late in the afternoon when most trades have finished their workday and left the job. A walkthrough at this time will probably not interrupt other workers or trade foremen, and the project superintendent can concentrate on preparing tomorrow's subcontractor meeting.

With a copy of the last meeting minutes in hand, the superintendent can check whether certain production goals have been met, whether the defective work has been corrected as scheduled, whether various subcontractors have kept their areas clean and relatively free of debris, and whether any other old business items have been attended to. Any items for discussion under new business can also be prepared during this walk-through. With a clear picture of the current status of the project, the superintendent will be prepared to respond to scheduling and quality issues, if raised at the subcontractors' meeting.

The topics to be covered in the subcontractors' meeting are discussed in Chap. 6, but we will address the actual format for these meeting minutes in this chapter.

Figure 5-5 is a typical subcontractor meeting minutes format and contains the basics:

1. Project name, number, site phone, site fax, and (if applicable) site e-mail address.
2. Date of meeting.
3. List of attendees, with their address, phone and fax numbers, email address, or all the above. (A sign-in sheet is passed around at the beginning of each meeting, and this forms the basis for filling in the attendee list.) Don't forget to add carbon copies in case you want to send copies to someone who did not attend the meeting.
4. Two main topics of discussion are highlighted—old business (which usually comes first) and new business (for some reason this particular set of minutes has the reverse).
5. Some method is needed to indicate that an old business item was not completed and was carried over to another meeting. This can be done as shown in Fig. 5-5, which is the 36th meeting. Under new business, item 1 is identified as 36.1, item 2 as 36.2, and so forth. If any of these items appear in meeting 37, they can be quickly identified as having first appeared on meeting 36. A look at the old business section of Fig. 5-5 shows that the topic of no smoking in the building was discussed at meeting 21.
6. Each item requiring action must be identified as such, and the person or persons responsible to provide that action must be indicated. Look at item 29.9 in Fig. 5-5. An individual has been assigned the responsibility to create the punch list, and that person's name would be indicated in bold type. Other superintendents will add an "Action by" column to the minutes in, say, the far right margin. In the case of item 29.9, someone would be designated to assume this task, and that person's name would be inserted in the column under the heading "action by."
7. The last item in the meeting minutes, following a statement establishing the next meeting, is a disclaimer of sorts. Quite often the person taking notes for the minutes will have missed an important point or misunderstood an important point. Because at some future date these minutes could be introduced as evidence in a court trial, their accuracy may be challenged. By adding a disclaimer similar to the one in Fig. 5-5, all attendees are given the opportunity to take exception to any item in the minutes; and if they do

Company Letterhead

Oriole Construction Company
566 Southway
Baltimore, Maryland 21200

A/E Collaborative
888 Airport Road
Towson, MD 21240

Re: The Academy
Project No. 5732

Attention: Mr. Arch Teck

Dear Mr. Teck:

As of (date) we have completed the work to the point where, in our opinion, the preparation of a punch list is warranted. We would appreciate your scheduling an inspection at your earliest convenience for the purpose of creating a punch list while most of the trades are still on-site to address any items relating to their trade.

When the date of your inspection arrives, (name of superintendent or foreman who will walk through with the A/E) will accompany you on your tour of the project.

With best regards,

Will Spencer
Project Superintendent

Letter 13 Requesting a punch list while subcontractors are still on the job.

Company Letterhead

Oriole Construction Company
566 Southway
Baltimore, Maryland 21200

A/E Collaborative
888 Airport Road
Towson, MD 21240

Re: The Academy
Project No. 5732

Attention: Mr. Arch Teck

Dear Mr. Teck:

On (date of initial letter requesting a punch list) we requested that your office conduct an inspection to create the official punch list. To date we have not had a response to our request. We had hoped to receive the punch list before most of the subcontractors have left the site in order to achieve a more rapid completion of any required work.

Your prompt response to our request will not only expedite the completion of the punch list but will also assist us in achieving a more rapid close-out of the project.

With best regards,

Will Spencer
Project Superintendent

Letter 14 Second letter requesting a punch list.

Company Letterhead

Oriole Construction Company
566 Southway
Baltimore, Maryland 21200

A/E Collaborative
888 Airport Road
Towson, MD 21240

Re: The Academy
Project No. 5732

Attention: Mr. Arch Teck

Dear Mr. Teck:

During your (date) inspection of the status of the official punch list dated (date), several items were added to the original list. We assume that this will be the final punch list, and once these additional items are completed, final payment will be authorized by your office (assuming all other close-out documents have been submitted).

If you still have other close-out documents to submit, substitute the following for the second sentence:
We assume that this will be the final punch list, and once these additional items are completed, your office will officially sign off on this list.

With best regards,

Will Spencer
Project Superintendent

Letter 15 The architect/engineer adds more items to the original punch list on a subsequent punch list inspection.

Company Letterhead

Oriole Construction Company
566 Southway
Baltimore, Maryland 21200

A/E Collaborative
888 Airport Road
Towson, MD 21240

Re: The Academy
Project No. 5732

Attention: Mr. Arch Teck

Dear Mr. Teck:

While we were completing items contained in the punch list received from your office, we received another punch list from the (Owner/Owner's Representative) adding items that did not appear on the list prepared by your office.

Although we wish to accommodate the Owner, we are of the opinion that the Owner should discuss these items with your office so that only one official punch list is prepared and submitted to the General Contractor.

With best regards,

Will Spencer
Project Superintendent

Letter 16 Owner sends the punch list to the contractor and the contractor wants to stop this practice.

Company Letterhead

Oriole Construction Company
566 Southway
Baltimore, Maryland 21200

A/E Collaborative
888 Airport Road
Towson, MD 21240

Re: The Academy
Project No. 5732

Attention: Mr. Arch Teck

Dear Mr. Teck:

We have received a letter from your office denying our request for final payment because certain punch list items have not been completed and accepted. We refer to (list the items that are in question).

These items, in our opinion, represent *warranty* work, not incomplete or unacceptable *punch list* work, and would therefore be covered by our 1-year builder's warranty/guarantee certificate.

We will, of course, expedite the repair or replacement of this work, but question whether this is cause to withhold final payment.

We would appreciate your response to our interpretation of this matter.

With best regards,

Will Spencer
Project Superintendent

Letter 17 When final payment is withheld because of "incomplete" punch list work when the work is really covered by the warranty.

not, they in effect have agreed to their content. A typical disclaimer is as follows:

> The attached meeting minutes represent the writer's interpretation of items discussed at this meeting. Any comments to the contrary are to be submitted, in writing, prior to the next scheduled meeting.

Monthly Project Reviews

In many companies, at least once each month the project manager is required by upper management to prepare a review of the entire project, including

1. Costs to date and costs to complete each item in the schedule of values
2. The status of change orders to the owner and to subcontractors or vendors
3. Actual schedule of progress versus the baseline or adjusted schedule to complete
4. Quality issues
5. Safety issues

PROJECT: WALKER AVENUE JOB NO.: 1324 SITE PHONE: SITE FAX:	MEMORANDUM FOR THE RECORD SUBCONTRACTORS PROGRESS MEETING #36

Date of Meeting: November 18, 1999

Attendees:	Phone	Fax
Carol Gailey- Napoli-Cover		
Rusty Burkhead- Winchester Drywall		
Ron Asbury-		
Clyde McKinney- Business Flooring		
Butch Spangler-		
Bob Akins- My Electrician		
Rob Andrieux- Winchester		

ITEMS OF DISCUSSION:

The Progress Meetings will be held weekly on Thursdays at 9AM at the construction site/office The next meeting will be **Thursday December 2, 1999.**

All subcontractors are required to attend the subcontractor meetings every week If you are unable to attend please call Butch or Ron at the site **before the meeting is to occur and let them know that you will not be there.

NEW BUSINESS:

36.1 All subs are to get their punch lists completed or the punch out crew will be hired to finish it for you.

36.2 Electric work needed for Napoli-Cover: AHU-4 starter mounted and wired, pump 1,2 heating starters mounted, CP-2 wired up, Domestic boiler CP-1 pump needs 120v feed with switch, 120v feed to ATC panel, all cabinet heaters wired up, AHU 3 starter wired, permanent 120v feed to boiler panel

36.3 Window glass for the first floor here on Monday the 22nd Nov.

36.4 Maria will be here on December 1,2 to punch the units so get the work ready now.

36.5 Napoli-Cover needs the heat trace so they can insulate at the chiller

OLD BUSINESS:

1.6. All invoices should be sent to by the **Monday before the last Thursday** of each month in order to process the invoices to get paid for the work completed that month. Checks are available on Fridays after 4pm at Charles Street after the subcontractors have been notified by Lien releases will need to be signed. Please do not call accounting regarding the availability of checks. anticipates that we will get paid 30 days after the requisition meeting has occurred. **Due to the Thanksgiving holiday the next draw meeting with the owner will be scheduled for Thursday December 2, 1999. All requisitions from subs are due on Monday November 29, 1999.**

21.1 **There will be no smoking in the building.**

29.9 Columbia Roofing needs to finish their punch out list. Columbia is working to complete it and will be done in the next few days- by October 14th. **Columbia has finished the preliminary punch list and he will complete the final punch by the end of the job. Butch will make up the punch list by November 19th.**

28.2 The two condensing units at the back of the building need to be piped in and wired ASAP. .Napoli-Cover to pipe the units in ASAP and then Bob Akins will follow up to wire the units. Carol is to make a date to do the work and Butch will schedule Bob Akins for the same day.

30.6 would like D&H to fix the door frame in the electric room in the basement. **is to talk with Maria to see if there might be other alternatives.**

34.1 My Electrician is to man the job with 6 men or more in order to meet the commitments that have been made to finish the job.

34.7 Business Flooring has asked about the runner and the binding for the first floor main stair- what color? to discuss with Maria. Clyde to get Harriett a color sample for the binding.

34.8 Tom Newell has commented on concerns about the "ridging" in the walls on the third floor corridor. to discuss. to get with Winchester to discuss and fix 7 or 8 of the problem areas.

35.1 Bob Akins is to start the site lighting on Saturday November 13. The concrete is finished and Bob is to finish pulling the wire and setting the steel posts.

Figure 5-5 Typical subcontractor meeting minutes.

6. Any potential or actual disputes or claims involving the owner, subcontractors, or vendors
7. Any other problems that have surfaced on the project

If your company employs such a monthly analysis of each project under construction, you as the superintendent will need to prepare for a meeting with the project manager to assist him or her in the assembly of this report. The project superintendent should keep notes on the following topics during the month for which the report will be prepared, in anticipation of this project update meeting:

Costs

1. Have any unanticipated or unforeseen conditions occurred on the project during the period? If so, will the associated costs be an extra to the contract, or will they be absorbed by the general contractor?
2. Have any other costs been incurred that will result in a change order to the owner? Is the owner in agreement as far as scope and costs are concerned?
3. Have any subcontractors or vendors requested additional costs because their subcontract agreement or purchase order did not adequately cover the scope of work?

Change orders

1. If any change orders are being prepared by the project manager, does the project manager need additional cost information from the project superintendent?
2. Will the contract completion time be extended or reduced or remain the same if the change order work is authorized?
3. Does the project superintendent have any knowledge of impending scope changes?

Schedule

1. Are any adjustments to the schedule warranted during the monthly update?

35.2 Winchester has said that there are still 3 units where the wall tile is not finished- Units 212, 224 and 313. Ron to check drywall work and call Ace Tile.

35.3 Business Flooring is to get the base done on the first floor by Monday November 15 and the second floor would be done by Wednesday November 17. told he has to finish the second floor ASAP.

35.4 The punch lists are attached to the doors of the units. All subs are to go back and designate a crew to finishing the punch lists. This needs to be completed now so that the remaining punch work can be completed. The punch out company needs to start immediately. Subcontractors need to finish their work in the units by Monday the 22nd of November or will go back and do the work and backcharge each sub for their that is not done.

35.5 has asked Carol to complete the chiller work so we can get it started and checked. Carol asked if the electrical wiring was completed.

35.6 has stated that it will take ten to fifteen working days to complete the generator work when it arrives. **NEW:** The generator will be arriving at the site on Wednesday November 17th. The transfer switches arrived at the site on Monday November 15th.

35.7 John from D&H Carpentry said he would have ten men working at the site on Monday November 15th. He would start the punch work and the caulking in the units. There have not been ten people at the site as of Nov. 18th. to call

The sequence for the punch out process for the building will be as follows:

Rough clean
list
finish list
Like- Nu- tub refinishing
IV Construction Punch- punch out company
Final Clean
TAT and CSI punch
IV Construction- 2nd
Touch up clean
Acceptance- Turn over to Owner

These minutes are a summary of the writer's interpretation of what transpired at the meeting. If there are additions or corrections or if you disagree with the interpretation, contact me at 332-1352 before the next scheduled meeting.

Submitted by:

Figure 5-5 *(Continued)*

2. Has the baseline or adjusted schedule extended or shortened the contract completion time?
3. If delays have occurred, what is required to prepare a recovery schedule and what are the anticipated costs to implement this recovery schedule?
4. Have any subcontractors advised the project superintendent of any potential delays in any delivery of equipment or performance of work?

Quality issues

1. Are there any quality issues that need to be discussed with the architect/engineer involving constructability or conformance with the plans and specifications?
2. Are there any subcontractor quality issues that need review and resolution?

Safety

1. Were any accident reports filed during the period that involve lost workdays or injuries?
2. Are there any safety issues to be addressed?

Claims

1. Is the project superintendent aware of any disputes that may be in the offing involving either the owner or a subcontractor?
2. Are there any unresolved issues that could escalate into a claim or dispute if not resolved during the period?

The Request for Information (RFI) and Request for Clarification (RFC)

Referring to the base contract and the general conditions document attached to it, generally AIA A201, we see that it is the contractor's responsibility to review the plans and specifications, and "any errors, inconsistencies or omissions discovered by the Contractor shall be reported promptly to the Architect as a request for information in such form as the Architect may require." The accepted format for such requests for information or clarification is the RFI form (Fig. 5-6) or RFC form (Fig. 5-7).

Although the completion of these types of forms appears to be rather straightforward, there are several entries that merit attention. Figure 5-6 has a preprinted ASAP in the "Date Needed to Be Answered" blank—this is not good practice. Just as the little boy shouldn't cry wolf, don't overuse ASAP. Give the architect or engineer ample time to respond to your RFI; ASAP should be used only when as soon as possible is actually required. By alerting the architect/engineer via email or phone, to an RFI in the works that requires prompt review and comment, such requests will generally be treated expeditiously.

Achieve Clarity in Requesting Clarity

It is important to be clear about the nature of the request being made. If at all possible, refer to a specific drawing and/or detail or specification section, using column lines or the north arrow to pinpoint the area in question if a drawing is involved.

For example, when you request clarification or further information regarding a conflict between the door size and location shown on an architectural floor plan and that item listed in the door schedule on another drawing, identify the door by number, say, door 201, and its location, say, second-floor electrical closet. Be specific as to the conflict: e.g., second-floor drawing indicates door 201 to be a 3070 hollow-metal door but item 15 on the door schedule, drawing A-20, states that a 3068 door is required. Please clarify.

Because it is likely that a number of RFIs or RFCs will be issued during the life of a project, and the status of each outstanding request must be tracked promptly, an

PROJECT:
JOB NO.:
SITE PHONE:
SITE FAX:

REQUEST FOR INFORMATION

To:	RFI#:
	DATE:
	DRWG REFERENCE
	SPEC SECTION

DATE NEEDED TO BE ANSWERED: ASAP
We request an answer to the following condition or problem:

Signed:

Answer:

Signed: *Date:*

Cc:

Figure 5-6 Request for Information (RFI) form.

REQUEST FOR CLARIFICATION

RCI NO.__________

DATE:__________

Project name: __________

Location: __________

RFC Directed To:__________

Respond To:__________

REQUEST FOR CLARIFICATION AS FOLLOWS:

Sender's Signature

Figure 5-7 Request for Clarification (RFC) form.

Company Letterhead

Oriole Construction Company
566 Southway
Baltimore, Maryland 21200

A/E Collaborative
888 Airport Road
Towson, MD 21240

Re: The Academy
Project No. 5732

Attention: Mr. Arch Teck

Dear Mr. Teck:

On (date) we were advised by your office that (explain which major changes are being considered by the architect/owner).

If sufficient information was supplied to quantify the change, continue as follows:
Based upon our initial review of this added scope, we anticipate that an extension of contract completion time of (days, weeks, etc.) will be required in order to add this work to the baseline schedule. We will be submitting our change order proposal, which will include the adjustment to the contract completion date, within (number of days, weeks required).

If insufficient information was supplied, substitute this paragraph for the one above:
Upon receipt of sufficient information from which to prepare a change order proposal, we will include not only the cost of the work, but the extension of the contract completion time required to incorporate this extra work into the project.

With best regards,

Will Spencer
Project Superintendent

Letter 18 When a major change in work is contemplated by the architect/owner and it will extend the contract completion date.

Company Letterhead

Oriole Construction Company
566 Southway
Baltimore, Maryland 21200

A/E Collaborative
888 Airport Road
Towson, MD 21240

Re: The Academy
Project No. 5732

Attention: Mr. Arch Teck

Dear Mr. Teck:

On (date) we were advised that a change was being considered by the owner in (describe the area in the building or on-site—be as exact as possible).

We have advised our subcontractors to delay (or stop) work in the area under consideration, as of this date. If further direction on how to proceed with the change is not received by (date), we can either commence work in the designated area or continue the "stop work" order, which may result in a delay to the project.

Please provide us with your decision in this matter.

With best regards,

Will Spencer
Project Superintendent

Letter 19 Architect requests that the contractor slow down work in an area where changes are being considered.

Company Letterhead

Oriole Construction Company
566 Southway
Baltimore, Maryland 21200

A/E Collaborative
888 Airport Road
Towson, MD 21240

Re: The Academy
Project No 5732

Attention: Mr. Arch Teck

Dear Mr. Teck:

On (date) we were advised by (owner, architect/engineer) that design changes were being prepared for (describe the area). Upon receipt of information from the (architect or engineer), we will prepare a proposed change order for the work in a lump sum amount for your review. Any changes in the contract completion time will also be included.

With best regards,

Will Spencer
Project Superintendent

Letter 20 Owner/architect advises of an impending change in scope.

Company Letterhead

Oriole Construction Company
566 Southway
Baltimore, Maryland 21200

A/E Collaborative
888 Airport Road
Towson, MD 21240

Re: The Academy
Project No. 5732

Attention: Mr. Arch Teck

Dear Mr. Teck:

On (date) we were instructed by (name, title, organization of owner's representative or architect) to proceed with (describe the work) on the basis of time and materials. We will commence work on (date) and will prepare daily work tickets for your representative to sign each day while this extra work is being performed.

It is important to have your representative available to sign these tickets since duplicates will be attached to our invoice for this extra work.

With best regards,

Will Spencer
Project Superintendent

Letter 21 Proceeding with extra work on time and materials basis, represented by daily work tickets.

Company Letterhead

Oriole Construction Company
566 Southway
Baltimore, Maryland 21200

A/E Collaborative
888 Airport Road
Towson, MD 21240

Re: The Academy
Project No. 5732

Attention: Mr. Arch Teck

Dear Mr. Teck:

On (date) we sent you a letter indicating that we were proceeding with (describe the extra work) on the basis of time and materials.

We indicated at that time that we would be preparing daily work tickets to be presented to your representative (you may name him/her if known) for (his/her) signature for each date that this extra work was performed.

On (date or dates), your representative was not on-site to sign these tickets, and we assume that there will be no disagreement as to the work performed on those days after our invoice has been submitted.

With best regards,

Will Spencer
Project Superintendent

Letter 22 Performing time and materials work when the owner's representative or architect is not available to sign the daily work tickets.

Company Letterhead

Oriole Construction Company
566 Southway
Baltimore, Maryland 21200

A/E Collaborative
888 Airport Road
Towson, MD 21240

Re: The Academy
Project No. 5732

Attention: Mr. Arch Teck

Dear Mr. Teck:

On (date) we were instructed by (name, title, organization of owner's representative or architect/engineer) to proceed with (describe the extra work); however, no agreement could be reached on the method to be used to establish the cost of the work.

We would recommend that you authorize us to proceed with this extra work in accordance with the procedures outlined in Article 7.3.3 of AIA Document A201, General Conditions, relating to the Construction Change Directives (CCD).

Note: If you are not using the A201 document, you can cite the basic provisions of this article as the method by which you want to document your costs.

Upon receipt of your authorization, we will commence with these changes.

With best regards,

Will Spencer
Project Superintendent

Letter 23 Proceeding with change order work when an agreement on costs cannot be reached.

Company Letterhead

Oriole Construction Company
566 Southway
Baltimore, Maryland 21200

A/E Collaborative
888 Airport Road
Towson, MD 21240

Re: The Academy
Project No. 5732

Attention: Mr. Arch Teck

Dear Mr. Teck:

On (date) we submitted (PCO#?, RFI#?, RFC#?) for your review and comment. As of this date we have not received your response.

If you have any questions regarding our submission, please give me a call; if not, your prompt review and return will be appreciated.

With best regards,

Will Spencer
Project Superintendent

Letter 24 Inquiring about the status of a recently sent Proposed Change Order (PCO), Request for Information (RFI), or Request for Clarification (RFC).

Company Letterhead

Oriole Construction Company
566 Southway
Baltimore, Maryland 21200

A/E Collaborative
888 Airport Road
Towson, MD 21240

Re: The Academy
Project No. 5732

Attention: Mr. Arch Teck

Dear Mr. Teck:

On (date) we inquired about the processing of (PCO, RFI, RFC) initially submitted on (date). We have received no response to our follow-up letter of (date),

If PCO add:
and, at our option, we may need to reprice this proposed change if your response is not received by (date).

If RFI, add:
and we may determine that additional costs will be required to complete this work unless we have received your response by (date).

If RFC, add:
and we may determine that additional costs will be required to complete this work unless we have your response by (date).

With best regards,

Will Spencer
Project Superintendent

Letter 25 When there is no response from the architect/engineer to a letter inquiring about the status of a PCO, RFI, or RFC.

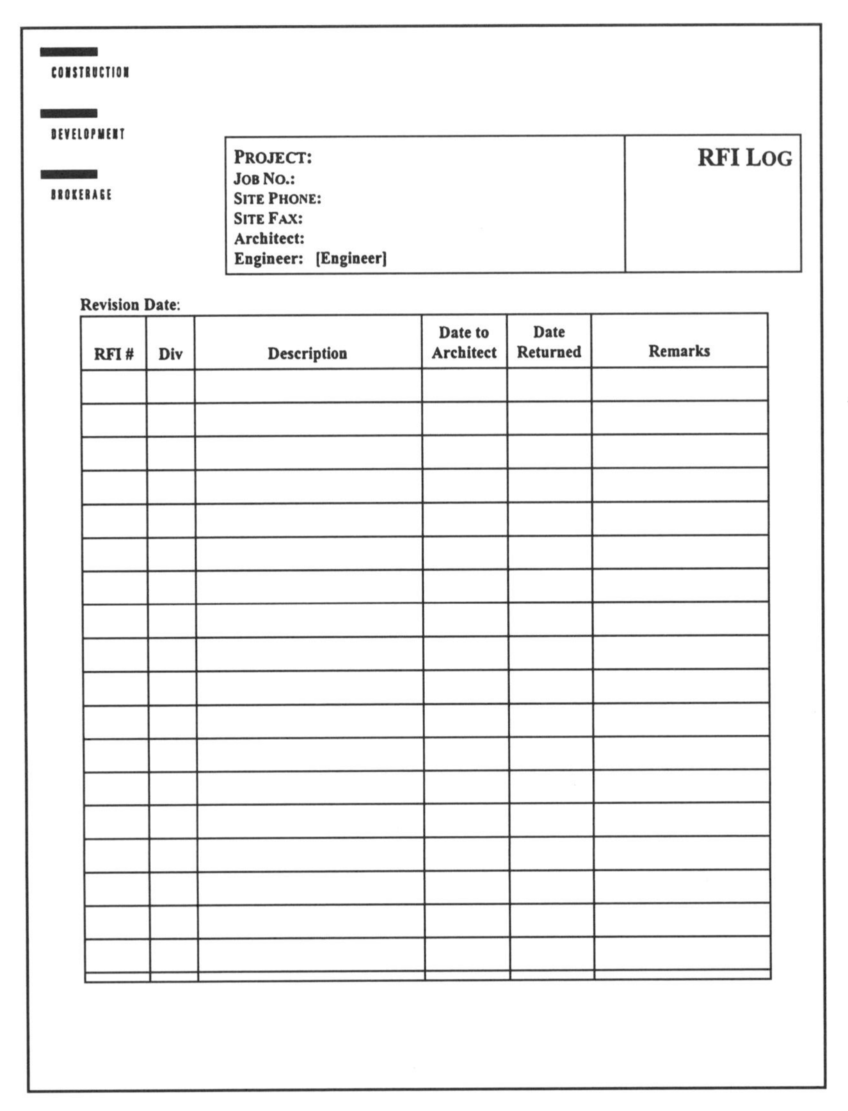

CONSTRUCTION
DEVELOPMENT
BROKERAGE

RFI LOG

PROJECT:
JOB NO.:
SITE PHONE:
SITE FAX:
Architect:
Engineer: [Engineer]

Revision Date:

RFI #	Div	Description	Date to Architect	Date Returned	Remarks

Figure 5-8 RFI log.

RFI or RFC log should be prepared similar to the one shown in Fig. 5-8. This log should be reviewed weekly to determine the status of all issued RFIs and RFCs. This update ought to be a topic of discussion at each project meeting where the owner and/or architect is in attendance. When response to an RFI or RFC has not been received by the date requested, the architect/engineer should be notified that this late response may have time-related consequences.

Preparing for Project Closeout

Preparing for the project's closeout ought to begin at the first subcontractor's meeting, where these requirements should be discussed. Each subcontractor should be requested to read the project's general and special requirements as they relate to closeout procedures in general and to read each subcontractor's applicable specification section for more definitive requirements. During the closing weeks of the current project, all eyes are turned toward that next project. It is at this point that the project superintendent needs to inject some urgency in the pursuit of closeout documents.

A checklist similar to the one displayed in Fig. 5-9 prepared at the beginning of the project can prove helpful in collecting all the closeout requirements on one document.

PROJECT: JOB NO.: SITE PHONE : SITE FAX:	MEMO Re: Project Closeout Procedure

	Task	Responsibility	Comments/completion date
1.	Review plans and specs for close out requirements		
2.	Prepare the table of Contents for the General Contractor's O&M early enough to track the requirements as needed throughout the project. See attached examples for reference.		
3.	Review contract for closeout requirements		
4.	Send letters to subs outlining procedures		
5.	Verify subcontractors have provided all extra material as stated in the contract documents		
6.	Officially transmit this extra material to the owner.		
7.	Receive and submit to owner as–built drawings and O&M manuals		
8.	Notify accounting department of completion to activate the Confirmation Form process		
9.	Generate and send statements to subs to confirm contract amount,		
	changes, claims and payments		
10.	Final lien releases signed, notarized, dated and returned		
11.	Lower tier obligations verified and paid		
12.	Provide subcontractors with preliminary punchlist before they leave the project		
13.	Ensure that no retention is released until punchlist work is completed		
14.	Release retention commensurate with submitting closeout materials and payment by owner		
15.	Schedule the deactivation of all services including phone, electric, water delivery, security, trailer return, dumpster, alarm monitoring, and routine supply deliveries		
16.	Properly pack and label for storage historical records from project		
17.	Return all excess office supplies, equipment, tools and storage, signage, etc.		
18.	Leave site in acceptable condition		
19.	Consolidate at PM location all critical contract, change order, submittal, permit, inspection (including occupancy), and other frequently requested project documents.		
20.	Maintain an accurate list of storage box numbers and contents		
21.	Complete download at date of substantial completion		
22.	Initiate Cost Certification based on Substantial Completion budget		

Figure 5-9 Project closeout checklist.

23.	Submit and track billing for retention as allowable by contract documents.		
24.	Prior to completion prepare a written close out plan that identifies known issues that would hinder completion per the schedule		
25.	Receive and distribute architect's punch list		
26.	Confirm Subcontractor and completion of all punch work		
27.	Obtain all necessary signatures from architect, owner, funding representative, etc. agreeing to completion of punch work		
28.	Establish process with owner for warranty communication		

Figure 5-9 *(Continued)*

The "Responsibility" column should be filled out to ensure that both the project superintendent and the project manager are aware of their individual obligations.

Since the company's final payment hinges on the completion, submission, and approval of all closeout requirements, this task, while tedious, nevertheless remains an important part of the construction process.

Problem Solving

Don't bring me problems—bring me solutions! How many times have you heard this from the boss. As the first line of defense in the problem-solving area, the ability to anticipate a problem is the mark of a good superintendent. Often problems can be averted or resolved if caught early and the situation is carefully thought through. But time is the key factor, and the superintendent's day is filled with answering questions from subcontractors and vendors, inspecting work for compliance with contract documents, documenting important events on the job site, and dealing with a multitude of documents that seem to show up on the site everyday. Without organization in the field office and in the superintendent's daily routine, the task of supervising a project can be overwhelming at times. So if it takes a little longer to organize the field office properly and establish the procedures that need to be followed on a day-to-day basis, these preplanned procedures will pay off during the life of the project.

Learning Effective Time Management

With all the demands placed upon a superintendent's time, learning how to productively manage that time is a key element in the effective management of the construction project. Pulled between answering subcontractor questions, reviewing important documents, receiving phone calls, and chasing down materials, a project superintendent's day often seems unending.

How often are you interrupted in the midst of tackling a problem? That equipment rental salesperson who shows up at the field office with a bunch of pads and pens requires you to stop what you're doing and hear what he has to offer. A subcontractor calls to inquire about submitting a bid for work and asks where she can pick up a set of plans and specifications for estimating. Some of these interruptions may be beneficial, but knowing when to politely turn them away may be just as beneficial.

With some time management tricks, learned and implemented, that seemingly endless day may be shortened. Experts in the field of time management have distilled problems relating to effective use of time down to 10 basic blunders.

Blunder 1—Inability to Deal with Drop-in Visitors

One of the problems with an open-door policy of being available to everyone all the time is that you allow drop-in visitors to interrupt your preplanned schedule. When these unanticipated guests arrive at the job site, ask whether this visit is important, what the purpose of the visit is, and how much time is needed to discuss or resolve the matter. Tell them how much time you can spare. "I can't talk to you now, but if you wait a half-hour (10 minutes) (until later this afternoon), I can meet with you."

Blunder 2—Lack of Priorities

Some superintendents create a daily A list and a B list. The A list contains those items that positively, absolutely must get done today. Obviously, this list must be short. The B list contains work items that will be dealt with once the A list has been completed. The key to preparing these priority lists is to only include items that truly are critical and, with respect to the A list, are achievable. Although the B list items can be carried over to the next day, if the A list items are not achievable in the day in which they have been created, rethink the priorities. Did they really belong on the top-priority list, or could they have been included on the secondary B list?

Blunder 3—Inability to Control Telephone Conversations

At the start of the project, is the flooring contractor calling for a meeting to discuss the scope of work? Is it necessary to carry on a long telephone conversation with this subcontractor whose work may not be required for 6 months or so? When these types of phone calls occur, be curt and tell the caller to get back to you in 5 months (3 months) or whatever time frame is appropriate.

If it is your nature to socialize during phone calls, think about how much more time you could free up if you limited the social talk. Could you make an additional three phone calls if you didn't review the weekend's football scores with that last four callers? Twelve calls of 10 minutes each computes to 2 hours on the phone! Can you afford to spend that much time on the telephone without affecting other work tasks?

Blunder 4—The Electronics Trap

With the widespread use of electronic assistants, are you spending more time reprogramming, updating, and revising this device than the time savings benefits it will provide? With the cell phone, adding frequently accessed phone numbers to the phone's electronic directory is a time saver, but is entering the job-site temperature three times a day into an electronic database more efficient than jotting down the temperature in the morning and at midday on paper and then accessing the electronic daily log at the end of the day to record all pertinent data? Don't invest in electronic devices solely to have the most up-to-date, or cutting-edge, product. Use devices that are time savers, not just flashy, cool gadgets.

Blunder 5—Reluctance to Delegate

How often have you said to yourself, "It's quicker to do it myself than to give it to Joe Hamfist" or "If I do it, it will be done right"? Both of these statements may be true, but your function is to manage, and managing means delegating—*not* doing the work yourself. It is up to you to select those assistants you can trust or to train them to perform their assigned tasks adequately. Although it may seem like too much time is spent instructing or supervising these delegated functions, in the long run you will have created an effective team that can be relied upon to do the job, thereby relieving you of the necessity to deviate from your core work responsibilities. You are also helping other team members grow in their job by taking on additional responsibilities.

Blunder 6—The Cluttered Desk Syndrome

Does your field office look like a tornado picked it up, turned it upside down, and set it back down right-side up? Neatness and orderliness in a field office achieve several goals. This displays an orderly approach to work which will be reflected and required in the construction taking place on the site. A neat office displays a certain amount of discipline on the part of the superintendent that is easily transferable to discipline in the construction process. And just as important, when the filing of documents is done promptly, plans and specifications are stored properly, and the field office in general is neat and orderly, it is so much faster to locate items, which translates to time savings.

Blunder 7—Procrastination

It is probably human nature to put off disagreeable tasks and promptly perform those items of work that are pleasant. Putting off important tasks because they are unpleasant or demand aggressive action usually results in significantly more time to perform these tasks and often is done in under urgent conditions that do not allow the sufficient amount of time and attention that should have been devoted to resolve these matters properly.

Consider dealing with those more difficult or unpleasant tasks first and leaving the easier, more pleasant tasks until later.

Blunder 8—The Need to Achieve Perfection

Allotting equal time to all tasks may not be the best use of time. Although the goal of perfection is admirable, as in the case of delegating work, a delicate balance between devoting equal time and energy to simple tasks when more time and energy are required for critical tasks is something that a time-stretched superintendent must deal with. When a limited amount of time is available to deal with problems, time must be directed to those tasks having the greatest impact on the project.

Blunder 9—Attempting to Do Too Much

A mental state of paralysis can develop when you attempt to do too many tasks at once. Where to start, how can I prioritize when everything I have to do is equally

Company Letterhead

Oriole Construction Company
566 Southway
Baltimore, Maryland 21200

A/E Collaborative
888 Airport Road
Towson, MD 21240

Re: The Academy
Project No. 5732

Attention: Mr. Arch Teck

Dear Mr. Teck:

I refer to our (meeting at the site, telephone conversation) on (date) and your verbal instructions to

Select one:
proceed with the work as follows: (describe the work).

do not proceed with the work (describe) until further notice.

We are proceeding as directed and would appreciate receiving your written confirmation of these instructions.

With best regards,

Will Spencer
Project Superintendent

Letter 26 Confirming verbal instructions received from the architect/engineer.

Company Letterhead

Oriole Construction Company
566 Southway
Baltimore, Maryland 21200

A/E Collaborative
888 Airport Road
Towson, MD 21240

Re: The Academy
Project No. 5732

Attention: Mr. Arch Teck

Dear Mr. Teck:

We have been requested to turn over a portion of the building by (owner/owner's representative/architect on (date).

We request that an inspection be made of that area and adjacent areas by your office before releasing the space for partial occupancy of the building. Our (superintendent or foreman) will accompany you on this inspection, which should include the following:

1. A clear delineation and definition of the space to be occupied.
2. A punch list for this space.
3. The status of work remaining to be completed in this area and adjacent areas.
4. Delineation of egress and ingress areas to be used by the occupant during this partial occupancy period.
5. An inspection of adjacent areas to document any existing damage or lack thereof.

We will establish an apportionment of all utility costs for the partially occupied portion of the building and forward to your office within the next (10/7/fewer) days.

With best regards,

Will Spencer
Project Superintendent

Letter 27 Responding to the owner's request for partial occupancy.

Company Letterhead

Oriole Construction Company
566 Southway
Baltimore, Maryland 21200

A/E Collaborative
888 Airport Road
Towson, MD 21240

Re: The Academy
Project No. 5732

Attention: Mr. Arch Teck

Dear Mr. Teck:

On (date) the Owner took partial occupancy of the building, occupying (identify the area — second-floor offices, computer room, Rooms 105 to 107, etc.). In order to apportion utility costs such as electricity, gas, and water, we determined that the area occupied at this time represents (X) percent of the total building area. The current monthly utility costs are (whatever the latest utility bills show — and you might be required to submit copies!), and so we have allocated (the same percentage as the space the Owner occupies) of the total monthly bills.

We will therefore be submitting a monthly invoice in the amount of ($$$$$) to the Owner for its share of the utility costs.

With best regards,

Will Spencer
Project Superintendent

Letter 28 Notifying the owner of apportioned utility costs for partial occupancy.

Company Letterhead

Oriole Construction Company
566 Southway
Baltimore, Maryland 21200

A/E Collaborative
888 Airport Road
Towson, MD 21240

Re: The Academy
Project No. 5732

Attention: Mr. Arch Teck

Dear Mr. Teck:

We wish to establish the date of (date) as the date of Substantial Completion per the requirements set forth in the contract documents (or you can include the section within the specifications that defines Substantial Completion; if none exists, refer to AIA Document A201, General Conditions, Article 9).

It would be appreciated if you could arrange a visit to the site, at your earliest convenience, to confirm that Substantial Completion has been achieved.

With best regards,

Will Spencer
Project Superintendent

Letter 29 Requesting a visit to establish the date of Substantial Completion.

Company Letterhead

Oriole Construction Company
566 Southway
Baltimore, Maryland 21200

A/E Collaborative
888 Airport Road
Towson, MD 21240

Re: The Academy
Project No. 5732

Attention: Mr. Arch Teck

Dear Mr. Teck:

We have received your Certificate of Substantial Completion establishing that date as (date). All utility companies servicing the project have been advised to submit billings to the Owner for all utility costs incurred after this date.

All applicable guarantees and warranties will commence on (date of Substantial Completion).

With best regards,

Will Spencer
Project Superintendent

Letter 30 Notifying the architect/engineer of the switch of utility costs to the owner after Substantial Completion has been established.

Company Letterhead

Oriole Construction Company
566 Southway
Baltimore, Maryland 21200

A/E Collaborative
888 Airport Road
Towson, MD 21240

Re: The Academy
Project No. 5732

Attention: Mr. Arch Teck

Dear Mr. Teck:

Due to the extended period of time during which severe weather (you may also name the severe weather — heavy rains, flooding, heavy snowfalls, sleet/hail) has occurred, the progress of the project has been seriously delayed. We are in the process of preparing a formal claim for delay due to weather and will be presenting that claim within the next (10 days, 2 weeks, or whatever).

Note: When weather delays are prepared, you need to obtain extended weather patterns from the local or national weather bureau to establish the fact that this weather is unusual and could not have anticipated.

With best regards,

Will Spencer
Project Superintendent

Letter 31 Notification of intent to file a weather-related delay claim.

Company Letterhead

Oriole Construction Company
566 Southway
Baltimore, Maryland 21200

Re: The Academy
Project No. 5732

A/E Collaborative
888 Airport Road
Towson, MD 21240

Attention: Mr. Arch Teck

Dear Mr. Teck:

On (date) we will have a pencil copy of Application for Payment No. (X) for the period (month) available for your review at the jobsite. Please advise when you would be able to meet (name of project manager or superintendent) to review this requisition.

With best regards,

Will Spencer
Project Superintendent

Letter 32 Requesting a review of a pencil copy of a requisition.

Company Letterhead

Oriole Construction Company
566 Southway
Baltimore, Maryland 21200

Re: The Academy
Project No. 5732

A/E Collaborative
888 Airport Road
Towson, MD 21240

Attention: Mr. Arch Teck

Dear Mr. Teck:

On (date) we submitted (number) copies of Application for Payment No. (X) to your office as requested. As of this date, we have not received payment. Please advise if you require additional information or documentation or have any questions regarding this requisition. If not, when may we expect the payment to be processed?

With best regards,

Will Spencer
Project Superintendent

Letter 33 Inquiring about the status of payment of a requisition.

Company Letterhead

Oriole Construction Company
566 Southway
Baltimore, Maryland 21200

A/E Collaborative
888 Airport Road
Towson, MD 21240

Re: The Academy
Project No. 5732

Attention: Mr. Arch Teck

Dear Mr. Teck:

On (date) we submitted Application for Payment No. (X) in the amount of $(XXXXX). When we inquired about payment on (date of previous letter or telephone call), we were advised that a check would be forthcoming on (date). As of this date, we have not received payment.

Our subcontract agreements contain the standard "pay when paid" clause and delays in payments from the owner will affect the project schedule if not received by (date).

Please advise when we may expect payment.

With best regards,

Will Spencer
Project Superintendent

Letter 34 When previous requests to the architect/engineer for payment go unanswered.

Company Letterhead

Oriole Construction Company
566 Southway
Baltimore, Maryland 21200

A/E Collaborative
888 Airport Road
Towson, MD 21240

Re: The Academy
Project No. 5732

Attention: Mr. Arch Teck

Dear Mr. Teck:

We have received (X) complete sets of plans and specifications to date. The contract for construction indicates that the owner is to furnish the contractor such copies of drawings and project manuals as are reasonably necessary for execution of the work; accordingly, we are requesting (X) additional sets.

Note: In case you need to refer to this portion of the contract, look at Article 2.2.5 of AIA Document A201, General Conditions.

With best regards,

Will Spencer
Project Superintendent

Letter 35 Requesting additional sets of plans and specifications at no cost to the contractor.

Company Letterhead

Oriole Construction Company
566 Southway
Baltimore, Maryland 21200

A/E Collaborative
888 Airport Road
Towson, MD 21240

Re: The Academy
Project No. 5732

Attention: Mr. Arch Teck

Dear Mr. Teck:

We wish to have you consider and approve a substitution for the following product:
(list the product/equipment and the drawing or specification section where it is contained).

We were recently advised that the delivery time required for the specified product is (8, 10, 12) weeks after approval and this extended delivery date will seriously impact our schedule. The (substituted product) is available within (1, 2, 3) weeks. Your consideration in this matter will be greatly appreciated.

For your review we are attaching a shop drawing for both the specified product and the substitution.

Note: You need a reason for substitution of a product, and delivery time is usually a good one. If the substituted product is of lower cost, you might offer a small credit.

With best regards,

Will Spencer
Project Superintendent

Letter 36 Requesting a substitution of a specified product.

important? That is the time to stop and reevaluate your work load. Don't attempt to do more than you reasonably think you can. Going back to blunder 2—rethink your priorities. Is it absolutely essential that I tackle all these issues this morning, or on this day? If I rush through all these tasks, will I be unable to spend sufficient time on the most important ones? I will therefore shorten my A list, I may have to reduce the number of items on my work list today, but what I do complete will be done thoroughly and done well. One expert advises that a planned work list should occupy only 50 percent of your available time, leaving the other 50 percent for additional time to complete any of these tasks that take longer than expected while also leaving time for unexpected events. Avoid the mental paralysis that comes from trying to do too many things at once, and rethink your daily goals.

Blunder 10—Inability to Say No

Sometimes it is difficult to say no to your supervisor or boss when you are asked to take on an assignment or two in addition to your already crushing workload. Does saying no suggest an unwillingness to cooperate or inability to take on more work? Will this extra workload seriously detract from your ability to perform your primary responsibilities, so that one or the other will have to be compromised? Possibly the best way to say no is to explain the pressing activities now occupying your daily workload, with the caveat that if there is time left to handle these new assignments, you would be more than happy to accommodate your manager. Or you can ask for a temporary assistant to help with your present-day assignments so you can free up some time to devote to the added tasks. This should send a signal that you are willing and able to assume these added responsibilities, but only if time allows or extra help is offered. If these assignments are that important, your supervisor will either offer to relieve you of some current activities or, more likely, look elsewhere for help.

Chapter

6

Working with Subcontractors

Rare is the project that does not employ subcontractors, and working with subcontractors encompasses the entire spectrum of management skills, from speaking softly to carrying and applying a big stick.

Although they are independent business people, subcontractors should be welcomed as team members and treated as such, unless proven otherwise. Subcontractor relations begin during the negotiation process when the bid proposals are being reviewed and analyzed by the project manager and the project superintendent in order to select the most qualified, competitively priced bidder for the project at hand. Problems often arise when one party or the other, or both, fails to adequately communicate what is expected of each other in the scope of work or performance of the work. Has the general contractor been specific in defining the work to be included or merely stated the scope of work as "plans and specifications"? With all the vagaries surrounding what is meant by plans and specifications, disputes and arguments can and do arise during the administration of a subcontract agreement unless a more definable scope is established and accepted by all parties.

A thorough review of the required work to be included in the subcontract agreement is the first step in this communication process. And a firm agreement and acceptance of that scope and the corresponding cost of work (subcontract amount) is the first step in establishing a professional relationship between the general contractor and the subcontractor. Unless the terms and conditions of the proposed subcontract agreement are fully understood and accepted by both parties at the time, it is merely a question of when disagreements will surface—and they will arise when the subcontractor may announce that the plans did not include a certain item or items of work, or when the superintendent demands certain tasks from the subcontractor only to be advised that this work was not included in the proposal and will not proceed as directed. And these disagreements always come at the most inopportune time.

Negotiating with Subcontractors to Avoid Disagreements

There are many ways in which to define the scope of work as completely as possible. Reviewing the appropriate plans and specification sections prior to meeting with the subcontractor and making notes of specific items to be discussed are an important first step.

Remember that there may be one or more "related" specification sections or drawings that ought to be reviewed to determine if they contain additional items of work to be included in the subcontract agreement. And don't forget to review the general and supplementary conditions for information regarding such things as change order procedures, project closeouts, requisition preparation, and backup documentation for both change orders and requisitions.

Some general contractors use a Subcontractor's Interview Form (Fig. 6-1) when they meet with the subcontractor to review the bid proposal. This preprinted form can be customized to include items or work specific to the project at hand. As the interview progresses and items are included or excluded from the work, addressed and checked off, at the conclusion of the meeting there should be no confusion as to what is and what is not included in the subcontract agreement. When the interview has been concluded, the subcontractor will be asked to sign the form, thereby evidencing agreement to the terms and conditions discussed during the meeting. When the subcontract agreement is prepared for execution and references this interview form, there is little room for disagreement about its content.

Other contractors use an interview form requiring greater subcontractor participation. This form is transmitted to the subcontractor in advance of the meeting so that he or she can be prepared to discuss this extensive list of work at the forthcoming get-together. Figure 6-2, which shows a preconstruction meeting checklist, is divided into six sections: an introduction and sign-in section, an overview discussion of the project by the general contractor (section 2), and a presentation by the subcontractor (section 3). Section 4 is a review by the general contractor to verify that the subcontractor is in possession of all required documents, and in section 5, the general contractor acquaints the subcontractor with many of the forms to be used in the administration of the project. Section 6 concludes the interview.

If the superintendent has not participated in any subcontractor interviews or negotiations and must therefore rely on the terms and conditions of the subcontract agreement, this definitive list of work becomes a critical part of the agreement. Although a scope definition of plans and specifications may suffice in many instances, as we discussed earlier in this book, errors, omissions, or inconsistencies in the contract plans or specifications, or both, often lead to confusion over what these plans and specifications *really* encompass. Quite often the clarifications and modifications to the basic scope of work will be added to the subcontract agreement as separate attachments, say, as exhibit A or B or attachment A or B, and so on, and that is all that's needed.

Review of the Subcontract Agreement to Become Familiar with Its Contents

Although the project manager, in many construction companies, is assigned the responsibility of administering the subcontract agreement, the project superintendent acts in a supporting role by advising the project manager in the managerial role. The project superintendent needs to read and fully understand the scope of work included or excluded in each subcontractor's agreement and must be familiar with the boilerplate provisions within the subcontract agreement, primarily those provisions dealings with performance, notices to correct, schedule compliance, and backcharge procedures. This review will alert the project superintendent to those specific subcontract provisions that are to be invoked if and when subcontractor-related problems arise. The most frequently referred to provisions in standard subcontract agreements are

1. Compliance with schedule requirements
2. Notice of nonperformance

SUBCONTRACTOR NEGOTIATION FORM Page 12 of 16

CONCRETE - SCOPE OF WORK (Cont'd) Including but not limited to the following:

ITEM	YES	NO	EXPLANATION AND/OR COMMENTS
GENERAL CONDITIONS 1. Superintendent - Submit resume for approval			
2. Line & Grade - Define			
3. Teamster Shop Steward			
4. Master Mechanic - proportionate share			
5. Maintenance Engineer			
6. Until such time as contractor is able to provide temp. light & power, subcontractor will provide his own with portable generators, etc.			
7. When temp. services (light, power, water) are available they will be provided with maintenance, without cost to subcontractor during regular working hours. If services including maintenance are required outside of regular working hours, subcontractor will pay for same.			
8. Furnishes hoisting for own work			
9. Pumping concrete			
10. Form Drawings Designed by Engineer to conform to OSHA requirements,			
11. Reinforcing shop drawings			
12. As built drawings			
13. Insurance as specified			
14. Hold harmless insurance			
15. Retained percentage per Prime Contract			
16. Cost of performance & payment bond included in contract sum and in unit prices.			
17. Licenses and permits for own work			
18. Payment schedule & retainer			

Figure 6-1 Example of a subcontractor interview form, this one for cast-in-place concrete.

SUBCONTRACTOR NEGOTIATION FORM Page 13 of 16

CONCRETE - SCOPE OF WORK (Cont'd) Including but not limited to the following:

ITEM	YES	NO	EXPLANATION AND/OR COMMENTS
19. Percentage for added work			
20. All work in accordance with local, state laws & regulations			
21. Restoration of damage to property			
22. Equal Employment Opportunity provisions			
23. Scaffolding for own work			
24. Personnel hoist for own work			
25. Contract amount shall include a minimum of two (2) moves of shanties field offices, sheds, etc.. Hook-up of electric services to be at subcontractor's expense.			
26. Working hours			
27. Stand-by requirements			
28. Perimeter rails - provide - maintain			
29. Rails at interior openings - provide - maintain			
30. Kick boards - install - maintain			
31. Conform to all OSHA requirements not elsewhere assigned to others			
32. Remove stripped lumber from floors daily			
33. Off-site disposal of rubbish and firewood			
34. Load rubbish onto containers provided by			
35. Project Work Rules.			
36. CGL Insurance to include X,C & U coverage.			
CLEAN UP & PROTECTION 1. Broom clean slabs prior to placing fill and finish			

Figure 6-1 *(Continued)*

SUBCONTRACTOR NEGOTIATION FORM — Page 14 of 16

CONCRETE - SCOPE OF WORK (Cont'd) Including but not limited to the following:

ITEM	YES	NO	EXPLANATION AND/OR COMMENTS
2. Broom clean slabs after stripping and removing forms			
3. Clean concrete from inserts, slots, reglets etc.			
4. Cut off nails, ties, etc.			
5. Concrete spillage, drippings, etc. to be removed immediately from adjacent surfaces of stairs, walls, bricks, etc.			
6. When floors are being added to existing building, protect existing areas from splashing concrete.			
7. Protect stair nosings during concrete placement			
8. Flush forms with water or air before placing concrete.			
9. Provide protection above & below men working in shafts			
WINTER PROTECTION 1. Heated concrete			
2. Provide temporary enclosures & heat			
3. Snow removal inside building site			
4. Protect sub-grade against freezing			
5. Covering & protecting foundation concrete against freezing			
6. Curing blankets - furnish, place, and remove			

Figure 6-1 *(Continued)*

SUBCONTRACTOR NEGOTIATION FORM

Page 15 of 16

CONCRETE - SCOPE OF WORK (Cont'd) Including but not limited to the following:

ITEM	YES	NO	EXPLANATION AND/OR COMMENTS
7. Wind breaks			
8. Additional formwork allowed for			
9. High early cement. Additional cost ______			
SCHEDULING & PROGRESS 1. Contractor has visited jobsite			
2. To facilitate jobsite mobilization and access to areas of the building by other trades, the subcontractor must agree to sequence his work as directed			
3. Participate in CPM planning			
4. Anticipated date concrete foundations will start			
5. Duration of foundation work			
6. Ready for structural steel by______			
7. Anticipated date concrete superstructure will start			
8. Duration of superstructure			
9. Amount of contact surface lumber included for foundations			
10. A) If additional foundation forms are used for acceleration what will cost be? (specify amount) B) Schedule will be accelerated by______ working days			
11. Amount of floor forms included for superstructure			
12. Subcontractor to submit shop drawings schedule within 15 days			

Figure 6-1 *(Continued)*

SUBCONTRACTOR NEGOTIATION FORM — Page 16 of 16

CONCRETE - SCOPE OF WORK (Cont'd) Including but not limited to the following:

ITEM	YES	NO	EXPLANATION AND/OR COMMENTS
13. A) If additional superstructure slab forms are used for acceleration, what will cost be? (specify amount) B) Schedule will be accelerated by______ working days			
14. Anticipated supply problems			
Subcontractor will submit a detailed schedule of values broken down for labor & material of the contract amount to the Project Manager			
within 10 days of the award date.			
BID DOCUMENTS & CONTRACT 1. Specification Sections Specify______________			
2. Drawings - Architectural - Foundation - Structural - Mechanical, Plumbing & Elec. - Site			
3. Invitation to Bid dated______________			
4. Supplement Nos.______________			
5. Unit prices submitted -			
6. Hourly wage rates submitted			
7. Contractor has read General Conditions and assumes responsibility for all requirements that pertain to his work.			
8. Subcontractor verifies he has included all escalation costs as required.			
9. Subcontractor must return signed copy of Subcontract Agreement form within three days of date of receipt. Should signed agreement not be returned within specified time, agreement may become null and void at General Contractors option.			
10. Subcontractor has reviewed and agrees to sign standard form of agreement, including Schedules "A", "B", "C" etc. at the time formal award is made.			"Boiler Plate" document to be accepted without change.
11. Subcontractor shall submit his Certified Financial Statement within seven (7) days of notification of award.			

Figure 6-1 *(Continued)*

1. Opening
- 1.1 Introductions
- 1.2 Sign-in sheet
- 1.3 Minutes and issues documentation
- 1.4 Project site meeting schedules

2. XXXX presents:
- 2.1 The project overview
- 2.2 The project CPM schedule
- 2.3 The project safety requirements

3. The subcontractor presents:
- 3.1 The subcontractor's team
- 3.2 The scope of subcontract work
- 3.3 The subcontractor's submittals and submittal schedule
- 3.4 The subcontractor's plan to accomplish the work
- 3.5 The subcontractor's workforce plan
- 3.6 Special materials, processes, or equipment to be used
- 3.7 Material suppliers and subtier subcontractors
- 3.8 AAAA material supply requirements if any
- 3.9 The subcontractor's detail schedule and all subcontract milestones and completion dates
- 3.10 The work interrelationships with other job-site subcontractors and trades
- 3.11 The subcontractor's quality control and inspection approach and implementation plan
- 3.12 The subcontractor's test and commissioning plan
- 3.13 The subcontractor's punchout workforce plan
- 3.14 The subcontractor's safety approach and implementation plan
- 3.15 The subcontractor's MSDS list and data sheets
- 3.16 The subcontractor's mobilization and demobilization plan

4. YYYY reviews the above for:
- 4.1 Verification that the subcontractor understands the work scope, schedule, submittals, and requirements
- 4.2 Verification that the subcontractor understands the project safety requirements
- 4.3 Verification that the subcontractor has the full drawings and specifications available and in the subcontractor's possession
- 4.4 Verification that the subcontractor has the project master schedule available and in the subcontractor's possession

5. ZZZZ presents:
- 5.1 The look-ahead schedule process
- 5.2 The short list tag process
- 5.3 The notice to correct procedure
- 5.4 The notice to clean up procedure
- 5.5 The safety violation notice procedure
- 5.6 The change request procedure
- 5.7 The RFI procedure
- 5.8 The time and material ticket procedure
- 5.9 The backcharge procedure

6. Meeting conclusion
- 6.1 By proper communication of our expectations, problems can be avoided and the work can proceed successfully.

Figure 6-2 Preconstruction meeting checklist.

3. Notice to correct
4. Disputed work or interpretation of contract scope
5. Requests for information or requests for clarification
6. Safety issues—violations and enforcement
7. Cleanup
8. Punch list work and other closeout procedures

Verifying Agreement with the Subcontractor's Field Supervisor

Prior to the subcontractor's starting work at the job site, a short meeting with the foreman or field supervisor assigned to the project by the subcontractor is advisable, if this person has not previously attended any of the subcontractor interview meetings.

The following items should be reviewed with the subcontractor's on-site supervisor:

1. Do you have, or are you familiar with, your company's fully executed subcontract agreement?
2. Do you have any questions regarding the scope of your work?
3. Do you have any questions regarding the construction schedule?
4. Have you begun to obtain and submit all required shop drawings? If not, when can we expect to receive them?
5. Are you aware of the working hours on the site?
6. Do you know which areas on the site have been set aside for your office trailer, material storage, and parking of workers' automobiles?
7. Have you received all Material Safety Data Sheets (MSDSs) pertaining to those materials you will be storing or using on site—and have you sent them to my office?
8. Do you have a copy of your company's safety plan on site?
9. Are you aware of any mock-ups that you will be required to prepare, and, if so, do you have all materials readily available to do so?
10. If preinstallation conferences for your trade are required, do you know what is expected of you?
11. If inspections or testing is required during the course of your work, are you familiar with these requirements?
12. Are you aware of the closeout procedures (if applicable) such as submission of as-builts, and manufacturer's site visits to obtain certification on certain items of work? It may also be helpful to review the various types of forms that will be used during the course of the project when the occasion arises:
    a. Superintendent's notice of pending backcharges (Fig. 6-3)
    b. Notice to clean up (Fig. 6-4)
    c. Notice to correct (Fig. 6-5)
    d. Stamp to be used on all time and material tickets (Fig. 6-6)

Compliance with Schedule Requirements

At the beginning of the project, the subcontractors will have been advised of the overall construction schedule and the time frame(s) in which they are expected to perform their work. They will have been required, or should have been requested, to review and comment on or accept this baseline schedule. If they take issue with their allotted sequence of work or the time frame in which the work is to be performed, these comments need to be addressed and resolved. As schedules change, are modified and adjusted, the subcontractor must be made aware of these change(s) and asked to review and either accept the changes or respond appropriately if the changes affect the contract sum or contract time. Once the changes are accepted, the subcontractors will be committed to the revised schedule. If they are unable to maintain the schedule owing to lack of workforce or materials or equipment, after a verbal notice of noncompliance is issued and ignored, the superintendent should notify the project manager so that a written notice to comply is transmitted to the subcontractor, referencing the paragraph in the subcontract agreement pertaining to noncompliance.

Superintendent's Notice of Pending Backcharges

Referenced Project: ____________________

Project No.: ____________ Date: ____________

To: ____________________

Attn: ____________________

Project Manager copied to (name): ____________________

This is to notify you that backcharges are pending against your contract, on the project referenced above, for the following reason(s):

- O Damages
- O Delays
- O Noncompletion
- O Backcharges by others
- O Errors in work
- O Noncompliances to safety laws/codes
- O Technical problems
- O Poor workmanship

Description and Comments:

____

____

____

____

This notification does not necessarily mean that your company will be charged. However, failure to make corrections reimburse cost, or complete work, or willful disregard of this notice will result in charges. Actual backcharge amounts will be determined by and charged against your contract at the discretion of the Project Manager. Backcharges may also include administrative cost, interest or other carrying cost, legal fees, or any other cost incurred by due to the problems or violations referenced herein.

If you have any questions concerning this notice, call me at:

____________________ Signature ____________________ Printed Name

____________________ Signature ____________________ Printed Name

Figure 6-3 Superintendent's notice of pending backcharges.

NOTICE TO CLEAN UP

Project #:
Date:

To: x
x
x
Attn: x

Ref:

Gentlemen:

Please be advised that your company has failed according to the terms and conditions of our subcontract agreement to properly clean up debris resulting from your work on the above-referenced project.

You are hereby directed to take all corrective action necessary to satisfactorily clean up your work by ____________ under terms of our agreement.

If satisfactory cleanup is not completed by the above date then will directly perform the cleanup by whatever means necessary and will backcharge to your company all resulting costs pursuant to the terms and conditions of the subcontract agreement.

Specific reference is made to the following items and areas: ____________________

____

____

____

____

____

____

____

Please be advised that this constitutes formal written notice as specified in our subcontract agreement.

Very truly yours, Received by:

____

Project Superintendent For: ____________________

Date: ____________ Time: ____________

Figure 6-4 Notice to clean up.

NOTICE TO CORRECT

Project #:
Date:

To: x
x
x
Attn: x

Ref:

☐ **EMERGENCY - IMMEDIATE RESPONSE REQUIRED**
If corrective action is not undertaken immediately will correct the conditions at subcontractor's expense.

☐ **STANDARD - RESPONSE AS NOTED BELOW**
Subcontractor must complete corrective action by: ____________

☐ **FINAL - 48-HOUR CORRECTION REQUIRED**
Subcontractor has failed to meet the conditions stated in the STANDARD Notice to Correct and unless corrective action is completed within 48 hours will complete the work at subcontractor's expense.

Gentlemen:

An inspection of your work was performed at the above-referenced project and the below-listed conditions and deficiencies were noted:

____
____
____
____
____
____
____
____

You are hereby directed to take all corrective action necessary to remedy these deficiencies under the terms of our agreement.

Please be advised that this constitutes formal written notice as specified in our agreement.

Very truly yours, Received by:

________________ ________________

Project Superintendent For: ____________

Date: ____________ Time: ________

Figure 6-5 Notice to correct.

A typical subcontractor agreement schedule compliance clause reads as follows:

> Subcontractor agrees to complete the subcontractor's work by (date). Subcontractors agree to commence the subcontractor's work (date to be filled in by the general contractor) in the most expeditious fashion but not later than indicated on the general contractor's construction schedule.

or

> The subcontractor shall commence work upon receipt of a notice to proceed from the contractor and shall prosecute the work in a manner that will not delay the completion of the prime contract and in accordance with any schedule provided by the contractor, which schedule may be changed from time to time. The subcontractor shall be entitled to such extensions of time as the contractor shall receive from the owner and as applicable to the subcontractor's work and not other damages arising out of a delay apart from such damages the contractor may receive from the owner on behalf of the subcontractor.

The determination of whether the referenced work to be performed or already performed is in-contract or not in-contract is subject to further review by and/or our customer in accordance with the terms and conditions contained in our Subcontract Agreement with your firm. The signature of an authorized employee below provides verification only that the work needs to be performed or has already been performed. It does not establish or accept the validity, dollar value or time impact of a claim.

BY:

For

DATE:

Print Name

Figure 6-6 Stamp to be used on time and material tickets (fill in name of general contractor).

Notice of Nonperformance

The ability to control the performance of a subcontractor is crucial to the administration of a successful construction project. When a subcontractor fails to meet the reasonable performance requirements of a project and verbal notification fails to correct nonperformance, the superintendent and project manager must review the subcontract agreement to determine which provision pertains to performance, or nonperformance, and how it is to be conveyed to the subcontractor.

The subcontract agreement provision for nonperformance is very similar to the schedule compliance requirement; however, most nonperformance clauses include specific notification rules and a time frame in which the subcontractor is given an opportunity to correct the problem. A typical provision allows the subcontractor anywhere from 48 to 72 hours' leeway to increase the pace of work, and this clause also alerts the subcontractor to the penalties that will accrue if performance is not accelerated. A standard nonperformance provision is set forth below:

> Time is of the essence.
> Should the subcontractor fail to prosecute the work or any part thereof with promptness and diligence, or fail to supply a sufficiency of properly skilled workmen or materials of proper quality or fail in any other respect to comply with the Contract Documents, the Contractor shall be at liberty, after seventy-two (72) hours written notice to the Subcontractor, to provide such labor or materials as may be necessary to complete the work and to deduct the cost and expense thereof from any money then due or thereafter to become due the Subcontractor under this Agreement, and the Contractor shall be at liberty to bar subcontractor from the job and take possession for the purpose of completing the work included under the Contract Documents, all of materials, scaffolding, ways, works, apparatus, machinery, equipment and appliances thereon, and to employ or contract with any other person or persons to finish the same.

When nonperformance is an issue, either a letter or an appropriate form should be promptly sent to the subcontractor. Even though a verbal notice of poor or nonperformance has been transmitted to the subcontractor's field representative or project manager, written notification must follow promptly.

Company Letterhead

Oriole Construction Company
566 Southway
Baltimore, Maryland 21200

The American Steel Company
855 Industrial Circle
Owings Mills, MD 21240

Re: Waterfall Plaza
Project No. 6444

Attention: Mr. Jim Beam

Dear Mr. Beam:

On (date) we sent you a letter requesting your recovery plan for accelerating your work in order to meet our schedule requirements. To date, we have not received your response to this request.

Following the provisions of (paragraph or article in the subcontract agreement pertaining to performance) in our subcontract agreement with your company, dated (date), we are hereby giving you the required (48- or 72-hour, or the time frame stated in the subcontract agreement) notification to increase your performance, or we will provide any additional manpower and materials required to do so. All related costs in this matter will be backcharged to your account.

With best regards,

Will Spencer
Project Superintendent

Letter 37 Putting the subcontractor on notice regarding poor performance.

Company Letterhead

Oriole Construction Company
566 Southway
Baltimore, Maryland 21200

The American Steel Company
855 Industrial Circle
Owings Mills, MD 21240

Re: Waterfall Plaza
Project No. 6444

Attention: Mr. Jim Beam

Dear Mr. Beam:

Reference is made to our (date) letter in which we gave you the required notification to increase your performance at the job site.

Since you have failed to respond to our request, we have engaged another subcontractor (or vendor) to (describe the work that is underperforming, i.e., install the doors, paint the walls, etc.).

Upon completion and acceptance of the subcontractor's (or vendor's) work, we will issue a backcharge to your account for all costs associated with this work, including our supervisory and project management costs along with an appropriate amount for overhead and profit.

With best regards,

Will Spencer
Project Superintendent

Letter 38 Hiring another subcontractor to supplement the poorly performing subcontractor.

Company Letterhead

Oriole Construction Company
566 Southway
Baltimore, Maryland 21200

The American Steel Company
855 Industrial Circle
Owings Mills, MD 21240

Re: Waterfall Plaza
Project No. 6444

Attention: Mr. Jim Beam

Dear Mr. Beam:

Attached is a copy of the architect's (or engineer's) punch list dated (date). Please correct and/or complete all indicated items pertaining to your trade, not later than (date).

If you have any questions relating to any of the items on this list, please let us know; if not, you will be expected to complete all of your listed work.

Notify this office when your work has been completed so that we can obtain the architect's (or engineer's) acceptance.

With best regards,

Will Spencer
Project Superintendent

Letter 39 Transmitting the punch list to a subcontractor.

Company Letterhead

Oriole Construction Company
566 Southway
Baltimore, Maryland 21200

The American Steel Company
855 Industrial Circle
Owings Mills, MD 21240

Re: Waterfall Plaza
Project No. 6444

Attention: Mr. Jim Beam

Dear Mr. Beam:

On (date) we forwarded the architect's (or engineer's) punch list dated (date) to your office with instructions to complete and notify this office.

During a walkthrough on (date) we found that most (or all, or some) of your punch list work remains incomplete. You are requested to expedite this work and complete your portion of the punch list not later than (date).

Please advise when this work is ready for inspection and sign-off.

With best regards,

Will Spencer
Project Superintendent

Letter 40 When punch list work is ongoing, but incomplete.

Company Letterhead

Oriole Construction Company
566 Southway
Baltimore, Maryland 21200

The American Steel Company
855 Industrial Circle
Owings Mills, MD 21240

Re: Waterfall Plaza
Project No. 3444

Attention: Mr. Jim Beam

Dear Mr. Beam:

Letters to your office on (dates) regarding incomplete punch list items have failed to achieve completion of your punch list work. Therefore, we plan to engage another subcontractor to complete all incomplete items on the punch list.

When this work has been completed, inspected and accepted by the architect (or engineer), all related costs will be charged to your account.

With best regards,

Will Spencer
Project Superintendent

Letter 41 Subcontractor fails to complete punch list work as directed.

Company Letterhead

Oriole Construction Company
566 Southway
Baltimore, Maryland 21200

The American Steel Company
855 Industrial Circle
Owings Mills, MD 21240

Re: Waterfall Plaza
Project No. 6444

Attention: Mr. Jim Beam

Dear Mr. Beam:

You have failed to complete your punch list work even though we have requested you to do so on several occasions (or state dates of letters sent requesting completion). On (date), the architect (or engineer) established the following values for each incomplete punch list item pertaining to your trade:

If any of these items remain incomplete after (date), we will deduct its value from any monies due and owing your company.

With best regards,

Will Spencer
Project Superintendent

Letter 42 Deducting value of punch list items from subcontractor's account.

Company Letterhead

Oriole Construction Company
566 Southway
Baltimore, Maryland 21200

The American Steel Company
855 Industrial Circle
Owings Mills, MD 21240

Re: Waterfall Plaza
Project No. 6444

Attention: Mr. Jim Beam

Dear Mr. Beam:

We call your attention to our baseline schedule dated (date or revision date), according to which you were to have started your work on (date).

As of this date, you have failed to carry out the agreed-upon start of work. Please advise, in writing, when you plan to commence work and what steps you will take to meet your original completion date.

With best regards,

Will Spencer
Project Superintendent

Letter 43 Failure to meet start date of work per schedule.

Company Letterhead

Oriole Construction Company
566 Southway
Baltimore, Maryland 21200

The American Steel Company
855 Industrial Circle
Owings Mills, MD 21240

Re: Waterfall Plaza
Project No. 6444

Attention: Mr. Jim Beam

Dear Mr. Beam:

We call your attention to our baseline schedule dated (date or revision date), according to which you were to have completed (tasks to have been completed) by (date).

That work is still in progress and it appears that it will not be fully completed for another (day, week, etc.), that is, until (date). Your delay in completing your work is affecting other trades and must be improved upon.

We require an immediate written response with your proposed method of getting back on schedule.

With best regards,

Will Spencer
Project Superintendent

Letter 44 When it appears that the subcontractor will be late in completing work.

Notice to Correct

During an architect/engineer inspection, reports will be issued to the general contractor recapping the results of that inspection and whether certain items of work comply with the contract requirements or whether specific items require removal, modification, or replacement to meet the terms and conditions of the contract.

During a walk through the project by the project superintendent or project manager, inspections may also reveal corrective action required by a subcontractor or subcontractors. Items to be corrected generated by these types of inspections can be verbally transmitted to the subcontractor(s) involved, but these items also need to be confirmed in writing in the event that the verbal notification goes unnoticed. A simple form such as the one displayed in Fig. 6-5 is often adequate for the purpose. When such a deficiency or notice to correct report is produced by the A/E, the general contractor is put on notice that if the work is not corrected within a reasonable period of time, an appropriate sum will be withheld from a future payment. When this involves subcontractor work, a similar notification to the subcontractor is required, making reference to the appropriate provision in the article of the subcontract agreement.

A typical notice to correct provision in a subcontract agreement is similar to this:

> The subcontractor shall provide sufficient, safe and proper facilities at all times for the inspection of the work by the Architect and Owner or by the Contractor or their authorized representatives and shall within twenty-four (24) hours of receiving notification from the architect or the contractor to that effect remove from the grounds or buildings, all work or materials condemned by them, whether worked or unworked, and shall take down and remove all portions of the work which the Architect or Contractor shall, by notice, condemn as unsound, improper or defective or in any way failing to conform to the contract documents or to the instruction of the Architect or Contractor, and the subcontractor shall at once remedy, replace or make good all its work so removed and all work damaged or destroyed by such removal and the replacement thereof, provided, however, that no inspection or failure to inspect by the Contractor or Architect shall relieve the subcontractor of any obligations imposed by the contract documents.

Disputed Work or Interpretation of Contract Scope—RFIs and RFCs

Disagreement over what constitutes *contract scope of work* is not unusual when you are dealing with subcontractors who may be interpreting their contract requirements differently from the general contractor's or designer's viewpoint. When scope issues are involved and relate to interpretation of plans and/or specifications, the subcontractor should be directed to prepare a request for clarification that will allow the general contractor to pass it on to the "interpreter of the contract documents"—the design architect or engineer. If upon receipt of the response from the architect the subcontractor disagrees with the ruling, the subcontractor can issue a formal protest to the general contractor which will be processed in accordance with either the provisions in the contract with the owner or the subcontract agreement, whichever document contains resolution of such matters.

Most agreements between subcontractor and contractor link this agreement to the contract with the owner with a simple statement, such as

> The Subcontractor shall be bound to the Contractor by the terms and provisions of all of the contract documents and assumes toward the Contractor, with respect to the Subcontractor's work, all of the obligations and responsibilities which the Contractor, by the contract documents, has assumed toward the Owner.

Company Letterhead

Oriole Construction Company
566 Southway
Baltimore, Maryland 21200

The American Steel Company
855 Industrial Circle
Owings Mills, MD 21240

Re: Waterfall Plaza
Project No. 6444

Attention: Mr. Jim Beam

Dear Mr. Beam:

During an inspection of your work on (date), we rejected (indicate the specific items of work that were rejected and the reasons why). You are requested to remove and replace this defective work not later than (date).

Failure to do so promptly will not only result in additional costs being charged to your account, but may also delay any current payments due and owing your company.

With best regards,

Will Spencer
Project Superintendent

Letter 45 Rejecting defective/nonconforming work.

Company Letterhead

Oriole Construction Company
566 Southway
Baltimore, Maryland 21200

The American Steel Company
855 Industrial Circle
Owings Mills, MD 21240

Re: Waterfall Plaza
Project No. 6444

Attention: Mr. Jim Beam

Dear Mr. Beam:

On (date) we sent a letter to your office advising you of work rejected for (poor quality, nonconformance with the contract documents, etc.). As of this date this work has not been corrected.

Failure to comply with our request by (date) will result in our engaging another subcontractor to remove and replace this (defective, nonconforming, etc.) work at your expense.

Your immediate response is requested or else we will proceed with any required corrective action.

With best regards,

Will Spencer
Project Superintendent

Letter 46 Second request to correct rejected work.

Company Letterhead

Oriole Construction Company
566 Southway
Baltimore, Maryland 21200

The American Steel Company
855 Industrial Circle
Owings Mills, MD 21240

Re: Waterfall Plaza
Project No. 6444

Attention: Mr. Jim Beam

Dear Mr. Beam:

Your failure to comply with our previous requests of (dates of previous warning letters) to remove and replace defective work has resulted in our engaging another subcontractor to complete these tasks.

Upon completion and acceptance of this work, we will assemble all related costs and backcharge your account.

With best regards,

Will Spencer
Project Superintendent

Letter 47 Engaging another subcontractor to correct defective work.

Company Letterhead

Oriole Construction Company
566 Southway
Baltimore, Maryland 21200

The American Steel Company
855 Industrial Circle
Owings Mills, MD 21240

Re: Waterfall Plaza
Project No. 6444

Attention: Mr. Jim Beam

Dear Mr. Beam:

We are enclosing a copy of the (architect's or engineer's) (letter, field report, memo) dated (date) and call your attention to (that part of the letter, field report or memo dealing with the rejected or nonconforming work). Please comply with the (architect's or engineer's) findings and correct the work as required. Upon completion advise this office so that we may schedule a reinspection.

With best regards,

Will Spencer
Project Superintendent

Letter 48 Notification of an inspection by the architect/engineer and the rejection of work.

Company Letterhead

Oriole Construction Company
566 Southway
Baltimore, Maryland 21200

The American Steel Company
855 Industrial Circle
Owings Mills, MD 21240

Re: Waterfall Plaza
Project No. 6444

Attention: Mr. Jim Beam

Dear Mr. Beam:

We are in receipt of your (date) letter in which you take exception to (either our rejection of work or the architect/engineer's rejection of work).

Please provide a detailed explanation of your position in this matter so that we may (render our opinion — if rejected by the GC — or submit it to the architect/engineer for review); we will respond appropriately.

With best regards,

Will Spencer
Project Superintendent

Letter 49 Responding to a letter in which a subcontractor disagrees with the reason for rejecting work.

Company Letterhead

Oriole Construction Company
566 Southway
Baltimore, Maryland 21200

The American Steel Company
855 Industrial Circle
Owings Mills, MD 21240

Re: Waterfall Plaza
Project No. 6444

Attention: Mr. Jim Beam

Dear Mr. Beam:

Attached is the (architect's/engineer's) response to your (date) letter in which you take issue with (his/her/its) finding(s). If you wish to pursue this matter further, you may file a claim in accordance with the provisions included in your subcontract agreement.

In the meantime, you are directed to immediately correct the work as specified or we will engage other forces to do so and charge your account accordingly.

With best regards,

Will Spencer
Project Superintendent

Letter 50 Responding to a letter in which a subcontractor takes issue with the finding(s) of the architect/engineer.

Company Letterhead

Oriole Construction Company
566 Southway
Baltimore, Maryland 21200

The American Steel Company
855 Industrial Circle
Owings Mills, MD 21240

Re: Waterfall Plaza
Project No. 6444

Attention: Mr. Jim Beam

Dear Mr. Beam:

Enclosed is an inspection report dated (date) as received from the architect (or engineer or inspection service). We call your attention to (Section X, or sections of the document pertaining to the subcontractor) that indicates rejection of your work (or remediation required).

Please advise by return mail (or e-mail or fax) as to the process you plan to implement to correct the work rejected by the architect (or engineer) and when you anticipate the corrective work will be complete and ready for reinspection.

With best regards,

Will Spencer
Project Superintendent

Letter 51 Transmitting an inspection report requiring rework.

Company Letterhead

Oriole Construction Company
566 Southway
Baltimore, Maryland 21200

The American Steel Company
855 Industrial Circle
Owings Mills, MD 21240

Re: Waterfall Plaza
Project No. 6444

Attention: Mr. Jim Beam

Dear Mr. Beam:

We refer to the inspection report dated (date) submitted by the architect (or engineer or inspection service) and forwarded to you on (date) with a request to respond with your proposed corrective action.

As of this date we have not received your response, and in accordance with the provisions of (paragraph or article in the subcontract agreement that gives the subcontractor X days to correct) be advised that unless this corrective work is completed, inspected and approved not later than (date), we will proceed with whatever actions are necessary to obtain the architect's (or engineer's) approval.

When all related costs have been received and tabulated, the appropriate backcharge will be issued to your company.

With best regards,

Will Spencer

Project Superintendent

Letter 52 When the rework required by an inspection report has not been performed by the subcontractor.

Company Letterhead

Oriole Construction Company
566 Southway
Baltimore, Maryland 21200

The American Steel Company
855 Industrial Circle
Owings Mills, MD 21240

Re: Waterfall Plaza
Project No. 6444

Attention: Mr. Jim Beam

Dear Mr. Beam:

On (date) we sent you a letter requesting (whatever was requested; e.g., clean-up by a certain date, accepting the schedule by a certain date — any matter that requires a response of acceptance). If we do not have your written response by (date), this will be taken to mean acceptance of the referenced matter.

With best regards,

Will Spencer
Project Superintendent

Letter 53 Letter forcing a response.

The provisions of AIA Document A201, General Conditions of the Contract for Construction, also apply to the contract between contractor and subcontractor. Therefore, in the absence of any other language, Article 4, Administration of the Contract, is applicable specifically.

Articles 4.2.11, 4.2.12, and 4.2.13 assign the decision-making process with respect to contract document interpretation to the architect. This assignment of design interpretation to the architect seems a one-sided affair in that the designer who prepared the documents is given the authority to decide on their completeness or "intent." It is also appropriate that Article 4.3 which follows is entitled *Claims and Disputes.*

Safety Issues

One of the project superintendent's more important tasks concerns the administration and implementation of the company's safety policy. Each subcontractor on the site will be required to comply with the provisions of that program. The subcontract agreement, recognizing the importance of job site safety, will generally include an article devoted to this subject, such as

Company Letterhead

Oriole Construction Company
566 Southway
Baltimore, Maryland 21200

The American Steel Company
855 Industrial Circle
Owings Mills, MD 21240

Re: Waterfall Plaza
Project No. 6444

Attention: Mr. Jim Beam

Dear Mr. Beam:

We are in receipt of your letter (or e-mail or fax) in which you claim that the contract documents, as you interpret them, do not require you to (portions of work in dispute).

We disagree with your interpretation and are forwarding your letter to the architect (or engineer) for (his/her/its) ruling in this matter. We will advise you of the response when received.

Note: If you need further reinforcement of this issue, refer to AIA A201, General Conditions, Articles 4.2.12, 4.2.13, and 4.4 (Claims).

With best regards,

Will Spencer
Project Superintendent

Letter 54 When a subcontractor's interpretation of contract obligations differs from the general contractor's.

Company Letterhead

Oriole Construction Company
566 Southway
Baltimore, Maryland 21200

The American Steel Company
855 Industrial Circle
Owings Mills, MD 21240

Re: Waterfall Plaza
Project No. 6444

Attention: Mr. Jim Beam

Dear Mr. Beam:

We refer to your letter of (date) in which you took issue with your contract responsibilities to (disputed items of work). We forwarded that letter to the architect (or engineer) for review and comment, and enclosed is the response dated (date).

If you wish to pursue this matter further, you must file a formal claim in accordance with the provisions in the contract documents. Unless you choose to file a claim, we consider this issue closed and request that you perform the work per your contract obligations.

With best regards,

Will Spencer
Project Superintendent

Letter 55 When a subcontractor's interpretation of contract obligations differs from the general contractor's—follow-up letter with architect/engineer's response.

> The subcontractor shall take all reasonable safety precautions with respect to the work, shall comply with all safety measures initiated by the Contractor and with all applicable laws, ordinances, rules and regulations and orders of any public safety authority for the safety of persons or property in connection with its performance hereunder. Subcontractors shall take whatever precautions are necessary to properly protect the work of other trades from damage caused by any operations.

The general contractor's safety program usually includes a provision for weekly *toolbox talks* to be held, attended by the GC's own field workers and by subcontractor tradespeople. A specific safety topic is discussed at each of these weekly meetings, and all attendees are required to sign in as an acknowledgment of their attendance. Although brief, generally 15 to 20 minutes, these toolbox talks act as a refresher to experienced workers and as a learning experience for new hires.

The format for a typical toolbox talk is shown in Figs. 6-7 and 6-8. There are a number of companies specializing in offering or assisting a general contractor in the preparation of safety programs, and these companies often offer subscriptions to an entire program of toolbox talks along with the related forms.

TOOLBOX TALK NO. 12

Concrete Construction Safety

Concrete construction can take three distinct forms:

1. Cast-in-place concrete
2. Precast concrete
3. Tilt-up concrete

Hazards associated with concrete construction of all types:

1. Fails
2. Caught-in-between
3. Impalement on exposed reinforcing bars or welded wire mesh
4. Struck by falling objects
5. Spatters on skin and in eyes

General safety requirements:

- Obtain assurance that the structure can support the additional load to be imposed by the placement of concrete. This information is generally available from the structural engineer.
- Place caps on exposed rebar, or otherwise place guards around areas where the pour is to take place, if there is a chance that a worker may fall and be impaled on exposed rebar.
- Wear appropriate personal protective gear to avoid contract with skin and eyes.
- When using a crane to lift and place overhead buckets, route the travel so that the fewest number of workers are exposed to hazards associated with this method of placement. No one should be working under the bucket.
- Watch that no one is standing in the way of a ready-mix truck maneuvering or backing up on the site.
- When pumping concrete, ensure that the concete is flowing properly and no area in the hose to the delivery tube is blocked at any time.
- If posttensioning is being performed, erect signs and limit employee access to the posttensioning area. Do not permit any employee not essential to the operation to stand behind the jack during the posttensioning process.
- When lifting tilt-up panels, check to see if the lifting hook is located properly and is firmly embedded and the crane is able to maintain the center of gravity required.

Figure 6-7 Format for a concrete construction safety toolbox talk.

TOOLBOX TALK NO. 12

Concrete Construction Safety

Sign-in Sheet

This sign-in sheet documents that the undersigned employees of (Company) have taken part in the training session on concrete construction safety, held on (date) at (location).

This Toolbox Talk covered the following:

- Three types of concrete construction
- Hazards associated with concrete construction of all types
- General safety requirements

Employee signature	Print name and company affiliation here
____________	____________
____________	____________
____________	____________
____________	____________
____________	____________
____________	____________
____________	____________
____________	____________

____

Supervisor's Signature

Figure 6-8 Typical toolbox talk attendance sheet.

Job Cleaning and Subcontract Provisions to Enforce This Task

Possibly no event provokes greater controversy between subcontractor and project superintendent than period or progress cleaning of the job site and related removal of debris and waste materials from the job site. Subcontractors frequently promise to clean their areas by a certain date and fail to do so; or a subcontractor will argue that all or a major portion of the debris that he or she has been requested to remove was actually generated by other subcontractors. As the project superintendent's patience begins to wear thin, a review of the subcontract agreement to highlight the provision relating to cleaning should be made before a letter of notification is sent to the subcontractor to clean the area, remove all the trash off-site—or else. Quoting the exact provision in the subcontract agreement relating to cleaning and possibly another article relating to notice to correct in the letter to the subcontractor is effective. A typical cleaning provision will follow this format:

> The Subcontractor shall at all times keep the Project site free from rubbish, debris and waste, and/or surplus materials resulting from its operations and shall turn over the subcontractor work in such a condition as to permit the next succeeding or intervening work to be commenced without further cleaning. At the time of completion of the subcontract work, such work is to be clean and in a condition acceptable to the Owner. If the Subcontractor fails to comply with the provisions of this paragraph, after 24 hours written notification by the Contractor, the Contractor shall have the right itself or through others to perform such cleaning and to charge the cost thereof to the Subcontractor.

Company Letterhead

Oriole Construction Company
566 Southway
Baltimore, Maryland 21200

The American Steel Company
855 Industrial Circle
Owings Mills, MD 21240

Re: Waterfall Plaza
Project No. 6444

Attention: Mr. Jim Beam

Dear Mr. Beam:

Please provide the necessary labor and materials to clean your work areas per the provisions of your subcontract agreement. We request that this work commence immediately upon receipt of this (letter, e-mail, fax).

With best regards,

Will Spencer
Project Superintendent

Letter 56 First request to a subcontractor to clean work areas.

Company Letterhead

Oriole Construction Company
566 Southway
Baltimore, Maryland 21200

The American Steel Company
855 Industrial Circle
Owings Mills, MD 21240

Re: Waterfall Plaza
Project No. 6444

Attention: Mr. Jim Beam

Dear Mr. Beam:

In response to our repeated requests to clean your work areas, you now indicate that the trash remaining is that of another subcontractor.

We are proceeding to engage forces to clean your work areas and charge your account for all associated costs. If you disclaim responsibility for the generation of debris and trash in your areas, it is up to you to resolve this matter with the subcontractor you say did deposit the trash. In no way does this relieve you of your cleaning responsibility.

With best regards,

Will Spencer
Project Superintendent

Letter 57 Responding to a letter in which a subcontractor claims trash was generated by another subcontractor.

A Project Cleaning Checklist

1. Establish firm job cleanup rules and procedures at the first on-site subcontractors' meeting. Include this statement in the meeting minutes in that initial meeting, and include it in all subsequent minutes—as a reminder to all attendees.
2. Report both good and bad cleaning performance at each subcontractor's meetings.
3. When you are walking the site, comment on good cleaning activities and let the supervisor know that you appreciate the effort. A thank-you note to the subcontractor's office can achieve a great deal.
4. When poor performance or nonperformance is noted, don't delay in issuing a verbal notice to correct; and if that doesn't work, issue a written one.
5. Never threaten without being ready to carry out that threat. If no action results from the notice to correct or notice to clean up, within the time frame contained in the subcontract agreement, promptly have other forces clean the area in question.
6. Prompt submission of the costs to clean and dispose of debris (including the general contractor's supervision costs and applied overhead and profit) should be sent to the subcontractor's office to warn of the impending backcharge.

Change Orders and the Subcontractor

The change order process, as it relates to subcontractor involvement, is threefold:

1. Request for change in scope by owner, architect, or engineer for which the subcontractor will become involved
2. Request for change order by the general contractor for nonowner-generated changes
3. Request for change order by the subcontractor for perceived errors and omissions in the plans and specifications or in the scope of work, as outlined in the subcontract agreement

Given that changes to the contract scope can increase or decrease the scope of work, costs can either increase or decrease accordingly. When costs increase, they are generally scrutinized pretty closely; but often when credits are presented for scope decreases, that same amount of attention is not displayed. Change orders can also affect contract time, either increasing or decreasing it.

Each company has its own policy when it comes to change order authorization, allowing some superintendents limited authorization for some change order requests of a specific scope or dollar value, while other companies place this responsibility solely in the hands of the project manager. Whatever the case may be, the project superintendent can play an important role in this process.

Requests by Owner

Requests for changes to the contract originating from the owner are often presented by the owner's design consultants, and these requests are usually accompanied by a drawing and/or specification revisions, sketches, or a written description of the work to be modified. These directions or instructions are distributed to subcontractors and vendors affected by the change, and it is usually helpful if the project superintendent reviews these changes, when received, with the subcontractor or vendor. Clear understanding as to what is being requested, whether deleted work or extra work, how the associated costs are to be presented needs to be reviewed.

Company Letterhead

Oriole Construction Company
566 Southway
Baltimore, Maryland 21200

The American Steel Company
855 Industrial Circle
Owings Mills, MD 21240

Re: Waterfall Plaza
Project No. 6444

Attention: Mr. Jim Beam

Dear Mr. Beam:

Attached are (plans, sketches, specifications, letters, etc.) for your review and pricing. Please submit your cost proposal in the following format:

Labor: trade classification, hourly rate and number of hours
Materials: list separately
Equipment: rental rates as applicable — hourly, daily, weekly, monthly if an operator is required, list separately per Labor above

List any additional or miscellaneous costs separately, i.e., small tools, insurance, permits, etc. Attach proposals from subcontractors, if applicable. When estimates include adds and deducts, apply overhead and profit percentages on net amount. Refer to the contract documents for allowable overhead and profit percentages for your work and that of lower-tier subcontractors.

Please respond not later than (date).

With best regards,

Will Spencer
Project Superintendent

Letter 58 Requesting a cost proposal for change-order work.

Company Letterhead

Oriole Construction Company
566 Southway
Baltimore, Maryland 21200

The American Steel Company
855 Industrial Circle
Owings Mills, MD 21240

Re: Waterfall Plaza
Project No. 6444

Attention: Mr. Jim Beam

Dear Mr. Beam:

On (date) we sent you (plans, sketches, specifications, letters, etc.) relating to a proposed change in the work. We requested that you respond not later than (date); however, as of this date, we have not received your response.

Please expedite this cost proposal to our office.

With best regards,

Will Spencer
Project Superintendent

Letter 59 Follow-up letter requesting cost proposal submission.

Company Letterhead

Oriole Construction Company
566 Southway
Baltimore, Maryland 21200

The American Steel Company
855 Industrial Circle
Owings Mills, MD 21240

Re: Waterfall Plaza
Project No. 6444

Attention: Mr. Jim Beam

Dear Mr. Beam:

We submitted our proposed change order (if it has a number, include it) for (describe the work) to the architect (or engineer or owner), which included (dollar value of subcontractor's work as submitted in its cost proposal) pertaining to your work.

Upon review, the architect (or engineer or owner) questioned the cost of the work in your proposal.

Please provide further documentation justifying the costs and scope of work included in your proposal so that we may forward it to the architect (engineer or owner).

With best regards,

Will Spencer
Project Superintendent

Letter 60 When the architect/engineer questions a subcontractor's costs.

Particularly when some portions of work are to be deleted and other portions of work are to be added, a clear understanding of how credits and extras are to be presented will help process the request smoothly.

The contract with the owner often includes specific language regarding *adds* and *deducts,* stipulating that the cost of the deleted work is to be subtracted from the cost of the added work before any overhead and profit percentages are added. The contract may also specify the nature and extent of documentation to support the credit that is to be applied against the added costs.

The subcontractor ought to provide sufficient documentation for the changed conditions to allow the owner to reasonably evaluate the quotation being presented. At a minimum the subcontractor should include the following information:

1. Break down labor costs, including the division of labor employed, i.e., laborer, mechanic, supervisor. The number of hours for each worker proposed should be listed along with the applicable hourly rate including fringe benefits. Quite often an owner will request additional labor rate breakdowns—the basic hourly rate plus a listing of each fringe benefit that all together equals the *burdened rate.* So the subcontractor preparing the estimate should be able to back up any detail of its published labor rate, if requested.

2. Cost of materials is documented with invoices reflecting the actual purchase of the material(s) or invoices for similar materials recently purchased.

3. Rental equipment can take two forms: equipment from a recognized rental company supplying these items or subcontractor-owned equipment that will be billed at competitive hourly or daily rental rates.

4. When equipment is purchased specifically for the change order work, the owner may request that the equipment be turned over to her or him when the work has been put in place. Or the owner may require the contractor to issue a credit for the residual value of the equipment after the work for which it was purchased has been completed.

5. The small tool issue is often questioned by owners who see a dollar amount for "small tools" or a percentage of the total cost of work, less overhead and profit, included for this item. The subcontractor should offer an explanation for this small tool item if included in the quote, or at least be prepared to defend it, if and when questioned by the owner.

6. Last, but not least, the subcontractor's overhead and profit will be added to the costs. If the contract's general conditions or special conditions relating to change orders contain a limitation on second- and third-tier subcontractor fees, or contain a graduated fee structure based upon the value of the work, then these restrictions ought to be brought to the subcontractor's attention. And remember, if the change requires adds and deducts, a full accounting of the deleted work must be included.

These procedures should be followed for all change order work. *View the change order preparation from the perspective of the person or organization receiving the proposal. "Is there enough background, documentation, and breakdown of costs included that I can intelligently review this proposal and fully understand the changes that will take place and their related costs?"*

Requests by the General Contractor

There are at least two reasons why a general contractor will want to consider issuing a change order to the subcontractor or subcontractors:

1. To include items of work previously omitted from the subcontract agreement by either design or error.
2. To incorporate additional work activities into the subcontractor's scope of work to effect better coordination or single-point responsibility. Examples could include adding wood blocking into the drywall subcontractor's work or including masonry waterproofing in the mason contractor's scope of work.

Once again, the project superintendent's input can be valuable in reviewing the subcontractor's proposal to comment on the reasonableness of the costs, the impact the added work will have on the schedule, and whether any work had been deleted for which no credits had been included in the proposal.

Requests by the Subcontractor

Requests for change orders emanating from the subcontractor can encompass questions relating to his or her interpretation of the contract documents or the scope of work presented and negotiated into the subcontract agreement, or from damages, real or alleged, arising out of the general contractor's action or inactions.

In the first case, these requests for change orders relating to contract interpretation need to be passed through to the owner, and this may result in an increase in both the general contractor's and the subcontractor's scope of work. But when the owner rejects any claim for extra work because of the architect's ruling in the matter, the general contractor and subcontractor need to discuss the matter to determine if they wish to pursue the issue further or withdraw their request.

Subcontractor Claims for Extra Work Where There Is No Owner Reimbursement

When the scope of work in the subcontract agreement is not fully defined, it can often lead to a claim for extra work that cannot be passed on to the owner. A subcontract agreement solely defining the scope of work as *per plans and specifications* is ripe for disagreements to occur. Some subcontractors dismiss the *intent* of the contract and rely on the strict language of the plans and specifications to request extras. For example, the hardware specification section requires the subcontractor to install hardware on all interior and exterior doors. However, the specific hardware set for the aluminum entry doors is listed in another specification section, say, storefront work. Is the subcontractor entitled to an "extra" to install this hardware by stating that these items were not included in the finish hardware specification section? Some specifications sections include a paragraph entitled *Related Work,* and these types of disagreements may often be settled if such a paragraph was included in either specification section. By the strict interpretation of the subcontract agreement that references the hardware specification section but not the storefront specification section, they may have a point. But why wasn't the issue raised when they were awarded the contract? Does it seem unreasonable to hold the subcontractor responsible for this work? Many of these types of disagreements can be approached by using the argument of "reasonableness."

The Time and Material Trap and How to Avoid It

Time and material work is often the method of choice when you decide to proceed with some types of "extra" work. In many cases this type of authorization is considered the most equitable way to proceed with work in an emergency or when exact costs are difficult to determine beforehand. But in other cases, time and material work becomes the basis for arguments that escalate into disputes and claims. When what is more familiarly referred to as *T&M work* proceeds with established controls are in place, few problems occur. When controls are lacking, problems surface.

Control over T&M work involves the following:

1. *Documentation of all labor expended.* This is accomplished by having the subcontractor or vendor prepare daily labor tickets for verification by the contractor's on-site representative or the superintendent. Verification is to confirm the number of hours worked and the number of workers. The actual work performed that day, written right on the ticket, may prove to be invaluable when the scope of work is in question.
2. *Prior knowledge and agreement of the labor rates.* The subcontractor or vendor should be requested to submit a breakdown of labor costs for each type of tradesman to be used in the T&M work. Figure 6-9 is an example of a complete hourly labor rate that includes all costs and overhead and profit percentages.
3. *An agreed upon percentage for overhead and profit.* This should have been either previously established by negotiation with the subcontractor/vendor or included in the contract or specifications. If no specific percentage has been established, do so as the first cost proposal is being prepared, because this will serve as the precedence for all future change order work.
4. *Procedure for verifying materials used or equipment rented or leased.* Receiving tickets for materials, signed by the authorized subcontractor representative, is generally sufficient to satisfy the materials issue, but costs for equipment rentals may be somewhat vague. If the subcontractor is using her

BREAKDOWN OF HOURLY RATES

Worker's Title:	LABORER	MA DRILLING JOBS	
	Straight Time	$1^1/_2$-Time Premium	Double-Time Premium
Base Wage Rate	23.35	11.68	23.35
FICA 7.65%	1.79	0.89	1.79
FUTA .80%	0.19	0.09	0.19
SUTA 7.42%	1.73	0.87	1.73
Gen. Liability	1.03		
Workers' Comp.	2.63		
Welfare Fund	3.50		
Pension Fund	6.90		
Apprentice Fund	0.35		
Vacation Fund			
Ed. & Cult. Fund	0.95		
Deferred Income Fund			
Paid Holidays			
Bond Premium			
Incidentals			
Other: Umbrella	0.57		
Subtotal	42.99	13.53	27.06
Overhead and Profit (10%) + (5%)	6.66	2.10	4.19
Total	49.65	15.62	31.25

Figure 6-9 Complete hourly rate for laborer breakdown.

or his own equipment, it is important to verify hours in use. The "cost" of the rental equipment can be resolved if both parties agree that the hourly rate of any subcontractor- or contractor-owned equipment will not exceed the rental rates of an equipment rental company. Prior agreement to rates when the equipment is "idle," i.e., not working while stationed on the T&M project, can be established. This will generally apply only to rather large pieces of equipment such as bulldozers, backhoes, excavators, compactors, hoists, and cranes.

A Daily Ticket Checklist

1. Require the subcontractor to submit the daily T&M by the close of business each day, so the work can be reviewed and verified while fresh in everyone's mind.
2. Each ticket should be reviewed for correct data:
   a. Personnel are identified and applicable hours assigned.
   b. Equipment is identified—statement as to active or idle time is ascertained.
   c. Materials are itemized. If receiving tickets are available, attach a copy.
3. Description of work completed or in progress is noted on the T&M ticket. If area on site or in the building is easily definable, include the location.

4. On all accepted tickets, indicate acceptance. On all rejected tickets, indicate the reason for rejection along with a request to void or resubmit ASAP.
5. On all tickets where disagreement exists, note the area of disagreement and write *Disputed* on the ticket. Note the area of disagreement: personnel, hours, equipment, idle and active hours, materials, quantities, costs, portion of "contract" work included on T&M ticket.
6. Meet with the project manager promptly to resolve any disagreements.

The Superintendent's Limited Authority to Approve T&M Work

If the subcontractor is required to do work that may or may not be clearly defined as an extra but is proceeding with T&M work, what is your company's policy with respect to authorization to proceed? Generally this decision is arrived at in consultation with the project manager.

The project manager and the project superintendent should come to an agreement on the procedure to follow when engaging in time and material work:

1. The subcontract agreement is to be reviewed to determine whether any unit prices exist that could apply to the work being considered. If so, some or all portions of the proposed T&M work will not apply, but unit prices will be used.
2. Review the intended work with the subcontractor to determine if a lump-sum agreement can be reached as it applies to the work under consideration, eliminating the need to proceed on a T&M basis.
3. If time and material is the only equitable method to be used to accomplish the planned work, establish the ground rules for its implementation. What costs will be included and what costs will not? Subcontractors often want to include supervision costs or project manager costs. Will these costs be acceptable to the owner, who may disallow any on-site supervision costs, stating that this activity is included in the general contractor's overhead percentage?
4. Any tickets prepared by the subcontractor must be explicit enough to define the work being pursued, the number of workers employed, their status (foreman, laborer, trades person) and the hours each devoted to the work at hand. The project superintendent should establish the procedure with the subcontractor to verify that workers listed on the ticket were actually engaged in the T&M work and not "contract" work.
5. The material tickets presented for the T&M work must also be explicit, identifying the product and where it was incorporated into the work. A clear distinction must be made as to the limits of the superintendent's authority when signing daily work tickets and/or material and equipment tickets. Does the signature acknowledge only that the work was performed? If that is the case, each ticket signed by the superintendent should contain the following caveat: "The undersigned verifies only that the work contained in this ticket has been performed, but does not establish or infer any contractual obligation." A stamp can be made containing specific language addressing the definition of a superintendent's authority, or lack thereof, when signing a T&M ticket by referring to Fig. 6-6.

The project manager can resolve this matter of scope responsibility with the subcontractor within the overall context of the subcontract agreement.

Extra work often requires additional time to complete that work, time that may extend the contract completion date. The T&M process should make clear whether the contract time has been affected. Does it remain the same, is it decreased, or is it increased?

Limiting Exposure to Damages for Delay Claims

There are actually two types of damages that can be created by delays in the project: direct losses and consequential damages. Direct losses or costs include such items as extended office trailer rental, additional salaries paid to on-site staff, increased ongoing utility costs, and other similar time-related costs. Consequential damages are not so clearly defined and are more subjective, but could include extended corporate costs—accounting, legal, and office administrative expenses required to continue servicing the extended project. Claims for damages for delays may be presented by the subcontractor when he or she is of the opinion that extra costs other than direct costs are being incurred or will be incurred because of delays created by the general contractor for which the subcontractor was not responsible. These would be probably fall into the consequential damages category. As discussed previously, if there is a "no damages for delays" provision in the subcontract agreement, the subcontractor is contractually blocked from instituting such a claim. If no such protective clause has been included in that agreement, there may be some way in which the delay for damages claim can be presented, if the facts do justify that action.

The superintendent should never agree with the subcontractor that the subcontractor is being delayed in the performance of the work, or else the superintendent will be quoted later on as having agreed to these delays. All such claims by the subcontractor should be forwarded to the office so that the project manager can formulate a response. There is more on this subject in the chapter on disputes and claims.

Third-Party Subcontractors and the Lien Waiver Problem

The project superintendent needs to be aware of every subcontractor working on the site. Often subcontractors will employ other subcontractors (referred to as second-tier or even third-tier subcontractors) to perform various specialized work on the project. The most common situation involves an HVAC subcontractor hiring an insulation contractor, or an air and balancing subcontractor or even a firm that fabricates the metal ductwork. It is important for the superintendent to be advised of these hires for any number of reasons—safety and security, to name just two. But there is also another important reason, and it involves the lien waiver process.

Subcontractors are generally required to submit a lien waiver along with their payment application request, signifying that monies previously received were used to pay for all labor, materials, and equipment employed during that period covered by that prior payment. If the subcontractor had engaged another subcontractor (a second-tier subcontractor) during the previous pay period, a lien waiver from this second-tier subcontractor should also be submitted—but quite often this is not done. If, at a later date, this second-tier subcontractor provides evidence of not having been paid, she or he can place a lien against the property, requiring the general contractor to satisfy (pay) the amount of the lien. If the general contractor cannot backcharge the prime subcontractor for the cost of removing the lien and must pay this lower-tier subcontractor, the GC will have, in effect, paid twice for the same work—once when the subcontractor requested payment, received payment, but did not honor the second-tier subcontractor's invoice, and payment a second time to remove the lien.

A provision inserted in the subcontract agreement can alleviate this situation to a degree, if enforced. The subcontractor can be required to notify, and obtain approval from, the general contractor for any lower-tier subcontractors the subcontractor intends to employ.

Company Letterhead

Oriole Construction Company
566 Southway
Baltimore, Maryland 21200

The American Steel Company
855 Industrial Circle
Owings Mills, MD 21240

Re: Waterfall Plaza
Project No. 6444

Attention: Mr. Jim Beam

Dear Mr. Beam:

We are in receipt of your current application for payment in the amount of ($$$$$) covering the period (whatever the period is). This requisition cannot be processed since you did not include lien waivers for your second-tier (or third-tier or both) subcontractors. At your option, you may deduct the amounts for these lower-tier subcontractors from your request for payment, or furnish the required lien waivers by (date).

Any payment applications received after this date will be processed during the next pay period.

With best regards,

Will Spencer
Project Superintendent

Letter 61 When payment for lower-tier subcontractors is requested, but their lien waivers are not included.

> The subcontractor shall not subcontract or delegate all or any portion of its work nor shall it assign any amounts due or to become due or any other claim or right arising in connection with this Subcontract Agreement without the prior written consent of the Contractor. In the event consent is granted to the Subcontractor to delegate or further subcontract any portion of its obligations hereunder, the subcontractor shall require that such delegee or subcontractor bind itself to the terms of the Contract Documents insofar as they pertain to its work and subcontractor shall remain fully responsible for all work performed by its Subcontractors.

Subcontractors do not often notify the general contractor either verbally or in writing when they hire a lower-tier subcontractor, so it is important for the project superintendent, walking the job daily, to note any unfamiliar faces and determine to which subcontractor they are assigned.

Damage to the Subcontractor's Work or Damage to Work of Others by the Subcontractor

How many times has one subcontractor damaged the work of another, claiming it was necessary to do so, but denied responsibility to repair the damage? This often occurs when one subcontractor must penetrate a block wall or a drywall partition to install the work. The penetration can be performed neatly or without any regard to the work already in place. Arguments then arise about the extent of cutting required and who should be responsible for patching. A provision in the subcontract agreement directing the subcontractor to protect his or her work and be responsible to repair the work damaged by other subcontractors will be helpful. Such a provision would be similar to the following:

> The Subcontractor is responsible for the protection of the subcontract work, including all materials contained therein or stored at the Project site until final completion and acceptance thereof by the Owner. The Contractor shall not be responsible for damages to the Subcontractor's work caused by other subcontractors. The Subcontractor warrants and guarantees the workmanship and materials covered by this Subcontract Agreement and agrees to make good, at its own expense and at the convenience of the Owner, any defect in material or workmanship which may occur or develop prior to the Contractor's release from responsibility to the Owner.

The Subcontractor Quality Control Process

Quality control as it relates to the subcontractor begins with the subcontractor's *knowing the job.* What administrative responsibilities are assumed by the subcontractor and what are required of the general contractor relating to such quality items as inspections and testing? What installation procedures and quality control measures are included in the subcontractor's work in question? A quick reading of the subcontract agreement will define the basic parameters of the subcontractor's obligations to the general contractor as they relate to performance and the plans and specifications. Taking note of any exceptions or qualification to the scope of work as defined by the plans and specifications is certainly a start to understanding what obligations the subcontractor has undertaken. Were there any addenda or bulletins issued that affect the subcontractor's work, and are they included in the subcontract agreement?

A review of the specification section or sections covered in the agreement is the next step in defining the subcontractor's responsibilities.

We know that the phrase *plans and specifications* is not always all-inclusive, so a review of each contract drawing with the appropriate subcontractor is one way to determine whether both parties agree on what is included what is not included, and what requires further investigation.

The Weekly Subcontractor Meeting

Some companies require the project superintendent to hold weekly or biweekly subcontractor meetings while other companies assign this task to the project manager. These meetings are important for a number of reasons:

1. They introduce the subcontractor team members to one another and define the responsibility of each team member.
2. They provide a forum for construction schedule reviews and updates and a method by which to document subcontractor performance commitments.

3. They address field-related issues that affect one or more subcontractors.
4. They disseminate information received from the owner, architect, engineer, or general contractor.
5. Safety concerns can be addressed, and the meeting offers a place where accident investigations can be reviewed.
6. Materials, equipment schedules, and deliveries can be reviewed and addressed along with other site logistics concerns.
7. Specific problems involving one or more subcontractors can be aired and resolved.

Preparing Meaningful Meeting Minutes

Accurate meeting minutes are an essential element of every subcontractor meeting. Meeting minutes can be extremely helpful in providing additional documentation to prepare or defend against potential disputes and claims. Timely dissemination of these meeting minutes is also important. Meeting minutes are divided into four basic components:

- List of attendees
- Old business—review of topics discussed at the previous meeting
- New business—new items or topics initiated at the current meeting
- Closing statement of writer's interpretation of events that transpired and the date and time for the next scheduled meeting

Key elements of any set of meeting minutes will include the following topics:

1. After the first meeting, establish a time and date for the next meeting or meetings. The specifications may require biweekly meetings, but confirmation of the actual time and date of the subsequent meeting still needs to be listed.
2. List all persons attending the meeting; their company affiliation, position, phone/fax number, or email address. This can be accomplished by distributing a sign-in sheet at the start of each meeting.
3. Review important documents.
   a. The construction schedule and a two-week look-ahead schedule. Each affected subcontractor should be prepared to participate in this discussion, and her or his remarks and response noted in the minutes.
   b. The shop drawing submittal schedule
   c. The delivery schedule for materials and equipment that have been approved by the architect and engineer
   d. Issuance of change orders or proposed or impending change orders
   e. Review of outstanding Requests for Information (RFIs) or Requests for Clarification (RFCs)
4. Old business—a review of topics discussed at the previous meeting but not concluded at that time
5. New business—any new items of discussion
6. A closing statement requiring that any objection to the writer's interpretation of the meeting's content be submitted in writing within a certain time period

During the course of the meeting, various individuals or companies will be asked to perform certain tasks or to commit to certain schedule requirements. These commitments must be documented in the meeting minutes so that it is very clear which party assumed responsibility for the specific event and the time frame in which this event is to be concluded. For every action item or event, the responsible individual or company must be listed. As an example, let's look at one event presented in a vague manner and the same event presented in very specific fashion:

Indecisive. The concrete slab in the hallway is to be flash-patched prior to the installation of carpet.

Specific action. Concrete subcontractor is to flash-patch corridor 105 in a manner acceptable to the flooring contractor, not later than September 14, 2003.

There will be no further need to send a letter unless the subcontractor fails to perform this work and another subcontractor is to be engaged to do the work.

The meeting minutes should be prepared as quickly as possible and distributed to all attendees promptly. Other interested subcontractors or individuals who receive informational copies should be included on the distribution list. A sample meeting minute format is shown in Fig. 5-5.

Backcharges—The Right and Wrong Ways to Deal with Them

Backcharges can originate due to a subcontractor's failure to perform specific items of work in a timely manner, or from his or her refusal to perform what is considered "contract" work, or when the subcontractor damages the work of other subcontractors and fails to make the necessary repairs. In either case, prompt verbal and written notification will avoid the many disputes over the justification and legitimacy of the general contractor's backcharge.

How many times have you heard this? "Why didn't you tell me it was important for me to clean up my area on the third floor by Friday? If I had known it was that important, I would have brought in some laborers to do so. I don't think this cleaning backcharge is proper." Or, "What damage to the drywall in room 105? My guys weren't even in that room on Monday." And there is always, "That's not my stuff in that trash pile, so I'm not going to accept a backcharge for cleaning."

A backcharge should not be considered in a situation where prior notice to correct was not presented to the subcontractor. The backcharge should be administered only after one or more efforts to have the subcontractor remedy the problem have been ignored. These prior notices could have been verbal, but written documentation will provide the superintendent with more ammunition should a backcharge be decided upon.

Then and only then should the superintendent notify the subcontractor's supervisor on the job site that a backcharge will be forthcoming, stating the reason why and even advising the subcontractor how the remedial or repair work will be handled. For example, "Because you failed to repair the damage (be specific about the type and location of the damage) I'm going to perform the work with my company's own forces on Saturday at overtime rates," or "I'm having the drywall subcontractor repair the wall you damaged." Whichever method is used, the superintendent should send a note to the project manager, advising of the decision to proceed with the remedial work. When the work has been completed, a detailed labor and material cost report or an extra work ticket signed by the superintendent should be forwarded to the office. The nature of the work and why it was required will be helpful in forcing the charge back to the subcontractor. Developing a team approach laced with a little bit of give and take is an effective, harmonious, and productive way to work with subcontractors. But when the team approach fails, knowing when to invoke those provisions of the subcontract agreement that afford control and promptly issuing the proper notices are often the only avenue open to the project superintendent.

Company Letterhead

Oriole Construction Company
566 Southway
Baltimore, Maryland 21200

The American Steel Company
855 Industrial Circle
Owings Mills, MD 21240

Re: Waterfall Plaza
Project No. 6444

Attention: Mr. Jim Beam

Dear Mr. Beam:

Please be advised that all requests for payment must be received in our office not later than (date). Any requests received after that date may be delayed until the next requisition period. Please carefully review the requirements for all accompanying material such as (certified payroll, lien waivers, etc., or you can merely state, "Refer to the contract specifications to insure that all accompanying documentation is provided with your application for payment.").

With best regards,

Will Spencer
Project Superintendent

Letter 62 Advising a subcontractor of the due date to apply for payment.

Company Letterhead

Oriole Construction Company
566 Southway
Baltimore, Maryland 21200

The American Steel Company
855 Industrial Circle
Owings Mills, MD 21240

Re: Waterfall Plaza
Project No. 6444

Attention: Mr. Jim Beam

Dear Mr. Beam:

Your current application for payment includes a request for payment for materials (or equipment) stored off-site. No such payment can be honored without a previous agreement with the architect (or owner). If you wish to requisition for off-site storage, you must comply with the following requirements:

Submit a request to do so and include a detailed description of the materials (or equipment).
Provide insurance certificates including coverage for transport to the site.
Supply a bill of sale that will transfer title to the owner upon payment.
At the architect's option, storage in a bonded warehouse may be required.
At the architect's option, reimbursement for all expenses involved in traveling to the site and inspection of the stored materials (or equipment) prior to payment may be required.

If these conditions are met, and with the architect's prior approval, you may include off-site materials (or equipment) in your next application for payment.

With best regards,

Will Spencer
Project Superintendent

Letter 63 Responding to a subcontractor's request for payment for materials stored off-site.

Company Letterhead

Oriole Construction Company
566 Southway
Baltimore, Maryland 21200

The American Steel Company
855 Industrial Circle
Owings Mills, MD 21240

Re: Waterfall Plaza
Project No. 6444

Attention: Mr. Jim Beam

Dear Mr. Beam:

During review of your current application for payment dated (date) in the amount of ($$$$$), the architect requested (select one of the following: additional documentation to support the value of your work in place, or additional documentation to support the value of materials stored on-site, or additional documentation to support the value of materials stored off-site).

The architect, however, has rejected the amount of your requisition as submitted and reduced the value to ($$$$$). If you disagree with this revaluation, we will arrange a meeting with the architect to discuss the matter more fully.

With best regards,

Will Spencer
Project Superintendent

Letter 64 Architect doesn't agree with a subcontractor's requisition amount.

Company Letterhead

Oriole Construction Company
566 Southway
Baltimore, Maryland 21200

The American Steel Company
855 Industrial Circle
Owings Mills, MD 21240

Re: Waterfall Plaza
Project No. 6444

Attention: Mr. Jim Beam

Dear Mr. Beam:

We have received your current application for payment in the amount of ($$$$$) covering the period (date to date, month, etc.). The processing of this requisition cannot be completed until a lien waiver for the previous pay period is submitted. Upon receipt of this lien waiver we will continue processing your application for payment.

Note: When second- and third-tier subcontractors have been engaged by the prime subcontractor, all waivers will be required and you should so state in your letter.

With best regards,

Will Spencer
Project Superintendent

Letter 65 Requesting lien waiver from subcontractor.

Company Letterhead

Oriole Construction Company
566 Southway
Baltimore, Maryland 21200

The American Steel Company
855 Industrial Circle
Owings Mills, MD 21240

Re: Waterfall Plaza
Project No. 6444

Attention: Mr. Jim Beam

Dear Mr. Beam:

We have been advised by (name of person at subcontractor's office) that unless you receive payment on your current requisition by (date) you will substantially reduce your manpower (or pull all workers off the project).

Please refer to your subcontract agreement, and particularly to the "pay when paid clause." As of this date we have not received payment from the owner and therefore cannot remit payment to you.

We remind you that any action to reduce or remove manpower from the project at this time is a direct violation of your contract obligations and will be dealt with accordingly.

With best regards,

Will Spencer
Project Superintendent

Letter 66 Responding to a subcontractor's threat to reduce manpower unless paid.

Chapter

7

Rehabilitation and Renovation of Older Buildings

Working in urban areas often exposes the project superintendent to construction projects involving the rehabilitation of older, often previously abandoned buildings. The recycling of a building often results in creating an upscale office building from a derelict waterfront warehouse or saving an historical structure from the wrecking ball.

These types of projects can result in either a rewarding experience or a frustrating, painful, and costly one. And rehab and renovation projects have their own idiosyncrasies and require an entirely new way of looking at how the work is to be administered. Often new drawings for these types of projects are prepared without the benefit of a partial or complete set of the original building's plans. The building in question may have been constructed decades ago and changed hands several times over the years with many unrecorded improvements or changes made to the property. The original drawings may have been misplaced or destroyed, and "as built" drawings are rare.

Architects hired by the owners of the property to prepare drawings for the current building's function may be limited in their ability to perform a thorough investigation of the standing structure by time, monetary restrictions, or lack of experience in similar projects. Trying to determine the skeleton of some older buildings without removal of substantial portions of wall, ceiling, and floor finishes is a very difficult, time-consuming, and costly proposition. But without exposing the existing structural system, it will be difficult to determine how new building components and finishes can be installed with any degree of accuracy.

Recently, some architects, via the use of digital camera technology, are including photographs in their bid documents to alert bidders to key existing conditions or to further augment specific instructions or directions contained in those documents. Figures 7-1 and 7-2 are examples of digital photographs included in the bid documents of a rehab project where a local defunct brewery complex was to be converted to subsidized housing.

There are two basic types of rehabilitation projects and numerous shades of gray in between. At one end of the spectrum will be found the "gut" or total rehab project, where demolition work will remove all interior finishes back to the structural system, including removal of all existing electrical, mechanical, and plumbing systems.

Figure 7-1 Bid document photos of existing conditions.

Figure 7-2 Other bid document photos of existing conditions.

The other end of the rehab spectrum is the partial rehabilitation project, where more of the finishes will be left intact and patched or repaired as required. This type of renovation or rehab work is more prevalent in projects of historical value or where important architectural features of the older building are to be retained and refurbished. Sound wood flooring, interior partitions, and ceilings may be left intact to be patched and refinished. In many instances exterior walls, ornamental cornices, and stonework and intricate carvings are to be repaired, some being used as molds for newly fabricated pieces to replace missing ones.

The reuse of existing systems and building components is limited only by building codes, life and safety issues, and financial and design considerations. In fact, some of the old building's dormant or abandoned manufacturing equipment may be refurbished in place, to add charm and character to the new structure.

More, rather than scant, investigation of an existing structure is crucial to the proper design and dimensioning of the new architectural and mechanical layouts. Often the structural system in an old building is found to differ materially from its assumed design, and if that is the case, the potential for significant dimensional differences in architectural, mechanical, and electrical drawings will present problems for the owner, designer, and contractor.

Before That Wrecking Ball Arrives

If a significant amount of demolition is required to remove old finishes or even parts of the structure, a thorough and complete review of the drawings must be undertaken to fully comprehend the extent of demolition—what is to remain and what is to be removed—and whether the initial scope of work is clearly represented by the contract plans and specifications.

After a comprehensive review of the drawings that includes making mental and written notes of questionable items, it will be very beneficial to take a slow walk through the building with drawings in hand (preferably a reduced set) and a flashlight to check on the following:

1. Check various elements of demolition indicated on the drawings, existing surfaces to remain, and the quality of the structure. If any portion of the demolition scope of work is unclear or inconsistent, the architect should be contacted immediately for clarification and possibly requested to visit the site to walk through the building as questions are asked and answers received. Any such walk-through should be followed up with a comprehensive list of items discussed, resolved, or open to further resolution. If a list like this is not going to be prepared by the architect, it is important that one be written by the contractor and sent to the architect for "review and comment."
2. What is involved in removing existing finishes from walls, ceilings, and other parts of the structure, as required by the contract documents? Will these finishes be easy or difficult to remove, and will the extent of removal required be achievable? Will more or fewer areas require removal due to the presence or absence of existing sound finishes?
3. Can any existing utilities be used during construction for temporary light, heat, or power, or will all new temporary utilities have to be provided? And will it take only a phone call to the local utility company to reactivate an existing service, or will a new service have to be installed from some distant point?
4. Determine whether that which is indicated on the drawings to exist does, in fact, *exist* and that which is *not* supposed to exist *does not* exist.
5. Do the contract documents correctly indicate the accurate composition of the items designated to be removed? That is, do the drawings indicate that a

steel stud and drywall partition is to be removed, for example, when, in fact, the partition in question is plaster on wood lath on rough wood framing members or even possibly terra cotta block?

6. A note should be made of the condition of a surface, subsurface, or area to be refinished or to be retained. Will it be possible to apply a new finish over the existing surface as is, or will it be possible to laminate gypsum board over the existing surface, if that is what the drawings indicate?

7. Do any of the items or materials removed have any appreciable salvage or scrap value? If some items are labeled "remains property of owner and is to be removed and stored at (location)," this should also be noted and all appropriate subcontractors reminded of their obligation to remove for storage by whichever party is assigned that responsibility. In some historical buildings this is not an inconsequential item.

An October 13, 1998, *Washington Post* article reported on the construction contract for the restoration of the Washington Monument. The contractor, by contract, was required, among other items of work, to remove and replace several damaged marble stones that formed the exterior of the Monument, including one 1200-pound slab. The contractor, by virtue of the contract language, claimed ownership and removal of all debris generated in the restoration process, including any damaged materials slated for replacement. The one 1200-pound slab removed from this historic building, cut into 1- or 2-inch squares and sold as a memento or collector's item, would reap a tidy sum. As one can see from this example, in many instances it will be important to determine ownership of debris before demolition commences.

8. Approximately how long will it take to demolish enough of the building components so that the new work can start? This kind of information will be invaluable during the preparation of the initial, or baseline, construction schedule.

9. Does there appear to be any hazardous material in the building—asbestos, old transformers or light fixtures that could contain PCBs, or peculiar odors that could portend environmental problems? Although most contracts hold the contractor harmless from dealing with such materials if the owner is unaware of their existence, hazardous materials require the services of a remediation contractor for removal, and this could seriously impact the contractor's construction schedule.

Knowledge of the types of businesses of previous tenants or owners can be useful in determining the potential for hazardous materials being discovered at the project. But quite often this information is either not known or not made known. The writer, working on an inner-city project where demolition had been performed by the city agency involved in the development of the area, was provided documentation from the designers of the new project which spelled out all known "existing" below-grade obstructions. While excavating in one corner of the razed site at about 3:00 p.m. on a Friday evening, the big Komatsu excavator pulled the top off a 2000-gallon grease interceptor that was not shown on the contract drawings. Not only the top was torn off, but also the sides of this cement structure were split open, allowing some 2000 gallons of foul-smelling, congealed grease to ooze into the adjacent excavations. Some frantic telephone calls were made to get a tank truck to the site on a late Friday to pump out this sludge and cover over the remains, until final cleanup could be completed on Monday morning. This type of work is never without its surprises.

On-Site Inspection Tips

During this inspection, be on the lookout for areas in the building where water may have been infiltrating—through the roof, walls, and basement. If that has occurred, an inspection of the exterior walls may reveal missing or loose mortar joints, caulking joints that have failed, cracks in the exterior wall system, structural or otherwise, loose or missing flashings, or spalled masonry or concrete surfaces. Review the contract documents carefully to determine whether the contractor is required to repair any such items. Obscure notes placed strategically on drawings for new work take on added importance in rehab or renovation projects. The notes may contain detailed directions for various types of demolition work, instructions for preserving existing finishes, or directions on how to proceed if a whole host of unknowns arise. General notes, often skipped over as boilerplate on other types of construction projects, should be scrutinized carefully for rehab or renovation projects.

Look carefully at the interior surfaces of the building to see if there are any water stains, old or new. Are there rotted materials in the building that will need to be replaced, and if so, whose responsibility is it to do so? Can this be claimed as an extra to the contract? Review the drawings very carefully, and look for obscure notes that may direct the contractor to *include, exclude, leave as is, replace entirely, remove back to sound surfaces,* and so forth.

Although these conditions might not be specifically included in the contract scope of work, they might be required by the exculpatory language in the special, general, or supplementary conditions on the plans or in the specifications. Therefore these sections of the specifications book ought to be read, and important sections highlighted for quick future reference, if needed.

Cutting and Patching

Another item that should be thoroughly investigated while walking through the site in the early stages involves how much and what kind of cutting and patching will be required. If all-new mechanical and electrical systems are going to be installed and the building is multistoried, the cutting and patching may require large openings or chases for these systems. These openings and chases may require some sophisticated structural supports along with fire-safe penetrations and rated enclosures once installed.

Penetrations through thick stone or old concrete walls and foundations can be quite costly. Cutting through masonry partitions or walls so that new openings can be prepared for doors and windows will also be expensive, often requiring temporary shoring. If unsound finishes are disturbed when these penetrations are made, large chunks of lath or plaster could very well fall off the existing wall or ceiling being penetrated, and the contractor may be obliged to restore those surfaces. Generally this is done at no additional cost to the owner, unless a fully documented claim including photographs of conditions before, during, and after these operations can be presented to the owner along with a logical explanation of a request for scope increase.

Discuss these types of issues with the project manager prior to the award of major subcontracts so that some responsibility and related costs are properly assigned to the various specialty contractors performing work involving cutting and patching. It may be more economical for the general contractor to assume all the risks associated with these cutting and patching operations by performing this work himself or herself. When subcontractors prepare the estimate for work such as this, they invariably include a contingency in case unforeseen conditions are encountered. If the entire dollar amount of the contingency is not absorbed (and it is generally so liberal that it is most often not used completely), the total costs when this type of work is performed by the general contractor should be less than the subcontractor's quote.

Prior to the Start of Demolition

Some demolition crews are like the proverbial bull in the china shop. They come on the job prepared to race through their work, sledge hammers and pry bars in hand, frequently without a full understanding of the scope of their work. Often they are required to work in a dimly lighted building, making it difficult to read the drawings that they brought with them. Prior to the start of work, review any demolition drawings carefully, and read *all notes* on the architectural, structural, mechanical, and electrical drawings for both existing conditions and new work. Often there will be small notes in out-of-the-way places on drawings, such as "If any item is shown on the existing drawings and does not appear on the new work drawings, it shall remain or be relocated as shown." This statement means that if some electrical work, for instance, is to remain, it must be identified so that it is not removed by the demolition crews. The same would hold true for architectural and mechanical items.

Fluorescent spray paint is useful in denoting all items scheduled for removal or demolition. If that procedure is followed, all crews must be made aware that anything *not* marked with bright paint is to *remain*. But make certain that there is no confusion among demolition crews about what spray-painted surfaces mean. There may be a misunderstanding between one contractor's concept of items to remain and what is to be removed; so when this work is being subcontracted, inform the subcontractor's foreman accordingly.

The Preconstruction Survey

But even before any demolition takes place, the building should be closely inspected to expose any significant or suspicious cracks, either structural or aesthetic, that existed prior to start of construction. This inspection is best performed and documented in the presence of the architect, and lots of photographs should be taken to establish the condition of the building prior to start of any work. Only when questions arise later on concerning the condition of an existing wall, floor, or finish will the project superintendent say, "I wish I had taken photos of that wall, floor, shaft, etc., before we started work in that area."

Particular attention should be paid to exterior wall surfaces that may have developed cracks or may have deteriorated from age, freeze-thaw cycles, or possibly minor structural failures over the years. Any significant areas that have cracked, spalled, or deteriorated should be identified and reported to the architect by letter as soon as these conditions are discovered, so there is no misunderstanding later as to when these cracks or failures occurred. Once again, take photographs to document the condition.

This *preconstruction survey* is often conducted not only on the building in question, but also on any nearby or adjacent structures, particularly if any heavy equipment or extensive demolition will be employed in close proximity to those buildings where excessive vibration could cause damage. It is not unheard of for occupants or owners of the building next door claim that some damage to their property was caused by excessive vibration or shock arising out of the demolition activities in the building under construction, but the "damage" they report had existed for months or years before construction actually started. These types of claims should be resolved as they arise and not left for resolution later on. A close inspection of the "damage," if on the exterior, may reveal mold or moss growing in the "new crack," indicating that this condition existed long before any construction began. Close inspection of interior "damage" may also show signs that the cracks or loose plaster were existing and not new as alleged. Documentation by high-resolution photographs may prove to be invaluable. Be sure to orient the photographs so that their exact location can be pinpointed at a later date, if necessary.

Interior Demolition Tips

When plaster walls and/or ceilings are to remain, tap the surfaces to determine whether these surfaces are sound. Any investigation revealing loose plaster should be reported to the architect, because the surface might loosen further or might fall off completely while demolition is taking place in adjacent areas. Before you send a letter to the architect, scan the drawing carefully to determine whether there are instructions for dealing with situations like this, such as a note stating, "When a portion of an existing surface is to be removed, but adjacent surfaces are not sound, those removals must extend to sound surfaces." These types of instructions often appear in out-of-the-way places on the drawings. Make certain that no such instruction exists before firing off that letter to the architect.

However, even if such a restrictive clause exists, consider whether this condition would have been *apparent* during the prebid site inspection or whether this condition could have *reasonably* been assumed. Just because restrictive clauses do exist, don't preclude the possibility of claiming an extra if extensive additional work was required that could not have been *reasonably* anticipated.

When portions of exterior walls are to be demolished and the building is located in or near densely populated areas, a water source should be available to wet down portions of the building to be demolished. Reducing dust problems before neighbors complain will create good public relations and possibly avoid future problems.

Fire extinguishers should be available in areas where open-flame torches are used to cut metals inside the building. Remember that fires can smolder for hours undetected in walls behind areas that have been exposed to open flame; therefore, a fire watch should be established when such cutting takes place.

Safety Concerns

Safety hazards abound in rehabilitation work. In addition to the obvious dangers from falling debris during demolition, there are many other areas in which accidents can occur. Floor openings created during demolition should be either secured by covering with temporary plates or barricaded. Air compressor hoses and electric power cables snake through the demolition area like spaghetti and can easily become tripping hazards. Other tripping hazards are created by failure to cut off protruding pipes, conduits, or other projections flush with the floor level when these types of removals are taking place. Water used in conjunction with cutting tools or dust control can also create slippery floors.

Cutting through active electric cables is always a danger, especially when some panels must remain active to provide temporary lighting and power for tools. And in older buildings not all lines are easily traceable to their respective panels, so they can be deactivated. Always err on the side of safety. If you are unsure whether an electric cable is alive or dead, assume it is active until the electrician determines otherwise.

The tendency to allow debris to accumulate before removal can also be the cause of many accidents. Boards ripped off walls or floors may have exposed nails, and these nails should be either removed or bent over to prevent injury to feet or hands. Prior to the start of demolition it is a good practice to assemble all workers and their supervisors and to conduct a safety meeting, pointing out the dangers inherent in rehab and renovation work and stressing the fact that proper procedures must be followed. At this safety meeting the use of appropriate personal safety equipment is stressed. The use of hard hats, safety goggles, and ear protection, when necessary, will be enforced, and the penalties for failure to comply should be firmly established—and enforced. The use of proper footwear is also important; thin sole shoes, sneakers, running shoes, or work shoes, without safety toes are strictly forbidden.

Shirts of synthetic fiber are not to be worn by anyone using open flames or cutting torches since these materials have a tendency to continue to burn and melt when ignited. Cotton shirts, if set afire by sparks, tend to extinguish themselves rather rapidly.

A "get tough" safety policy ought to be initiated and strictly enforced. Violators must be directed to leave the work area immediately if they lack the proper personnel protection equipment and must not be allowed to return to work until they are in full compliance with the safety program.

Problem Areas during Construction

In the early stages of the demolition process, the job will require a great deal of close supervisory attention. Questions concerning what stays and what is to be demolished will arise hourly and daily, even though the areas to be demolished have been spray-painted. Questions will arise when conditions different from those specified in the contract documents are discovered and immediate resolution is required to maintain job progress.

When partial electrical and/or mechanical systems are to remain in the building and portions of each type are to be removed, demolition can be accomplished in one of two ways. The mechanical or electrical contractor can disconnect the conduit and cable or piece of equipment, so that the demolition contractor or general contractor can complete the removal; or the mechanical or electrical contractors can perform the entire removal process themselves. Since they will use a combination of mechanics and helpers, it may be more cost-effective to have the higher-paid mechanics make the disconnects only and leave the removal and disposal of the debris to laborers earning lower hourly rates.

Electric cables and conduits will probably be found where they are not supposed to be, and the electrical subcontractor should be readily available during the entire demolition process to deal with these unexpected events. This may be true of the plumbing contractor as well when some existing water lines are to remain active for fire and dust control or in the event that a gas line is discovered where none was indicated.

The problem with active versus inactive electrical conduits usually can't be resolved by disconnecting all electric lines in a rehabilitated building project. Electricity will be needed for power tools during demolition and will be required for temporary lighting; to disconnect *all* power would mean that another temporary service, independent of the existing one, has to be brought into the building. On the other hand, if the removals are complicated and safety is the key factor, another electrical service may be the best solution.

Existing Conditions—The Problem Area

Many project contract documents for these rehab and renovation projects will designate existing conditions or systems on a separate set of drawings frequently entitled "Existing Conditions" or some other words to that effect.

The portions of the building that are to remain will generally be shown on the applicable "new" architectural, structural, mechanical, and electrical drawings. To somewhat ease the supervisory burden on the project superintendent, the project manager should include a stipulation in the appropriate subcontract agreement that requires each subcontractor to monitor the work for that trade, especially those items that are required to remain and those that are scheduled for demolition. The appropriate subcontractor can tag, paint, or otherwise identify these items in some way that is easily recognizable. To carry out this responsibility, the subcontractor(s) may have to have a mechanic on the job to assist the demolition crew in identifying which items are which. No matter how well the work slated for removal is marked, invariably the demolition crew will have many questions throughout the entire process.

Verify the Dimensions for New Work

If a great deal of interior demolition is to take place, stop and check dimensions once demolition has progressed to the point where some new work layout can begin. Do the dimensions of the exposed areas agree with the dimensions on the contract drawings? Are they greater or smaller? If the mechanical and electrical drawings were based upon dimensions that are no longer valid, all drawings referring to new work must be carefully reviewed, and if major discrepancies are uncovered, the architect must be notified immediately, first verbally, then followed up by a written memo.

Too often the architect, in the early stages of design, either can't commit forces to conduct a thorough investigation of the existing structure or has not included a detailed investigation due to an owner's budget constraints. But absent such a complete and thorough investigation, which includes considerable poking around, how can the architect define and create the new space?

A Case Study of a Severe Dimensional Problem

The author once had an experience that vividly dramatizes this point. He was the project manager on a project involving the recycling of a late-1920s hotel into units for subsidized elderly housing. The old hotel was a cast-in-place concrete structure, reportedly one of the oldest of its kind in New England. The building was 14 stories high with four step-backs in the structure and façade as the building progressed upward. The exterior walls were brick, and the interior walls were terra cotta with wood lath and plaster finishes. The ceilings on the underside of the concrete decks were plaster.

This project was a gut rehab, and all interior finishes were to be removed back to the concrete structure. All concrete slabs, columns, and beams were to be stripped of finishes. All existing mechanical and electrical systems were to be removed along with two huge coal-fired furnaces currently under 10 feet of water in the subbasement. The two old traction elevators were to be removed, and new elevators fitted into existing shafts. All windows were to be replaced with new insulated, glazed aluminum double-hung units. Apparently a set of original structural drawings existed, but the new owners of the building refused to purchase them from the previous owner who, assuming he had a good thing, was asking a fortune for them. Instead, the drawings prepared for the project were based on an *assumed* system of columns, beams, and slab thickness, and the dimensions of these structural components were determined by the design team by poking through plaster surfaces around selected beams and columns on only a few floors. Needless to say, many of the design assumptions arrived at by this cursory investigation proved to be wrong, but were only verified when the actual structural components had been fully exposed during the demolition process.

Demolition progressed from the basement up, with one crew, while another demolition crew worked from the penthouse elevator down. The plan of attack was to get to the basement and subbasement areas, pump out the boiler room, cut up and remove the old boilers, remove the coal chute, and in general clean out the entire area so that new equipment locations and piping layouts could commence as quickly as possible, allowing the appropriate long lead time for equipment to be ordered. In the meantime, demolition starting from the top down would allow room layouts to proceed and mechanical and electrical risers to be located. This, in turn, would permit concrete floors to be cored or saw-cut so that mechanical piping and duct risers could be plumbed down on a floor-to-floor basis.

After the three top floors had been stripped back to the structure, centerlines were established along the north-south and east-west building axes. From these centerlines, partition layouts in accordance with the dimensions shown on the architectural drawings began. The centerline of one floor was transferred to the floor below by coring a 2-inch hole in the slab and dropping a plumb bob on piano wire through this opening to the lower floor, thus marking the centerline on that lower floor.

Suddenly it became apparent that dimensions were just not adding up, and all kinds of dimensional discrepancies came to light. A thorough investigation of the three uppermost floors began by comparing actual dimensions with those shown on the architectural drawings. It appeared that, among other things, the structural columns and beams on the top three floors were smaller in section than those which had been exposed on the lower floors. This seemed logical since the upper floors were carrying less dead load than the floor below. However, the architectural drawings did not reflect the dimensional differences in the structural beams and columns because, as it was later determined, the architectural team did not probe these upper-floor members when they made their initial, spotty investigation. All columns and beams on the new drawings, with just a few exceptions, were shown to be the same size. The actual size of the beams and columns, when stripped completely of their plaster coating, changed dimensionally every other floor.

Also, some structural columns in the exterior walls were found to be wider and deeper than the new drawings indicated. The mechanical drawings indicated hot water baseboard heat to be installed on the exterior walls, but did not take into account the space occupied by the columns that protruded into the room. The minimum square footage of each apartment had to be maintained in order to comply with the Department of Housing and Urban Development's (HUD's) minimum property standards, and these dimensions were very tight to begin with. It appeared that some of these structural columns would have to be notched at the base to allow the baseboard heat piping to be installed, in order to maintain minimum clear dimensions in the room. However, the structural engineer issued a directive *not* to notch the columns, and a chase wall had to be created, taking more valuable inches away from the bedroom areas, which were already dangerously close to the minimum size allowed by HUD.

And these exterior columns varied in size on every other floor, so that any new plans being developed had to be created for the seven apartment configurations on the floor, which in turn affected bathroom and kitchen dimensions and, of course, all mechanical and electrical riser sizes and locations.

Except for demolition, all other work stopped while a systematic review was made of each apartment on each floor by the project manager, project super, and the owner's entire design team. Subcontractors whose field experience would be called upon from time to time were also invited to join the group.

Most apartments had to be completely redesigned, and the uncovering of a number of junior beams, which were never indicated to exist in the structural system, added more problems since many were located directly above areas designated for mechanical and electrical risers and chases.

Drawings were revised on the spot and cost estimates for the changes hurriedly assembled for both architect's and owner's review and approval so work could start. The entire process was both time-consuming and costly. Since there was a liquidated-damages clause in the construction contract, the documentation of all delays placed another burden on the project superintendent and project manager.

The job finally got back on track, but a lot of crisis management was involved in getting it there. The lesson to be learned is that when this type of construction is undertaken, time must be spent early on to check all dimensions and wall thicknesses for even such things as hollow metal door frames and replacement window sizes. If variances are discovered and there are conflicting dimensions, at least these disparities can be picked up and corrected at an early stage via shop drawings.

Varying Conditions

When conditions are at variance with the contract documents or differ substantially or materially from those normally encountered, the architect should be notified immediately with a written confirmation follow-up. For example, the contract drawings may have indicated that a specific wall is to be removed, and the wall was noted to be nonbearing; however, upon investigation during construction, the wall was determined to be a bearing wall requiring substantial structural modifications before it could be removed. The project superintendent, after alerting the architect, should request in writing that the structural engineer visit the site as soon as possible and should issue a sketch reflecting the necessary structural modifications. All costs for the modifications outlined in the engineer's directive need to be submitted for approval quickly; better yet, the project superintendent should request authorization to proceed with the work on a time-and-materials basis.

All such delays should be reported to the project manager and documented promptly. Remember, the delay clock starts ticking when the discrepancy is discovered and may not stop until the modification work has actually been completed.

When corrective action or changes are required due to field conditions, they can be reported quickly by phone; but they can be just as quickly documented by issuing a Request for Information (RFI) or Request for Clarification (RFC), or simply a memo by email or via fax.

If the architect's response is slow in coming and work must cease in the affected area, send another email or fax, notifying the receiver of this condition and its potential impact on job progress. But if authorization does not immediately flow, possibly because costs are being reviewed, request authorization to start the work and determine final costs based on the basis of a construction change directive concept in Article 7.3 of the General Conditions Document A201.

All parties need to work closely together in a rehabilitation project. The relationship that must be established with the design team and the owner is that of a team effort. And all subcontractors must also join the "team" to gain their full cooperation for what will doubtless be a daunting project.

Because of the vagaries almost always present in the contract documents when you are dealing with these kinds of projects, the project superintendent/project manager must clearly communicate with the designers this team concept and that the contractor is not embarking on a campaign to generate change orders. The sole purpose of these change orders is to assist the architect in thoroughly investigating the conditions of the existing structure and to help solve problems quickly and in the most cost-effective manner.

If there are extra-cost items of work to be undertaken, perhaps options to change other items of work and effect tradeoffs can be considered. Because of the nature of rehabilitation and renovation work and the problems that will inevitably arise, everyone must attack the problems as expeditiously, harmoniously, and equitably as possible.

The Use of Contingencies

Normally an owner will include a contingency sum in the estimate for the total project, in anticipation of encountering additional costs for conditions unknown or unanticipated. The owner may not wish to divulge the contingency, assuming that the general contractor may see this as an untapped source for change orders. If it is suspected that the owner, generally one who is dealing with the first construction project, has not included a contingency in the construction budget, the owner should be made aware of the need to create one because it will be needed. It is a rare rehab or renovation project that doesn't need additional funds to cover the cost of unanticipated problems and related costs. Depending upon the nature of the project, the contingency set aside by the owner should be at a minimum 5 percent of total construction costs or, better yet, 10 percent of the total cost of construction.

Water Leaks and Other Concerns

Prior to the installation of new drywall partitions in the rehab or renovation project, a walk around the exterior of the building and an inspection of the existing roof are in order. If the scope of the contract work does not include any exterior façade restoration work, the exterior should be inspected to determine whether any areas might have the potential for water infiltration. And even though a new roof is scheduled for completion at some future date, an inspection of the existing roof or skylight might point out some areas that need immediate attention and patching to temporarily keep out water.

Not only will any leaks that are prevented keep the new drywall work from becoming damaged, but also when new finishes are applied and new flooring work is started, the extra money spent making the building watertight will certainly pay off. Wet or moist conditions present prior to the application of any wall system, except perhaps concrete masonry unit (CMU) walls, may result in the growth of mold and mildew with their real or imagined concerns.

If any appreciable demolition work is to be performed requiring heavy equipment inside the building, the structural engineer should be contacted to determine whether shoring will be required under the areas where this equipment will be operating. If permission to proceed is granted, a letter must be received from the engineer stating the conditions under which certain types of equipment can be employed.

Job progress photographs have a special purpose in rehab and renovation projects: They can reflect the condition of various parts of the structure prior to the application of new finishes or new wall openings in progress or completed. When a photograph is used to show unusual or uneven surfaces, a method of measurement should be introduced into the photo; for example, a carpenter's ruler can be inserted to indicate the degree of variance or thickness of an existing surface. Other forms of documentation take on added importance in these types of projects. With the proper attention to detail and cooperation from subcontractors and consultants, the project superintendents can gain experience while enjoying a great deal of satisfaction by knowing that they have kept the wrecker's ball away from a building worth keeping.

Encountering Hazardous Materials

Unless explicitly included, most construction contracts exclude the contractor from responsibility to deal with hazardous materials discovered during construction. AIA Document A201, General Conditions, 1997 edition, requires the contractor to notify the owner and architect when hazardous materials are found, and the contractor is to immediately cease work in the suspected area.

The discovery of hazardous materials may occur when least expected, and a project superintendent ought to have some working knowledge of some common hazardous materials most likely to be encountered.

The Asbestos Problem

Even with the government-mandated program in 1997 directing the abatement of asbestos in public schools, not all these materials were physically removed. Some were encapsulated behind nonhazardous materials, and others were enclosed in partitions or chases. Thus the potential for encountering asbestos in older schools during construction work today cannot be dismissed.

As late as 1981, one-half of all asbestos consumption in the United States derived from the manufacture of roofing felts, felt-backed sheet flooring and tiles, asbestos-cement pipe and fittings (*transite*), as well as being woven into some types of clothing. A survey conducted by the Environmental Protection Agency (EPA) some years ago revealed that 733,000 buildings in this country contained this hazardous material. According to the EPA, buildings constructed in the 1960s are most likely to have sprayed on or troweled on fireproofing materials containing asbestos. Older buildings are more likely to have asbestos pipe and boiler insulation.

Although the project superintendent may never encounter asbestos materials in any rehab or renovation project, the potential for meeting up with some form of this material cannot be dismissed. Therefore some background information may prove helpful.

Friable and Nonfriable Asbestos Materials

Asbestos in building construction materials can take two forms: friable and nonfriable. *Friable* asbestos is the type that can be crumbled, pulverized, or turned into a powderlike substance by crushing in the hand. Pipe insulation is a common form of friable asbestos. *Nonfriable* asbestos consists of asbestos fibers and a bonding matrix whereby the fibers will not be disturbed or released into the air until the product is cut, sawed, drilled, or sanded. Examples of nonfriable asbestos are vinyl asbestos tile, which contains about 21 percent asbestos; roofing felts, which contain anywhere from 10 to 15 percent asbestos fibers; and siding shingles, which contain 12 to 14 percent asbestos.

When the presence of asbestos in older buildings is known, it is usually identified in the bid documents along with an environmental consultant report allowing the contractor enough information to prepare an estimate for its removal. Most contracts nowadays exclude hazardous materials from the contractor's responsibility, stating the precaution

> If asbestos, PCBs, other hazardous materials, are encountered on the site by the Contractor, the Contractor shall upon recognizing the condition, immediately stop work in the affected area and report the condition to the owner and architect in writing.

Other materials may resemble asbestos when viewed by the naked eye, but the presence of asbestos can only be confirmed by subjecting the sample to laboratory analysis. The spearlike fibers of the material can be positively identified under an optical microscope. In fact one study criticized the method of establishing fiber count solely by optical microscope inspection because that instrument picks up fibers larger than 0.2 micrometer, whereas potentially dangerous fibers can be as small as 0.02 micrometer. The latter are detected only by using an electron microscope.

Once it has been determined that asbestos is present, the material can be removed, encapsulated, or enclosed. Encapsulation can be accomplished by coating the asbestos with a bonding-type sealant that will penetrate and harden. The asbestos can also be covered with a protective material, which is placed over it, sealed over, and seamed around all edges. Enclosing asbestos involves constructing an airtight enclosure around all surfaces that contain asbestos. The enclosures are to be built of an impact-resistant material, and signs are required to be placed on the exterior of the enclosure, warning that asbestos is contained therein.

When removal of asbestos-bearing materials is to take place, environmental remediation firms contracted to perform the work will be required to file documents with local, state, or federal agencies prior to removal. The EPA regulates the removal and/or disposal of asbestos-bearing materials, and the Department of Labor's OSHA division has regulations on restrictions of worker exposure to the material; so both agencies will become involved in any removal process.

Lead-Based Paint

Prior to World War II, lead was a common ingredient in exterior and interior paints, adding luster, longevity, and durability to the product. Before 1940, according to a study by the Department of Housing and Urban Development (HUD), 18.9 million residences contained lead paint; and during the period from 1960 to 1979, this total increased to 22 million homes. Figure 7-3 lists the percentage of lead-based paints by year and application.

With the advent of fast-drying and durable latex, alkyd, and acrylic-based paints, coupled with the federal government's enactment of a law in the mid-1970s mandating the content of lead in paint to be reduced to 0.06 percent, lead-based paints faded out of the picture. Most of the structures still containing lead paint today can be found in older inner-city commercial structures and residential areas. Cases of brain-damaged children who had eaten flaking lead-based paints have been well documented and are testimony to one of the dangers created by this material. Cumulative exposure to lead damages the brain, blood, nervous system, kidneys, bones, heart, and reproductive systems and contributes to high blood pressure.

Component Category	Interior	Exterior
Walls/Ceiling/Floor		
1960–1979	5	28
1940–1959	15	45
Before 1940	11	80
Metal Components[1]		
1960–1979	2	4
1940–1959	6	8
Before 1940	3	13
Nonmetal Components[2]		
1960–1979	4	15
1940–1959	9	39
Before 1940	47	78
Shelves/Others[3]		
1960–1979	0	—
1940–1959	7	—
Before 1940	68	—
Porches/Others[4]		
1960–1979	—	2
1940–1959	—	19
Before 1940	—	13

[1] Includes metal trim, window sills, molding, air/heat vents, radiators, soffit and fascia, columns, and railings.

[2] Includes nonmetal trim, window sills, molding, doors, air/heat vents, soffit and fascia, columns, and railings.

[3] Includes shelves, cabinets, fireplace, and closets of both metal and nonmetal.

[4] Includes porches, balconies, and stairs of both metal and nonmetal.

Figure 7-3 Percentage of lead-based paint by year and application.

Because of its widespread use decades ago, older buildings being rehabilitated or renovated might well have been painted with lead-based paints. And in some cases the original lead-based paint is covered over by multiple coats of nontoxic paint, thereby hiding the real culprit. Being alert to the potential for uncovering lead-based paints will be another of the project superintendent's responsibilities when working in a rehab or renovation environment.

Vacuum Blasting to Remove Lead Paint

Various removal options are being offered by certified lead abatement contractors, as outlined in Fig. 7-4. The system of vacuum blasting is widely used on exterior metal surfaces. Use on masonry may require a test panel or two to determine whether this abrasive method of paint removal will erode too much of the surface of the masonry unit. This method of directing a high-pressure stream of an abrasive through the inner nozzle of a sand blasting type of machine, while applying a vacuum around the outer surface of the nozzle, is effective in both removing the lead paint and containing most of the hazardous dust generated by the process.

In many cases lead-based painted doors, windows, trim will simply be removed and disposed of off-site, in accordance with the appropriate local, state, and federal regulations. Power sanding to remove lead-based paints should be avoided since it releases a great deal of harmful dust that will find its way into every nook and cranny of the building and is virtually impossible to vacuum away.

Various types of chemical stripping compounds can also be used to remove lead paint, and this method is often used when only limited quantities of lead paint are to be removed. The residue stripped off the existing surfaces must be carefully contained and disposed of in accordance with prevailing federal and state rules and regulations.

	Containment	Relocation	Recommended Practices	Cleanup
Demolition	Use plastic sheeting to prevent airborne dust migration. Interior Worksite Prep. Level 4; Exterior Worksite Prep. Level 3	No residents in dwelling during any work.	Wet surfaces, use covered containers to move debris; best subcontracted to abatement contractor, or a demolition contractor certified for abatement.	HEPA vacuum, wet mop, and HEPA vacuum.
Repainting	Floors and ground covered with 6-mil plastic. Interior Worksite Prep. Level 4; Exterior Worksite Prep. Level 3	No entry into work area during interior work.	Wet scrape, wet sanding, HEPA-filtered vacuum power tools.	Daily cleanup with HEPA vacuum, wet wash, HEPA vacuum.
Floor Sanding	Full containment of rooms, negative air recommended if leaded dust hazard identified.	No entry into work area during work.	Sanding lead-containing floors should be completed by abatement contractor, or other contractor certified for abatement.	HEPA vacuum of entire house may be needed.
Plaster Repairs	Localized containment for walls, entire room for ceiling. Usually Interior Worksite Prep. Level 1 or 2 for small jobs	No entry into work area.	Wet prior to removing.	HEPA final cleanup.
Window Replacement	Localized containment around each opening.	No occupancy during removal and initial cleaning and sealing.	Seal interior with plastic. Remove window from exterior if possible.	HEPA vacuum all areas with replaced windows.
Carpet Removal	Do dust sampling to determine contamination level. Usually Interior Worksite Prep. Level 3 or 4.	No occupancy during removal and initial cleaning.	Carefully remove and package carpet and pad in 6-mil plastic with taped seams. Wet down carpet before removal or disturbance.	HEPA vacuum floor after carpet bagged and prior to removal.

	Method										
	Removal							Enclosure			
Attributes	HEPA Needle Gun	Heat Gun	HEPA Vacuum Blast	HEPA Sand	Remove/ Replace	Caustic Paste	Offsite Stripping	Plywood Paneling	Gypsum	Pretab Metal	Wood, Metal, Vinyl, Siding
Capital Required	High	Low	High	Moderate	Moderate	Low	Low	Low	Low	High	Moderate
Worker Protection Required	High	High	High	High	Moderate	High	Moderate	Low	Moderate	Low	Low
Finish Work Required	Tentatively high	Moderate	Tentatively high	Moderate	Low	Moderate	Moderate	Wide	Wide	Limited	Wide
Product Availability	Limited	Moderate	Limited	Limited	Wide	Moderate	Limited, strip shops decreasing	Moderate	Moderate	Long	Long
Durability	Long	Long	Long	Long	Long	Long	Long	Moderate	Moderate	Moderate	Moderate
Labor Intensity	High	High	High	High	High	High	Moderate	High	High	High	High
Overall Safety	Moderate	Moderate	Moderate	Moderate	Very high	Moderate	High-high	High	High	High	High
Surface Preparation	None	None	None	None	None	Minimal—adjacent areas	Minimal—hardware removal	Minimal	Minimal	Minimal	Minimal
Cost	High	High	High	High	High	High	High	Moderate	Moderate	High	Moderate

Figure 7-4 Various methods of lead abatement.

PCBs, VOCs, Hydrocarbons, and Other Hazardous Materials

The previous occupant of an abandoned or razed building can provide an alert project superintendent with clues to the potential existence of hazardous materials on the site.

PCBs

Was the previous owner or occupant engaged in the electrical manufacturing or electrical repair business? Are there old fluorescent light fixtures installed or discarded ones stored on the property? Polychlorinated biphenyls (PCBs) were banned in 1979, so the age of the facility will be a tip-off to their possible presence in the building.

Polychlorinated biphenyls can be found in old electrical transformers and in the ballasts of older fluorescent lighting fixtures. This chemical degrades very slowly and is considered a dangerous health hazard. PCB received a great deal of notoriety when, in mid-2001, the federal government, under pressure from the state of New York, began a $500 million cleanup of the upper Hudson River. Under the aegis of the Superfund Act, the state was to contract to dredge approximately 1 million pounds of PCB-laden silt from the riverbed. This material had been legally discharged from General Electric's Hudson Falls plant at Fort Edward, New York, from 1940 to 1977, and the battle over responsibility for its total removal lingers on.

VOCs

Volatile organic compounds (VOCs) include solvents such as acetone and methyl ethyl ketone (MEK). Hardware stores and paint shops that sell any of these products may, over the years, have dumped leaking containers down the drain; and those drainpipes may have broken, allowing hazardous materials to leach into the soil. Any previous occupants of the building that had a need to use these solvents, say, an autobody repair shop, should serve as a red flag and require some poking around to determine if any residues of these chemicals are present. A sniff test will prove effective.

TPH

These initials stand for *total petroleum hydrocarbons*—fuel oil, gasoline, and kerosene. Gas stations and automotive repair shops would have been users of these products. In 1984 a federal law, referred to as LUST for *l*eaking *u*nderground *s*torage *t*anks, was enacted to enforce the replacement of leaking underground fuel tanks. In 1988, the Resource Conservation and Recovery Act (RCRA) established a 10-year window on these replacements. Owners were given three options:

1. Replace older tanks with new, government-approved designs
2. Upgrade older tanks by adding spill and overfill prevention accessories and corrosion protection.
3. Abandon them.

According to the federal government, there are at least 1.5 million known underground tanks, but there are probably millions more of *unknown* tanks.

In July 2000, the writer was working with a contractor-developer on an abandoned city block parcel previously owned by the city of Baltimore. All buildings but one small one had been demolished, and the site had been leveled and graded. No mention was made of the existence of any underground tanks. Well, during the digging with a big Caterpillar 235 excavator, a 1000-gallon tank leaking fuel oil was pulled from the ground. As everyone scurried to contain the oil spill, the excavator continued to work, only to uncover another tank about 10 feet away, this one of 1500-gallon capacity, leaking oil and water as it was raised out of the ground. The state EPA was immediately called to the site. With a lot of fast talking and a demonstration that the leaks were minor in nature and swiftly contained, only a warning was issued by that agency.

This is a prime example of the principle that a project superintendent cannot discount encountering underground hazards or hazardous materials when working in an urban environment or on a previously occupied building site. So be alert.

Chlorinated Hydrocarbons

This family of hazardous liquids includes degreasing agents and paint strippers. Trichloroethylene and trichlorethane are metal degreasers used in machine shops and metalworking shops. Perchloroethylene was once a very common dry cleaning solvent, and methylene chloride is a component in many paint strippers. All four chemicals are to be avoided, and long-term inhalation is a serious health hazard. Once again, the presence of these chemicals may be detected by applying the sniff test.

Even buildings previously containing landscaping and gardening shops are not immune from scrutiny, because they probably stored and sold dangerous herbicides and insecticides such as DDT.

Health Hazards and Legal Issues Associated with Mold and Mildew

Today's project superintendent must be concerned about more than the potential for hazardous materials to lurk underground and the possibility of lead and asbestos above ground. As the twenty-first century began, a new culprit, to the delight of the legal profession, attained a great deal of attention—toxic molds.

Legislation is being proposed on national, state, and local levels dealing with mold and associated problems, and this represents a new challenge in the field.

Airborne mold spores are virtually everywhere, both inside and outside; but given moisture and moderate temperatures accompanied by humidity and a medium upon which to feed, toxic molds can grow rapidly in older and even new buildings.

The presence of toxic molds in buildings came to the forefront when Legionnaires' disease was so named and discovered at a convention center in Philadelphia, Pennsylvania, years ago. A rash of severe respiratory ailments were traced to airborne bacteria, and a new term was born—*sick building.*

Builders who had previously ignored or discounted the problems that could occur when mold or mildew was allowed to grow in certain areas of the project were now beginning to realize their tremendous exposure to legal action. This became some lawyer's new class action darling, and others were quick to follow.

An article about toxic mold that appeared in the August 5, 2001, edition of *The New York Times* reported that one lawyer in California had 1000 current cases related to mold issues, and, in Texas, mold claims had doubled between 2000 and 2001. In 2002, in New York City, 125 families joined together to file an $8 billion lawsuit against their landlord, claiming that toxic molds throughout the building caused residents illnesses and the death of a 7-year-old girl who suffered from asthma. Whether these types of claim have much credence is for the courts to decide, but the project superintendent should not treat any observations or discovery of mold and mildew lightly.

Cause and Effect

Older buildings are often left open to the elements, with water penetrating through leaking roofs or open areas left when windows or doors have been removed. New construction projects are not immune to conditions that foster mold and mildew when areas are not adequately protected against water infiltration. In the presence of moisture, humidity, and "food," toxic molds will grow with alarming speed—and the cellulose covering on gypsum board is the preferred "food" for these colonies of mold.

Contractors who had been accustomed to finding black mold on Sheetrock when certain areas of their building were exposed to water and moisture dismissed these conditions as an annoyance, something that had to be removed before painting. The cure was simple; mix a solution of household bleach and water, spray it on the wall, and kill the mold. On the next day, wipe down the affected areas, apply a coat of stain kill or clear shellac, and keep on painting.

That simple solution may not have worked then, and it probably won't work now. Although the surface area may have been rid of mold and mildew, colonies of these toxic "bugs" may have spread to enclosed areas within the drywall partitions.

In a recent situation where the writer had been called in to evaluate a mildewlike condition on several walls in an elderly housing project, a testing laboratory was brought in to take samples from various areas in a finished apartment, on the surface of the wall, within the wall cavity, and from the room's heat pump filters. Airborne mold colonies present on the walls had been picked up by the heat pump's filters and distributed by the unit's fan throughout the unit, eventually settling onto the carpeting. Not only was it necessary to destroy the mold colonies growing on and in the partitions, but also the heat pump filters had to be replaced after the heat pump and the carpeting were carefully vacuumed using HEPA (high-efficiency particulate) filters. These procedures were necessary in every apartment where traces of mold were in evidence.

Prevention versus Remediation

Architects are paying more attention to creating air barriers and rain-screen systems as they design exterior walls. Air barriers are constructed to build a barrier that is air-impermeable and can withstand loads from differential air pressure fluctuations. These air barriers are most commonly formed by the use of materials such as Tyvek (registered trademark of Dupont) or exterior-grade sheathing. Rain-screen systems are designed to block water from the exterior environment from penetrating the outer skin of a building, while taking into account that some small amount of water is likely to penetrate this rain screen. Effective rain-screen systems incorporate a vented air cavity that permits the system to "weep."

The old saying "An ounce of prevention is worth a pound of cure" certainly applies to the mildew and mold problem. Preventing excess moisture buildup in a rehabbed or new building begins with measures to keep out the elements and maintain some degree of airflow through the building. Many construction sequences introduce significant amounts of moisture into the building. Concrete slabs or extensive concrete repairs will obviously create moisture buildup as the concrete cures.

A leading contributor to mold and mildew results from the process of pouring gypsum floor toppings, either before or after drywall installation has begun. Gypcrete, as it is known, introduces substantial amounts of moisture in the areas where it is placed, and without a concerted effort to create air movement, mold and mildew are sure to grow.

Some Mold and Mildew Terms You Need to Know

Testing for mold generally involves taking air samples within the concerned unit and comparing them with outside air samples to evaluate the composition of indoor versus outdoor air quality.

It is likely that one or more of the following molds may be discovered when a testing laboratory is engaged to check an area for mold or mildew:

Aspergillus glaucus—a toxin-producing mold.

Aspergillus fumigatus—most common mold found on gypsum drywall surfaces; can result in persistent lower respiratory infections.

Penicillium—a live mold that can reproduce and affects persons with allergies.

Cladosporium—a common fungus that offers no particular hazard.

Basidomycetes—common mushroom type of fungus with no particular hazard except at high concentrations.

Stachybotrys atra—a fungus that has been linked to anything from sinus infections to brain damage.

CFU/m^3—test results are reported in this manner (colony forming unit per cubic meter of air sampled).

Environmental Audits

The Environmental Protection Agency has a responsibility to promulgate and enforce regulations regarding the management of hazardous materials. It has several pieces of legislation at its command to enforce actions. At the state level, additional laws and regulations may have been enacted to effect more stringent regulations.

The Resource Conservation and Recovery Act (RCRA) was designed to promote the protection of human health and the environment and to conserve valuable resources. This act is the backbone of the EPA. In response to the need to clean up and reclaim hazardous sites, Congress enacted the Comprehensive Environmental Response Compensation and Liability Act (CERCLA), more commonly known as the Superfund. CERCLA was enacted to provide a system for identifying and cleaning up hazardous material releases and to establish a fund to pay for the cleanup where no responsible parties could be found.

The Superfund Amendments and Reauthorization Act (SARA) was the first major revision of CERCLA. It established a fund to clean up leaking underground storage tanks, a program to maximize the safety of workers engaged in hazardous material cleanup, and a new radon gas and indoor air quality (IAQ) research program.

Protection of Archeological and Paleontological Remains and Materials

Contractors have to be alert these days to another problem—the potential of unearthing archeological, paleontological (fossils), or historical materials while excavating in urban, suburban, and even some previously undeveloped areas. If there is reason to believe that any of these materials might surface during the excavation of a site, they may be of significance in the recording of historic or prehistoric events, and the contractor may be required by local law to cease work immediately and notify the proper authorities. In the late 1990s when the protection of archeological remains was just beginning to attract wide attention, while constructing a road in New Mexico, the contractor unearthed some American Indian religious tribal remains and had to cease work until the tribe to which these remains belonged was identified. Work could not continue until such time as the rightful owners, once established, could have access to those remains. Other instances of encountering similar remains have halted construction work around the country.

When such instances do occur, the contractor is to contact the architect immediately, notify her or him of the possible historical value of the unearthed materials, and await further instructions. The contractor is required to make every effort to preserve the site until further direction is given and may, in fact, have to reschedule or redirect the work, or possibly abandon work until a final resolution is achieved. Since the potential for lengthy delays is evident, the contractor should request a "stop work" order which will stop the "construction" clock and start the "delay" clock with its related and forthcoming array of costs.

Saving an historic or architecturally important building can be a gratifying experience if the work is carefully planned; if open communications exist between owner, designers, and subcontractors; and if all parties are willing to assume some portion of the risk inherent in this type of work.

Murphy's law is alive and well in rehabilitation and renovation projects. Always expect the unexpected, and you will hardly ever be disappointed.

Chapter

8

Safety at the Job Site

From 1980 to 1995, some 17,000 construction workers died as a result of injuries suffered on the job—that is an average of 1133 deaths per year. In 2001 there were 1225 fatal accidents in the construction industry, an all-time high.

Everyone knows that a construction site is a dangerous place in which to work; 19.5 percent of all industrial fatal accidents occur on construction projects. And within the industry subcontractors have been the worst offenders. In the year 2000, of the 1154 industrial deaths, 673, or 58 percent, were attributed to these specialty contractors. For the influx of Latino workers during the 1999–2000 period, the industrial fatal accident rate increased from 725 to 815. Whether this could be attributed to a lack of safety training or to a language barrier problem, it behooves employers of Latino workers to pay extra attention to their training and supervision.

The Occupational Safety and Health Act (OSHA)

The United States Congress in 1970 passed the Williams-Steiger Occupational Safety and Health Act, referred to simply as OSHA. The rules and regulations established by this act are governed by the U.S. Department of Labor. Most states have adopted industrial safety provisions similar to those of the federal government, and project superintendents often have to deal with inspectors from both agencies.

In 1990, OSHA created a separate Construction and Engineering Division, and in November of that year President Bush signed the Omnibus Budget Reconciliation Act, which included, among many other provisions, significantly increased penalties for OSHA violations. An additional fine with a $70,000 maximum limit was to be imposed on contractors who *willfully* contributed to a job site accident.

OSHA's Most Frequent Paperwork and Job Site Safety Violations

Not all OSHA violations involve dangerous or hazardous job site conditions; many violations are issued because of a contractor's failure to comply with various reporting and posting requirements. Check your job trailer for the proper postings, and be aware of your reporting responsibilities. OSHA's top five paperwork violations are

1. Failure to provide the Log and Summary of Occupational Injuries and Illnesses forms properly updated

2. Failure to adhere to the General Duty clause of the OSHA act (a citation based upon no specific violation or a citation issued after a previous one had been ignored)
3. Failure to report a fatality or multiple hospitalization incidents
4. Failure to record occupational injuries and illnesses on the Supplementary Record form
5. Failure to record and report occupational injuries and illnesses on the required OSHA Log form

The most frequently reported job site accidents are due to one or more of the following incidents:

- Falls from elevated areas
- Being struck by an object or machine
- Being caught in between
- Electrical hazards

Figure 8-1 lists 26 of the top 100 frequently cited OSHA violations along with the corresponding section of the 29 CFR 1926 manual pertaining to the subject matter.

OSHA, though, is much more than an on-site safety inspector. It has always provided the construction industry with training materials, but it has expanded its mission to include what it calls "outreach trainer presentations." These presentations are available online at *www.osha.gov*; the user needs a Zip file utility, Microsoft PowerPoint, and the Adobe Acrobat reader to receive the free downloads. These training sessions, tailored to the needs of the construction worker, cover such topics as electrical safety, fall protection, excavations, cranes, materials han-

RANK	DESCRIPTION OF STANDARD		STANDARD (1926.____)
1	**Fall Protection**	Guarding open sided floors/platforms	500(d)(1)
2	**PPE**	Head protection from impact, falling or flying objects	100(a)
3	**Electrical**	Ground fault protection	404(b)(1)(i)
4	**Electrical**	Path to ground missing or discontinuous	404(f)(6)
5	**Trench/Excavation**	Protective Systems for trenching/excavation	652(a)(1)
6	**Scaffolding**	Guardrail specifications for tubular welded frame scaffolds	451(d)(10)
7	**PPE**	Appropriate PPE used for specific operation	28(a)
8	**Ladders/Stairways**	Stair rails required @ 30" change of elevation or 4 risers	1052(c)(1)
9	**Fire Protection**	Approved containers or tanks for storing or handling flammable or combustible liquids.	152(a)(1)
10	**General Provisions**	General Housekeeping	25(a)
11	**Trenching/Excavation**	Daily inspection of physical components of trench and protection systems	651(k)(1)
12	**Scaffolds**	Safe access for all types of scaffolds	451(a)(13)
13	**Electrical**	Ground fault circuit interrupters (GFCI's)	404(b)(1)(ii)
14	**Concrete/Masonry**	Guarding protruding steel rebars	701(b)
15	**Scaffolds**	General requirements for guarding	451(a)(4)
16	**Trench/Excavation**	Spoil pile protection	651(j)(2)
17	**Welding/Cutting**	Securing compressed gas cylinders	350(a)(9)
18	**Welding/Cutting**	Additional rules for welding as per ANSI Z49.1-1967	350(j)
19	**PPE**	Eye/Face Protection for operations which create exposure	102(a)(1)
20	**Fall Protection**	Guarding floor openings	500(b)(1)
21	**Ladder/stairway**	Ladder extended 3' above landings	1053(b)(1)
22	**Trench/excavation**	Access/Egress from trench/excavation	651(c)(2)
23	**Electrical**	Listed, labeled or certified equipment used in manner prescribed	403(b)(2)
24	**Electrical**	Flexible cords designed for hard or extra hard usage	405(a)(2)(ii)(j)
25	**Electrical**	Strain relief for cords	405(g)(2)(iv)
26	**Woodworking Tools**	Additional rules for woodworking tools as per ANSI 01.1-1967	304(f)

Figure 8-1 OSHA's top 26 most violated standards.

dling and storage, hand and power tools, personal protective equipment, scaffolds, and stairways and ladders. A newsletter called *Construction eTools,* also available online, contains safety tips, new developments relating to safety, and listings of other programs available from the Department of Labor.

OSHA should be looked upon as a valuable source of safety and safety training information and as a partner in the quest to provide the safest working environment in an industry that is constantly facing workplace dangers.

Eye Injuries—Another Frequent and Preventable Accident

Although they are not specifically included in the most frequent accidents and many are not life-threatening, eye injuries can be disruptive to production. Generally they are so easy to prevent—just use eye protection! More than 1000 eye-related injuries occur each day on construction sites. Each year about 100,000 of these injuries result in either temporary or permanent damage to the worker's vision.

How many times have you walked the site and observed a laborer cutting rebar with a cutoff saw, or someone cutting a CMU with a dry masonry saw, and neither laborer wearing eye protection? Generally one reason for lack of eye protection is, "Well, I was only going to cut one rebar (or brick or block)." But that one cut could be the one that blinds. Or another reason given is, "My goggles fog up, and I can't see what I'm doing" or "It's too hot to wear face protection."

None of these reasons are valid, and the project superintendent must immediately stop any operations or activities where there is a lack of proper personal protection.

What Do You Know about Proper Eye Wear?

1. *Are some types of safety glasses bad for the worker's eyes?* No. Safety eyewear is produced from optical-quality glass or plastics. Prolonged use of these types of glasses during the workday will not be injurious to the worker's eyesight.

2. *Do face shields really offer adequate eye protection?* Face shields offer proper eye protection only when used in conjunction with safety glasses or goggles. Harmful particles can rise up beneath the face shield much as in a chimney, and without safety glasses or goggles, eye injuries can occur.

3. *Can workers obtain industrial-grade safety glasses that match their own prescription glasses?* Industrial-grade safety glasses can be made to personal prescription standards.

4. *Are there differences between optical glass eye protection and polycarbonate eye protection?* Polycarbonate lenses are the strongest and most impact-resistant type of safety glasses. They are also lighter in weight and more fog-resistant than glass lenses.

5. *Do serious injuries to the eye occur only on the job site?* Give your workers a pair of safety goggles or glasses to take home with them, because lots of eye injuries occur at home—cutting tree limbs with a chain saw or grinding or cutting materials in the home workshop. Giving workers an extra pair for home use not only shows that you care about their safety, but also gets them in the habit of wearing eye protection when performing certain tasks requiring this type of protection. And it improves chances of workers for showing up on Monday morning in good health and ready to work.

OSHA Offers Assistance in Dealing with Four Dangerous Situations

In October 1994, OSHA announced its Focused Inspections Initiation Plan to assist contractors and subcontractors who have implemented effective safety programs but wished to have an OSHA inspection focus on the four leading causes of fatal injuries mentioned above. These four categories constitute 90 percent of all construction-related deaths. Upon request OSHA will conduct a focused inspection and provide a guideline (Fig. 8-2) for contractors requesting these types of inspections and a form (Fig. 8-3) for inspectors to follow when assessing a contractor's commitment to safety. A typical focused inspection guide sheet relating to falls from elevated areas is shown in Fig. 8-4 and a portion of a similar guide sheet for electrical hazards is shown in Fig. 8-5.

Women in the construction workforce present another reporting problem that probably skews accident statistics, making them lower than they should be. A study completed by the Advisory Committee on Construction Safety and Health (ACCSH) in 1999 revealed that many female construction workers don't report their injuries. Tradeswomen may not report injuries because of their need to "prove" themselves to their fellow workers and to their bosses. During the course of the study, ACCSH personnel heard the following comments: "I didn't ask for help. I ended up getting injured." And "I ended up almost hurting myself just to try to prove that I can do the job as well as he can."

The project superintendent should pay attention to the safety and welfare of company employees on the construction site, and he or she must pay equal attention to *all* workers employed on the project, including those of the subcontractors. Being aware of the demographics affecting accident rates, the superintendent can be more sensitive to the working environment of this increasingly varied and diverse workforce.

OSHA's Metrification Venture

The metric system of weights and measures is used nearly worldwide, but not in the United States. Even though federal legislation was enacted in 1988 to begin the metrification of America and an executive order was issued in 1991 to augment Public Law 00-418 on metrification, little has been done in that regard. OSHA's involvement in the metrification process is concerned with industrial hygiene matters since most biological, chemical, and physical sciences have long used the metric system. Time-weighted averages (TWAs), permissible exposure limits (PELs), and sampling and reporting forms all use the metric system.

Metric base units of measurement are as follows:

Unit measurement	Unit name	Symbol
Length	meter	m
Mass	kilogram	kg
Time	second	s
Electric current	ampere	A
Thermodynamic temperature	kelvin	K
Amount of substance	mole	mol
Luminous intensity	candela	cd

The metric system entails the use of multiples or powers of 10 to describe magnitude greater or less than the basic units of the meter, gram, ampere, and so forth. For example,

CONSTRUCTION FOCUSED INSPECTIONS INITIATIVE

Handout for Contractors and Employees

The goal of Focused Inspections is to reduce injuries, illness and fatalities by concentrating OSHA enforcement on those projects that do not have effective safety and health programs/plans and limiting OSHA's time spent on projects with effective programs/plans.

To qualify for a Focused Inspection the project safety and health program/plan will be reviewed and a walkaround will be made of the jobsite to verify that the program/plan is being fully implemented.

During the walkaround the compliance officer will focus on the four leading hazards that cause 90% of deaths and injuries in construction. The leading hazards are:

- Falls (e.g., floors, platforms, roofs).
- Struck by (e.g., falling objects, vehicles).
- Caught in/between (e.g., cave-ins, unguarded machinery, equipment).
- Electrical (e.g., overhead power lines, power tools and cords, outlets, temporary wiring).

The compliance officer will interview employees to determine their knowledge of the safety and health program/plan, their awareness of potential jobsite hazards, their training in hazard recognition and their understanding of applicable OSHA standards.

If the project safety and health program/plan is found to be effectively implemented the compliance officer will terminate the inspection.

If the project does not qualify for a Focused Inspection, the compliance officer will conduct a comprehensive inspection of the entire project.

If you have any questions or concerns related to the inspection or conditions on the project you are encouraged to bring them to the immediate attention of the compliance officer or call the area office at:

______________________.

______________________ **qualified as a FOCUSED PROJECT.**
(Project/Site)

______________ ______________________.
(Date) (AREA DIRECTOR)

This document should be distributed at the site and given to the Contractor for posting.

Figure 8-2 Construction Focused Inspections Initiative.

CONSTRUCTION FOCUSED INSPECTION GUIDELINE

This guideline is to assist the professional judgement of the compliance officer to determine if there is an effective project plan, to qualify for a Focused Inspection.

	YES/NO
PROJECT SAFETY AND HEALTH COORDINATION; are there procedures in place by the general contractor, prime contractor or other such entity to ensure that all employers provide adequate protection for their employees?	
Is there a DESIGNATED COMPETENT PERSON responsible for the implementation and monitoring of the project safety and health plan who is capable of identifying existing and predictable hazards and has authority to take prompt corrective measures?	
PROJECT SAFETY AND HEALTH PROGRAM/PLAN* that complies with 1926 Subpart C and addresses, based upon the size and complexity of the project, the following:	

____ Project Safety Analysis at initiation and at critical stages that describes the sequence, procedures, and responsible individuals for safe construction.

____ Identification of work/activities requiring planning, design, inspection or supervision by an engineer, competent person or other professional.

____ Evaluation/monitoring of subcontractors to determine conformance with the Project Plan. (The Project Plan may include, or be utilized by subcontractors.)

____ Supervisor and employee training according to the Project Plan including recognition, reporting and avoidance of hazards, and applicable standards.

____ Procedures for controlling hazardous operations such as: cranes, scaffolding, trenches, confined spaces, hot work, explosives, hazardous materials, leading edges, etc.

____ Documentation of: training, permits, hazard reports, inspections, uncorrected hazards, incidents and near misses.

____ Employee involvement in hazard: analysis, prevention, avoidance, correction and reporting.

____ Project emergency response plan.

* FOR EXAMPLES, SEE OWNER AND CONTRACTOR ASSOCIATION MODEL PROGRAMS, ANSI A10.33, A10.38, ETC.

The walkaround and interviews confirmed that the Plan has been implemented, including:

____ The four leading hazards are addressed: falls, struck by, caught in/between, electrical.

____ Hazards are identified and corrected with preventative measures instituted in a timely manner.

____ Employees and supervisors are knowledgeable of the project safety and health plan, avoidance of hazards, applicable standards, and their rights and responsibilities.

THE PROJECT QUALIFIED FOR A FOCUSED INSPECTION ______

Figure 8-3 Focused Inspection Guideline.

RANK IN FREQUENCY CITED	1926.	
#1	**500(d)(1)**	GUARDRAILS NOT PROVIDED FOR OPEN-SIDED FLOORS OR PLATFORMS

RULE: ***Every open-sided floor or platform 6 feet or more above adjacent floor or ground level shall be guarded by a standard ramp, or the equivalent , as specified in paragraph (f)(1) of this section, on all open sides, except where there Is an entrance to a ramp, stairway, or fix ladder. The railing shall be provided with a standard toeboard wherever, beneath the open sides, persons can pass, or there is moving machinery, or there Is equipment with which falling materials could create a hazard.***

INTENT:
Falls from elevations are the leading cause of fatalities in the construction industry. From 1985-1989, 33% of all construction fatalities [10] resulted from a fall from an elevation. One hundred-seventeen fatalities occurred when employees fell from open sided doors and through floor openings. This standard specifies that guarding must be provided for all open-sided floors and platforms 6 feet or more in height. It also specifies minimum requirement for the type of guarding. Paragraph (f) of the same section species the requirement of a standard guardrail system. TABLE 5.2-1 lists guardrail specifications for various materials. Where there is an open-sided floor/platform and there is a potential for a person to pass or a hazard is presented by machinery, toeboards are required. The intent is to contain any materials near the edge from inadvertently getting pushed over the edge where they may strike persons or machinery below. TABLE 5.2-2 lists specifications for toeboards.

HAZARDS:

- Falls from elevations: probable injuries range from death to fractures; Fall from lower elevations such as 4-6 feet have caused serious lost-time accidence and occasionally have been the cause of fatalities.
- Struck by: the lack of material containment (toeboards) has caused both fatalities and lost-time accidents when falling materials have struck employees below.

(AMONG OTHER) SUGGESTED ABATEMENTS:

- Whenever an employee must work at <u>any</u> elevated location, ask the questions: 1) Are they protected from a fall? and 2) What measures must be taken to protect the employee at the elevated work location?
- Fall prevention systems such as standard guardrail systems provide more positive means of protection than fall protection systems such a bodybelt/harness-lanyard-lifeline combination, except when workers are suspended, i.e. working on suspended scaffolds, work platforms, etc.
- Construct/maintain all guardrail systems according to OSHA requirements.
- An acceptable method to preclude the use of toeboards, would be to determine the fall radius of materials on an open-sided door/platform. Place positive physical barrier outside the potential fall radius to keep workers and machines outside the danger zone.

SELECTED CASE HISTORIES:
An employee taking measurements was killed when he fell backwards from an unguarded balcony to the concrete 9'6" below.

COMMENTS:

- Falls from elevations accounted for 14% of all lost-time accidents[6].
- This standard was cited in 103 fatality/catastrophe inspections conducted by OSHA over a 4 year period.

ADDITIONAL DOCUMENTS TO AID IN COMPLIANCE:
[1] Section 500 & Steel Erection - 750 & 752(k); [11]; [12]; [13]; [26] Part - 701(f)(2) - Concrete and Masonry Const.

OSHA COMPLIANCE LETTER
Date 5/22/84; From-Directorate of Field Operations to Regional Administrators; Synopsis - Clarification of 1926.750(b)(1)(iii) stating that ½ " wire rope or equivalent safety railing must be used around temporary planked or temporary metal-decked doom during steel erection operation. Raging also must be provided at leading edge if spreading stops for any significant time period. ½ " synthetic or fiber rope would not be acceptable as a required safety railing for steel erection operations.

OSHA COMPLIANCE LETTER
Date 1/13/81; From-Assistant Secretary to Int. Union of Bricklayers & Allied Craftsmen; Synopses - Standards 1926.28, 1926.104, 1926.105 & 1926.500(d)(1) do not apply to overhand bricklaying operations.

OSHA COMPLIANCE LETTER
Date 2/13/86; From-Directorate of Field Operations to Individual Company; Synopses - When structural steel assembly including decking has been completed and other trades are working on the deck while concrete is being poured on the deck, the door must be guarded in accordance with 1926.500(d)(1).

MAJOR SUBJECT COMBINED STANDARD TITLE	#[1]	DESCRIPTION OF EACH STANDARD FOR COMPARISON
SCAFFOLDING		
GUARDING SPECIFICATIONS	6 15 32	Guardrail specifications for tubular welded frame scaffolds General requirements for guarding Guarding specifications for mobile scaffolds
ACCESS	12 59	Safe access for all types of scaffolds Ladder/stairway affixed or built-in to mobile scaffold for access/egress
FOUNDATION SPECIFICATIONS	40 55 77 100	Sound, rigid, and load capable footings or anchorages for all types of scaffolds Plumb and sound base for mobile scaffold – casters locked Foundation specifications for tubular welded frame scaffold legs Locking or pinning legs to prevent uplift
		STAND ALONE STANDARDS **32,43,51,71,76,82,85,91,92**
LADDER/STAIRWAY		
STAIR RAIL REQUIREMENTS	8 74	Stair rails required @ 30" change of elevation or 4 risers Guarding or stairway edges and landings
DEFECTIVE LADDERS	45	Defective portable ladders tagged and taken out-of-service
LADDER ERECTION	75 21	Siting and securing ladders Ladders extended 3' above landings
		STAND ALONE STANDARDS **37,63,65,66,67,81,90**

Figure 8-4 Focused inspection relating to falls.

MAJOR SUBJECT COMBINED STANDARD TITLE	#(1)	DESCRIPTION OF EACH STANDARD FOR COMPARISON
ELECTRICAL		
GROUND FAULT PROTECTION	3 13 64	Ground fault protection Ground fault circuit interrupters (GFCI's) Assured equipment grounding conductor program
CORD SPECIFICATIONS	25 24 46	Strain relief for cords Flexible cords designed for hard or extra hard usage Flexible cords and cables made suitable for specific conditions
SPECIFICATIONS FOR TEMPORARY LIGHTING	29 72	Protection and grounding for temporary lamps. Temporary lights suspended from electrical conductor cords
COVERED FOR BOXES, UNUSED OPENING, ETC	38 41	Covering provided for pull boxes, junction boxes, outlets, etc. Unused opening in boxes must be closed and conductors entering boxes must be protected from abrasion.
CONTROLLED ACCESS AND GUARDING OF EQUIPMENT OPERATING >600 VOLTS	30 70	Controlled access to installations operating at over 600 volts. Guarding provided for temporary wiring operating over 600 volts.
		STAND ALONE STANDARDS **4,23,35,39,44,53,57,62,84**

Figure 8-5 Focused inspection relating to electrical hazards.

- The *kilo*gram is 1000 grams, and the *milli*gram is 1/1000 gram.
- There are 100 *centi*meters to 1 meter and 1000 meters to 1 *kilo*meter.
- There are 10 *deciliters* in 1 *liter*.

Age as a Safety Factor

Age is another factor affecting the rates of accidents. A study performed 5 years ago by *The Wall Street Journal* revealed that workers between the ages of 55 and 64 were nearly 4 times more likely to be killed by objects or equipment than younger workers. Although this study was an industrywide one and not devoted solely to construction, it presents another group of workers requiring closer scrutiny, from a safety standpoint, on the job site.

Safety Pays—In Many Ways

A good safety record rewards the general contractor in several ways. No one wants to see fellow workers injured, but somehow safety often takes a back seat to other seemingly important matters on the job site, such as scheduling workers and materials and proper scheduling. But safety pays in more ways than one.

Skilled workers are a shortage in today's marketplace, and when a productive crew has been assembled, a serious injury can affect not only crew size but productivity as well. The company's worker compensation insurance premiums are based upon accident and injury experience. When a company has a poor safety record and experiences multiple worker injuries, insurance rates can climb dramatically, to the point where corporate overhead can also increase by a percentage point or two. Since the cost of overhead is factored into every bid the company prepares, increased insurance costs and increased overhead costs may make the company noncompetitive in some of those hotly contested markets.

Developing a Safety Program

The development and implementation of an effective and comprehensive safety and health program require, first and foremost, management leadership and commitment. Both must be ongoing and active—not just lying dormant until the next accident. Employee involvement should be encouraged in such activities as problem resolution, hazard analysis, accident investigations, and safety and heath training.

The company safety program should cover the following:

1. A statement of company policy that simply expresses the company's commitment to protect its employees and those of its subcontractors.
2. The appointment, duties, and responsibilities of a safety director or safety coordinator.
3. Responsibilities of field supervisors in administering the plan and their relationship to the safety director or safety coordinator.
4. Procedures for reporting job-related injuries and illnesses.
5. Working rules and regulations of the safety program. This is the "nuts and bolts" of the program in which specific items of personal protection, procedures for operating power-actuated tools, and procedures for operating various pieces of equipment are set forth in detail.
6. A hazard communications (hazcom) program to conform with OSHA regulations.
7. Procedures for dealing with safety violations and violators

More on Hazard Communications

This program deals with the proper handling, storage, and use of products containing hazardous materials, as well as directions to follow if any of these are ingested or come in contact with human skin or eyes. Violations of OSHA's hazcom regulations rank in the top 25 list of citations issued by that agency.

And leading the list of hazcom violations is failure to have all MSDS materials on file. The materials safety data sheet (MSDS) is a document provided by the manufacturer of the material containing hazardous components. By law, each company manufacturing such a product is required to prepare a sheet listing its hazardous nature, proper handling and storage instructions, and in case of contact or ingestion by a worker, the necessary first aid or medical procedures to follow.

Prior to shipment of the product to the job site, an MSDS is to be sent by the manufacturer to the general contractor's office for dissemination to the field. When the product does arrive on the job site, the provisions of the MSDS with respect to handling and storage are to be followed.

The field supervisor must keep an accurate and orderly file of all MSDS documents and ensure that all such products delivered to the job site have a corresponding data sheet. During an OSHA inspection, the inspector will most likely request a review of all MSDS documents, so they should be kept in an easily retrievable file.

Although the safety program's rules and regulations may have been formulated in the office, it will fall to the superintendent to strictly enforce the program in the field.

MSDS and Dust Control

Five MSDSs submitted by drywall joint compound manufacturers were reviewed by the National Institute for Occupational Safety and Health (NIOSH), and each one warned workers to avoid generating dust when sanding taped drywall joints. Respiratory protection is recommended when dry sanding is used; four manufacturers recommended using "wet sanding" whenever possible, and the fifth manufacturer advised to cut dust exposure by providing ventilation.

Health experts estimate that thousands of construction workers are exposed to unhealthy levels of airborne dust and that at least 250 workers in all industries die each year of silicosis.

Dealing with Drywall Sanding

Drywall sanding can be accomplished by dry sanding, wet sanding, and pole sanding. Wet sanding is generally avoided because of concern for drying time and also for the effect water might have on the finish texture of the drywall. Pole sanding increases the space between worker and the sanding surface, and while frequently used on ceiling tape joints, is less frequently used on walls. Dust exposures can be cut, according to NIOSH studies merely by changing from hand sanding to pole sanding which reduces the amount of dust close to the sander's nose and mouth.

Several systems are now on the market that use portable vacuums to capture airborne dust. NIOSH engineers compared dust exposure to workers using the conventional hand sanding and pole sanding method to those workers employing one of the commercially available vacuum sanding methods; vacuum sanding reduced dust exposures by 80 to 97 percent.

Unacceptable dust levels are often created when laborers use pneumatic jackhammers, drills, and chippers. Masons using dry chop saws to cut brick and block create high levels of harmful dust and simply using a wet cutting method is sufficient to greatly reduce, if not totally eliminate exposure to hazardous dust. Painters are complaining about high levels of silica in the air, in some cases 225 times the permissible levels, when employed on bridge painting projects where sand blasting has been used to prepare the metal surfaces.

By enforcing better methods to control airborne dust and debris, the project superintendent will create a much cleaner work environment. For workers this cleaner environment will be not only more comfortable, but less irritating to the eyes, nose and throat and result in a more productive atmosphere where absenteeism, frequent breaks and clean up time will be much less.

The Weekly Toolbox Safety Talk

An overwhelming majority of contractors conduct weekly toolbox talk sessions at the job site. Relatively short in nature, ranging from 5 to 15 minutes, these weekly sessions are meant to highlight specific work tasks and remind attendees of the potential hazards associated with these tasks and safe working procedures to follow when performing these tasks. After the toolbox talk is concluded, usually each attendee inks a sign-in sheet, signifying she or he has attended this short training session and been informed about the safety procedures associated with the subject discussed. Refer to Fig. 6-7 for a sample of a typical weekly toolbox talk.

Enforcement of the Safety Program

A typical disciplinary procedure portion of a company's safety program will probably read as follows:

> Compliance with OSHA and company safety rules and regulations is a condition of employment at (company name). All employees working for (company name) will be trained in, and must familiarize themselves with, both OSHA and company safety rules and regulations before beginning work on a construction site. A copy of the company's safe work rules and procedures will be provided by your supervisor. Management personnel at all levels, including project managers, field supervisors, and foremen, are responsible for taking action when a violation is observed. If a violation is observed, they must take action immediately to correct the violation and enforce this disciplinary policy. Employees who fail to follow safety rules and regulations established to protect them and their fellow employees endanger themselves and others.

We have all been guilty of walking the job site during an important inspection or running late for a meeting and observing a safety violation. The tendency may be to continue on and make a mental note to stop back later to deal with the violation, but later may be too late to prevent a debilitating injury. When a violation is spotted, *stop*—bring it to the attention of the violator with a stern warning. The written report can wait until later in the day, but the verbal warning must be timely.

A Typical Infraction Warning Procedure

The company may elect to have a system of three or four warnings before strict enforcement is instituted. A four-warning notification system will be similar to this one:

First warning: The first time an employee is observed violating any safety rule, the employee shall be given a *first warning*. The first warning will be an oral one and will be so noted in the employee's personnel file.

Second warning: The second time the employee is observed violating any safety rule, the employee shall be given a *second warning*. The second warning will be an oral warning accompanied by a written safety violation notice. A copy of the written safety violation will be given to the employee, to the employee's union steward (if applicable), and to the company's safety director. A copy of the notice will be placed in the employee's personnel file. The employee will be required to meet with the safety director for counseling.

Third warning: The third time an employee is observed violating any safety rule, the employee shall be given a *third warning*. The third warning will be a written safety violation notice. A copy of the written notice will be given to the employee, to the employee's union steward (if applicable), and to the safety director. A copy of the written notice shall be placed in the employee's personnel file. A meeting with the employee, his or her immediate supervisor, the safety director, and a representative of top management will be held to determine why the employee has failed to comply with the company's safety program—for the third time. Top management must determine what action will be taken at this time. One option open to management, which must be included in the disciplinary section of the safety program, is suspension from work. For example, the program might include a provision that states, "Employees who have accumulated three warnings in a 12-month period may be suspended from work, without pay, for up to one week."

Fourth warning: The fourth time an employee is observed violating any safety rule, the employee shall be given a *fourth warning.* A fourth warning will be a written safety violation notice. Employees who do not follow safety rules, especially after being warned several times previously, are a threat to themselves and to their coworkers. Therefore, employees who receive a fourth warning may, at management's discretion, be terminated from employment or be subject to other disciplinary action deemed appropriate by management.

It is important that these warning notices, whether oral or written, be applied whenever a safety violation is observed. Consistency and fairness are key factors in administering any safety program. Employees will not take the disciplinary policy seriously if they know that safety violations have occurred, were observed by their supervisors, but no warning was issued.

What to Do When an OSHA Inspector Appears at the Job Site

When an OSHA inspector appears at the job site and requests permission to enter the site, the superintendent should be aware of the proper procedures to follow once the inspector arrives at the job trailer:

1. The inspector will make her or his presence known, and should be asked to present credentials, if not already offered. Make a photocopy of the inspector's credentials.
2. Ask about the nature of the inspection—is it routine or initiated by OSHA because of a complaint it received?
3. The inspector should be asked whether he or she plans to tour the entire site or just selected portions. The inspector should be confined to those areas relevant to the complaint.
4. The superintendent should call the office and advise the project manager or director of safety that an OSHA inspector is on site and the nature of the visit. Will either the project manager or safety director be available to walk with the inspector, or will the superintendent be designated to do so?
5. The superintendent should contact the crew supervisors for all subcontractors on site, tell them about an impending OSHA inspection, and request that they or their appointed representative accompany the inspector.
6. If the sole general contractor representative on the inspection tour, the superintendent should bring along pad and pencil to take notes. If a camera—digital, instant or otherwise—is available, bring that along as well.
7. If any OSHA-related documents are requested by the inspector, bring them along.
8. Do not volunteer information, and avoid making any statement that could be construed as a violation of any OSHA rule or regulation. This should be interpreted not as being uncooperative, but as being unfamiliar with all the OSHA regulations.
9. If the inspector takes samples, ask to take two of each so that you can retain one of each sample for your records.
10. If a violation is discovered and the inspector inquires about the time necessary to correct the situation, respond but without acknowledging the "violation."
11. If possible, videotape the entire inspection process.
12. If citations are issued during the inspection tour, don't voice an objection. Wait until you are officially notified of the violation. Then, after reviewing the "violation" and accompanying citation, respond, citing the company's position.

13. A closing conference with the OSHA inspector should be requested to ensure that everyone understands the findings of the inspector.

14. Citations will be issued by OSHA's area director, and when received, they are to be posted at the job site for 3 days. Any corrective work must be completed within the time frame indicated in the citation.

At the end of the inspection, there will be a closing conference; the inspector will review the results of the inspection and point out the violations observed. If any of the problems are minor, the OSHA inspector may request that your company submit an abatement plan within 10 days. If the problems uncovered are more serious, the inspector may simply state that time is needed to evaluate the findings and she or he will notify the company promptly.

It is likely that another inspection—a reinspection—will be scheduled, and it is important that detailed notes be taken all during this closing conference.

Safety Program Forms

1. Safety violation notice for a company other than the general contractor or subcontractor (Fig. 8-6)
2. Safety violation notice to be issued to an individual (Fig. 8-7)
3. An accident/incident report (Fig. 8-8, p. 182)
4. An accident/incident report for a minor accident (Fig. 8-9, p. 183)
5. A witness statement to an accident/incident report (Fig. 8-10, p. 183)
6. An emergency number list (Fig. 8-11, p. 184)

SAFETY VIOLATION NOTICE

COMPANY

To: ____________ Date ________ Time ________

____________ Job #

____________ Project

ATTN: ____________ Location

Company

The company identified above is not following safe practices as follows:

____

____

____

____

____

____

____

Heffner & Weber requires that all practical effort be taken to eliminate or minimize exposure of workers to unsafe and/or unhealthy conditions on our job sites. Subcontractors and visitors are obligated to cooperate with this policy.

FAILURE OF A WORKER TO COOPERATE FULLY IN CARRYING OUT OUR SAFETY PROGRAM IS CAUSE FOR DISCIPLINARY ACTION INCLUDING REMOVAL FROM SITE, PROBATION, SUSPENSION AND / OR DISMISSAL.

Very Truly Yours, Received By

____________ ____________

Project Superintendent For ____________

Date ________ Time ________

Figure 8-6 Safety violation notice for a company other than the general contractor.

SAFETY VIOLATION NOTICE

INDIVIDUAL

To: ____________ Date ________ Time ________

____________ Job #

____________ Project

ATTN: ____________ Location

Individual ____________ ☐ Heffner & Weber Employee

Employer ____________ ☐ Subcontractor Employee

Craft ____________ ☐ Site Visitor

The individual identified above is not following safe practices as follows:

____

____

____

____

____

____

Heffner & Weber requires that all practical effort be taken to eliminate or minimize exposure of workers to unsafe and/or unhealthy conditions on our job sites. Subcontractors and visitors are obligated to cooperate with this policy.

FAILURE OF A WORKER TO COOPERATE FULLY IN CARRYING OUT OUR SAFETY PROGRAM IS CAUSE FOR DISCIPLINARY ACTION INCLUDING REMOVAL FROM SITE, PROBATION, SUSPENSION AND / OR DISMISSAL.

Very Truly Yours, Received By

____________ ____________

Project Superintendent For ____________

Date ________ Time ________

Figure 8-7 Safety violation notice issued to an individual.

ACCIDENT / INCIDENT REPORT
Sheet 1

JOB NUMBER ____________ DATE ________ SHEET ______

JOB NAME ____________ SUPERINTENDENT ____________

1. WAS ANYONE INJURED? ☐ YES - ACCIDENT REPORT ☐ NO - INCIDENT REPORT

2. Enter Data for all injured parties. Use second sheet if additional space is needed.

Name	Employer	Injury	Treatment

3. WERE THERE WITNESSES? ☐ NO ☐ YES - Enter below. Use second sheet if additional space is needed.

Name	Employer	Job Function	Location

4. PROJECT SUPERINTENDENT - DESCRIBE OBSERVATIONS AND IMMEDIATE RESPONSE ACTIONS TAKEN.

Line	Time	Observation / Action
1		
2		
3		
4		
5		
6		
7		
8		
9		
10		
11		
12		

ACCIDENT / INCIDENT REPORT
Sheet 2 Cont'd

JOB NUMBER ____________ DATE ________ SHEET ______

JOB NAME ____________ SUPERINTENDENT ____________

10. Continuation from above sections. Enter Question Number and Line number being continued.

CONT'D	Time	Continuation

Figure 8-8 Accident report.

ACCIDENT / INCIDENT REPORT
MINOR

JOB NUMBER ______________ DATE ________ SHEET ______

JOB NAME ______________ SUPERINTENDENT ______________

1. WAS ANYONE INJURED? ☐ YES - ACCIDENT REPORT ☐ NO - INCIDENT REPORT
2. Name Of Employee Injured ______________ Employer* ______________

   Describe Injury. ______________

   Was Medical Attention Required? Describe. ______________

   Treated By ______________ Location ______________
3. Name Of Witness(es) ______________ Employer ______________
4. Describe Work Being Performed At The Time Of The Accident / Incident And Applicable Safety Rules
5. Were Safety Rules Violated? ☐ NO ☐ YES (If YES Describe Violation)
6. Describe What Happened In Sequence.
7. What Caused The Accident / Incident
8. Describe What Was Done Or Can Be Done To Prevent This From Taking Place Again.
9. Project Superintendent - Describe Observations And Immediate Response Actions Taken

Figure 8-9 Accident report for a minor accident.

ACCIDENT / INCIDENT REPORT
WITNESS STATEMENT

JOB NUMBER ______________ DATE ________ SHEET ______

JOB NAME ______________ SUPERINTENDENT ______________

A. Witness Name And address

Name ______________ DATE ________ SHEET ______

Address ______________ Home Phone ______________

City ______________ Work Phone ______________

State ZIP Employer

B. Witness Description Of Accident / Incident

Line	Time	Witness Statement
1		
2		
3		
4		
5		
6		
7		
8		
9		
10		
11		
12		
13		
14		
15		
16		
17		
18		

Witness Signature ______________ Date ________

Project Superintendent ______________ Date ________

Figure 8-10 Form to report witness statement relating to an accident.

EMERGENCY NUMBER LIST

Job # ______ Project ______ Location ______ Project Superintendent ______

Plat # ______ Tax # ______ Street Address ______ City ______ State ______ ZIP ______

	EMERGENCY NUMBER					ADMINSTRATIVE NUMBER			
	Type Of Emergency Service	Name Of Emergency Provider	Emergency Telephone Number	Limits On Hours Of Operation	Limits On Capabilities	Non-Emergency Telephone	Point Of Contact	Address Of Emergency Provider	Fees / Pre-Use Agreements Or Contracts
1	FIRE								
2	POLICE								
3	MEDICAL								
4									
5	Ambulance								
6	Hospital								
7									
8	Water Break								
9	Gas Leak								
10	Sewer Leak								
11	Elect - OH								
12	Elect - UG								
13	Tele - OH								
14	Tele - UG								
15	Cable - OH								
16	Cable - UG								
17	HazMat								
18	Environ								
19									
20									

Figure 8-11 Form to create emergency telephone number directory at the job site.

Chapter

9

Quality Control and Quality Assurance

Quality construction is what every builder wants to achieve, and quality of construction has become a key element in an owner's evaluation of a general contractor. The cost of construction and the ability to meet project completion dates will always be important factors in the selection of a general contractor. But quality of product has taken on a more important role in this highly competitive business, and many a contract hinges on the builder's proven track record of quality projects.

There is often confusion about the difference between *quality control* (QC) and *quality assurance* (QA), and there are distinct differences. *Quality control* can be defined as the standard used in the construction or assembly of a building component, while *quality assurance* is a process that verifies that these standards have been met.

The quality of workmanship on a project will be determined, in part, by the nature of the contract documents. Construction specifications define the qualitative requirements for materials, products, equipment, and installation procedures as well as the level of workmanship. Within these defining documents, tolerances are established, often by the various and many trade groups and trade associations referenced in the specifications.

As an example, quality standards for concrete work are established by the American Concrete Institute (ACI), and numerous references to ACI are made in Division 3 of the projects specification manual. ACI Specification 301 contains the following tolerances for cast-in-place concrete:

Plumb—columns, piers, walls

- In any 10 feet: ¼ inch
- Maximum for total height of structure (less than 100 ft): 1 inch

Level—slab soffits, ceilings, beam soffits (measured before removal of shores)

- In any 10 feet: ±¼ inch
- In any bay or in any 20 feet: ±⅜ inch
- Maximum total length of structure: ±¾ inch

Level—columns, walls, beams, partitions (deviation from dimensions on plans)

- In any bay: ±½ inch
- In any 20 feet: ±½ inch
- Maximum for the structure: ±1 inch

These are the tolerances required for cast-in-place concrete members when the ACI requirements are referred to in the contract specifications—this is a quality control standard.

When the superintendent places a level on the concrete beam, column, or wall to determine if it meets these standards, that is quality assurance.

And tolerances for a specific product can vary depending upon the end use of the product. Concrete slabs to receive pad and carpet in an office building will not be required to be finished to the same tolerances as a "superflat" floor for an automated warehouse project. The tolerances included in the specifications ought to consider another criterion—"constructability" tolerances. A wood-frame building will not meet the same level and plumb qualities as a steel stud or structural steel frame building, due primarily to the quality of the materials—wood studs with all of nature's imperfections versus the manufactured tolerances of metal studs or structural steel rolled sections.

Drawings that specify a masonry opening $^1/_8$ inch larger than the window that is designated to fit into that opening defy normal field allowable tolerances and do not take into consideration the space required to properly align the window in both horizontal and vertical axes; nor do they provide room for blocking.

When such constructability issues are discovered during the plan review, they should be brought to the architect/engineer's attention for a relaxation or modification of the tolerances required to install or build a specific component.

But quality control and quality assurance are not limited to construction products, assemblies, and components. Division 1 of the project specification booklet generally includes requirements pertaining to shop drawing submission and review, record documents such as as-built drawing preparation, and even the way in which monthly requisitions are to be prepared and submitted with accompanying backup documentation or substantiation. These are, in effect, quality control standards, and compliance with these standards will be an exercise in quality assurance.

So quality control and quality assurance span the entire spectrum of the construction process, from complying with the requirements to establish a site logistics plan, if included in the specifications, to complying with closeout procedures, to ensuring that the cast-in-place concrete complies with ACI standards.

First Steps First

To fully understand the general contractor's obligations and responsibilities, a read, or reread, of the contract with the owner is necessary. That which is included in the scope of work and that which is not included should be defined. What exceptions is the owner making to the terms and conditions of a standard AIA contract?

The project superintendent should be familiar with the terms and conditions of the contract with the owner that affect the supervision of the project. The next order of business is to begin the process of carefully reviewing the plans and specifications, making notes of important things to remember and any obvious problems, errors and omissions, or inconsistencies contained in the contract plans and specifications.

An in-house kickoff meeting is strongly recommended at this time, if this is not standard procedure in your company. This meeting should be attended by the estimator, the project manager assigned to the project, and the staff member who will be buying out the job, if it is not being done by the project manager. The intricacies of the project can be discussed as well as any problem areas uncovered during the estimating process, pointing out any constructability or quality concerns expressed by anyone attending the meeting. The estimator may have noted problems or concerns during the takeoffs, and these problems or concerns need to be passed on to the construction team. Quite often, during the bidding process, subcontractors or vendors submitting their quotes will qualify their bids, or take exception to the bid documents, stating concerns about specific product or installation procedures contained in either the plans or the specifications. Sometimes these subcontractors or vendors will actually offer suggestions to improve quality at no additional cost, and these suggestions should be passed on to the architect/engineer for review and comment.

Remember Article 3 of AIA Document A201, General Conditions of the Contract for Construction, states that any design errors or omissions noted by the contractor during the review shall be promptly reported to the architect as a request for information or clarification. The first pass at any errors and omissions issue should be made at this kickoff meeting. By doing so the contractor establishes an early relationship with the architect, one that says, "I read and understand my obligations and will continue to act as a professional all the way through this project—I hope you do the same."

A typical agenda for a kickoff meeting would include most of these items:

1. Review of the project superintendent's and project manager's responsibilities
2. The schedule included in the bid and the need to establish a baseline schedule as soon as possible
3. The budget and any potential shortcomings or problems with the scope of work assigned to a particular line item in the budget
4. Quality control and quality assurance
   a. Inspections required and who is responsible to provide them and pay for them
   b. Any preinstallation meetings required by contract
   c. Any required mock-ups
   d. Controlling the quality of subcontractor work and how inspection of the work is to be carried out. Will this be performed primarily by the project superintendent, who will report directly to the project manager or directly to the subcontractor with a copy to the project manager?
5. Safety
   a. Location of closest area medical center and fire station
   b. MSDS requirements and compliance with OSHA hazcom
   c. Toolbox talks
   d. Safety violation forms and their use
   e. Periodic inspections by the company's safety director
   f. Dealing with OSHA inspections, if and when they occur (Will project manager be required to meet with the inspector, or will this responsibility fall to the project superintendent alone?)
6. Company policies and procedures
   a. Review of company manuals
   b. Review of field report forms
   c. Issuance of Request for Information (RFI) and Request for Clarification (RFC)
   d. Subcontractor meeting procedures and preparation of meeting minutes
   e. Notice to correct and backcharge procedures
7. Review of the shop drawing submittal log
   a. Will the responsibility to conduct periodic reviews of the status of the shop drawing log lie with the project manager or project superintendent?
   b. Will the superintendent be given an advance copy of the shop drawing for review and comment? (It is always a good idea to do so!)

   c. Who will assume primary responsibility to notify a subcontractor or vendor if the shop drawing is late in being submitted?
   d. Who will follow up on delivery of supplies or equipment when an approved shop drawing has been returned to the subcontractor or vendor?
8. Status of buyouts and critical nature of specific items and subcontracts
   a. Establishment of a priority in buyouts, what trade or product is needed immediately, etc.
   b. Quality and performance records, if known, of the subcontractor and vendors under consideration for the current project
9. Role of general field superintendent in this particular project
10. Site logistics, availability of office and storage trailers, supplies and equipment
11. Availability of utilities at the site and need to supplement, if required
12. Review of closeout procedures and assigned responsibilities
   a. Preparation of as-builts by the general contractor
   b. Periodic inspections of as-builts required of subcontractors

Testing and Inspections

Probably the first encounter with QA/QC will occur in the site and excavation phase of the project when, in many cases, the owner's consultants will be engaged to inspect the excavation operations, including soils analysis, requiring specific tests such as a Proctor test with related reports. This test is used to determine the optimum moisture content of the soil removed from the excavation in order to achieve the desired compaction. The firm hired by the owner to perform these tests may also be retained to conduct cast-in-place concrete inspections and testing.

The Geotechnical Engineer—First Encounter with QA/QC

The determination of acceptable or unsuitable soils will test the project superintendent's knowledge of this subject, and control over acceptable levels of compaction requires close attention.

The owner frequently hires the geotechnical or civil engineer to observe, inspect, and perform the various tests required during the site work phase of the project. The geotechnical consultant may be hired to perform any or all of the following tasks:

1. On-site inspection of rough grading operations to verify quantities of cuts and fills and elevations of rough or finish grades
2. Testing of soils, either on-site materials or off-site (borrow) materials, to include compliance with soil composition and gradations, moisture content, and compactability
3. Visual inspection of any underground obstructions encountered, whether indicated on the contract drawings or not
4. Verification of quantities of unsuitable soils removed form the site and the import of acceptable replacement materials
5. Verification of any rock removal and quantities blasted and removed
6. Inspection of site utilities to ensure compliance with the materials and installation procedures contained in the contract documents
7. Inspection of foundation subsoils to verify proper depth and adequate bearing capacity
8. Inspection and testing of ready-mix concrete to ensure compliance with the mix design, other quality-of-materials issues, and inspections of concrete forming and placement procedures

9. Inspection of the placement of concrete in footings, foundations, and slabs, including proper size, fabrication, and installation of reinforcing steel, wire mesh.
10. Review of the contractor's request for extras relating to the scope of work being performed by the geotechnical engineer

The division of quality assurance and quality control is rather easy to apply in each of these geotechnical procedures because the standards are specific in nature and allow little tolerance in their application. For example, soils testing for compaction will either meet or fail the 95 percent compaction requirement.

The key to working successfully with the owner's geotechnical consultant is based upon good communication and recognizing the role of each player in the process; that is, the geotechnical consultant is there to confirm to the owner that the contractor is complying with all the contract requirements. The project superintendent's job is to ensure that the specifications are being met, and when the scope of the contract work is increased or decreased, that the owner is made aware of these scope changes and any resultant time and cost implications.

To adequately perform quality assurance procedures relating to site work, the project superintendent should

1. Reread the specification sections pertaining to site work
2. Read Division 1, which may include testing procedures required for this particular project and the contractor procedures pertaining to receipt of and responses to reports issued by the owner or the owner's consultants
3. Be present to observe all soil and concrete testing and inspection performed by the owner's geotechnical engineer and request "pencil" copies if available
4. Note and correct any deficiencies as the operation continues, and have the geotechnical consultant note that corrective action was initiated
5. Be familiar with the proper procedures to achieve "design" compaction requirements and visual recognition of suitable and unsuitable soil conditions

Field observations and reports and copies of test results prepared during this phase of work will be sent directly to the owner or the owner's representative and should include copies to the general contractor. After comparing the test results with the contract requirements, the architect or engineer will issue a report to the contractor with comments regarding acceptance of the test data or noting any deficiencies and corrective action required.

The project superintendent should maintain daily contact with the on-site geotechnician to be aware of all acceptable, marginal, or unacceptable earthwork operations.

Other Testing and Inspection Requirements

Other testing requirements may be listed in the specification section to which they apply, such as this one:

> Division 3—Concrete
>
> Testing inspections may apply to ensuring that the subgrade for foundations is at the elevation required and that the subsurface presents the specified bearing capacity. Inspection of the type of concrete forms and their proper installation may also be a function of testing and inspection procedures in Division 3, as would the proper fabrication and installation of reinforcing steel or welded wire mesh.
>
> Approved shop drawings containing concrete mix design and reinforcing steel should be close at hand so the project superintendent can refer to them from time to time to ensure compliance with the contract requirements.

And of course placement of concrete, specifically during "cold" or "hot" weather as well as with correct curing procedures, would qualify as quality control/assurance procedures. These instructions are often included in the specific section of Division 3 or contained in a separate section devoted to installation.

Division 4—Masonry

With the appropriate approved shop drawings in hand, the project superintendent is ready to review the QC requirements for masonry work.

Are the correct concrete masonry units (CMUs) on-site? Are they lightweight or regular weight, as required, and of the proper size? What about reinforcement, both vertical and horizontal? Does it comply with the approved shop drawing? Not only will inspection of these materials be required, but also the project superintendent should periodically check to ensure that they are installed properly.

Flashings installed improperly, or even missing, in masonry walls will be a source of trouble (spelled *cost*) for years to come. The project superintendent must be familiar with the project's flashing details; but if they appear to be inadequate, a conversation with the architect is in order to discuss any suggested improvements. *Note:* If any such conversation takes place and the architect directs the superintendent to proceed with the installation per the drawings, it is critical to confirm that conversation as soon as possible, so that the responsibility for any future leaks can be laid at the door of the designer. If revision to flashing details or installation procedures is suggested by the architect, request a sketch before proceeding.

Cavity wall inspections must be frequent to ensure that this space is kept free from mortar and will weep moisture as designed.

And often more than a casual inspection is required. The writer had such an experience while working at a campus setting where two three-story, 180,000-square-foot brick-veneer office buildings were under construction. The brickwork had lots of intricate details, and the limestone quoins and pediments were representative of the high-quality design and the owner's high-quality expectations. It was important for the masonry contractor to have quality issues at the forefront of the work so that close inspection as the first courses of brick were installed would be key to establishing the quality standards. As the brickwork proceeded to the height of about 4 feet, the writer inspected the cavity and found excessive mortar buildup, which would prevent any moisture in the cavity from weeping out through the pea gravel placed there for that purpose. He told the masonry contractor's crew supervisor that this condition was not acceptable and the excess mortar was to be removed, carefully, so as not to damage any installed flashings. About 3 hours later when this wall was reinspected with the masonry crew supervisor, the appearance of the cavity had changed dramatically. All that could be seen was a layer of clean pea gravel with very little trace of mortar. In fact, it looked too clean. The writer found a rebar and probed through about 2 inches of clean pea gravel before hitting a layer of hardened mortar. What the foreman had done was merely to place a clean layer of gravel over the mortar that had previously dropped into the cavity and hardened to the point where no moisture would be able to reach the weep holes underneath. After being instructed to tear down about 100 feet of constructed wall, the masonry crew supervisor finally realized that we would accept no substandard work, and the balance of the brickwork was installed properly.

And of course, just as in concrete work, cold weather and hot weather installation procedures are to be followed per the specification instructions.

Mortar Mix—Does It Meet Specification?

An often overlooked item in the inspection process is confirmation of the correct mortar mix. This is of particular importance if the specification does not require the mason to prepare and submit mortar cubes for testing. Mortar types generally fall into the following categories:

Type M—high compressive strength (2500 pounds per square inch average). This type is highly durable; therefore, it is generally recommended for unreinforced masonry walls below grade.

Type S—reasonably high compressive strength (1800 pounds per square inch average), having great tensile strength and recommended for reinforced masonry walls where maximum flexural strength is desired.

Type N—midrange compressive strength (750 pounds per square inch average), suitable for general above-grade masonry work for parapets and chimneys.

Type O—low compressive strength (350 pounds per square inch average) and suitable for interior non-load-bearing masonry walls.

Type K—very low compressive strength (75 pounds per square inch average). Occasionally used for interior non-load-bearing walls where permitted by code.

The project superintendent, from time to time, ought to look at the full or even empty mortar bags to verify that the correct type of mortar is being used.

Division 5—Metals

Along with shop drawing approval, the engineer frequently requires *mill reports*—basically a chemical analysis of the steel used to form the rolled sections such as beams and columns. This report contains all the elements incorporated into the steel along with its certification. This report verifies that the exact type of steel, such as grade A572 or A992 (high-strength) or grade A242 (corrosion-resistant), that has been specified has been supplied.

A Word about Mill Reports

Requesting, receiving, and verifying information contained in the mill report obtained from the steel manufacturer can have serious consequences if not pursued diligently when this requirement is included in the contract specifications. These mill reports are prepared by the steel mill producing the product and contain the detailed chemical composition of each representative sized rolled section produced and shipped to the steel fabricator. They also contain a certification that the steel meets specific American Institute of Steel Construction (AISC) standards.

Another "war story" experienced by the writer will vividly reveal the importance of mill reports. The writer was the project manager on a 16-story, structural steel framed apartment building some years ago. In this particular project, Division 5 of the specifications included a requirement for the structural steel contractor to submit a mill report that contained the detailed composition of the steel being produced for this particular job along with the certification that the steel met specific American Institute of Steel Construction standards.

Repeated requests to the steel subcontractor for these reports went unheeded, even as steel began arriving on site. As erection began, the threat of nonpayment resulted in the structural steel subcontractor finally submitting a thick stack of mill reports, which were sent on to the design engineer. A week or two later as erection passed the 10th floor and concrete decks were being poured on the lower floors, the engineer began to question several of the mill reports, indicating that some of them applied to beam sizes that were not called for on the job. There were several mill reports for W 24 × 56 and W 18 × 36 steel when no such sized members were required. Further investigation revealed that the subcontractor had falsified the reports, and the engineer questioned whether the high-strength steel, as required, was actually furnished and installed. He directed that some coupons (small rectangles of steel) be removed from several beams and submitted for destructive testing to determine if high-strength steel, as specified, was or was not furnished and installed. When the writer attempted to contact the steel subcontractor, he found that the subcontractor's phones were disconnected and the office had shut down. So much for the low bidder! The outcome could have been much worse. A computer analysis of the structural steel calculations and design, by the structural engineer, indicated that a dozen or so fish plates throughout the building would suffice, and coupled with the cost of the design reevaluation, the final bill for this mishap was not so bad. But it did instill the importance of having mill reports prior to the delivery of steel.

Division 5 testing and inspection requirements can include inspection of metal deck welds, shear stud inspections, framed openings, edge stops, structural steel moment connection-bolted connections, and welds.

Structural steel QC/QA measures begin before one piece of steel is purchased or fabricated. Many specifications include bidder qualification requirements, such as the fabricator must demonstrate experience in the specific type of structural steel system being proposed.

The structural steel fabricator's shop may be required to meet American Institute of Steel Construction standards and be so certified, and if not, the design structural engineer may be required to visit the fabricator's shop and approve the facility (and the general contractor will have to pay for all such travel and inspection fees).

Field inspections nearly always require the following:

1. Inspection of field welding
2. Ascertaining that fit-up and alignment are proper
3. Inspection of placement of bolts with proper tension (when torsion-control self-indicating bolts are used, the torsion test is a visible one)
4. Inspection to ensure compliance with the approved shop drawings

And where welding occurs, the following are included:

1. 100 percent visual inspection
2. Fillet welds—one spot test per member (magnetic particle)
3. Partial penetration welds—one spot test per weld (magnetic particle)
4. Full-penetration welds in joints and splices—tested 100 percent by ultrasonic testing

Generally if more than 10 percent of tested welds are rejected, an additional 10 percent are to be taken.

Metal deck with shear studs will be tested by tapping each stud with a hammer to obtain the desired "ring," and even structural steel shop paint may be subjected to QA to ensure that all field touchup, as required, has been satisfactorily completed.

Don't assume anything when it comes to testing and inspections. Only a complete read of the specifications will uncover all the inspection and testing requirements and, in combination with any contract modifications, reveal whether the owner, general contractor, or subcontractor is responsible for the testing and/or inspection.

When the owner is responsible for testing, ample notice must be given to allow the owner to schedule a consultant to visit the site and not delay any of the general contractor's operations.

Mechanical and Electrical Testing and Inspections

Carefully read Division 15 (mechanical) and Division 16 (electrical) specification sections where significant inspection and testing requirements frequently appear. In fact one subsection of Division 15 may even be entitled "Testing, Adjusting, Balancing," often referred to simply as TAB. Many QC/QA requirements rest within the pages of Divisions 15 and 16, and one place to start to look for this type of work is in the General Mechanical and General Electrical Requirements section that prefaces each division.

A typical general mechanical requirement, as an example, could include the following:

1. Approval of the selection of the subcontractor by the engineer prior to the general contractor's awarding a contract to that subcontractor
2. Compliance with state and local authorities and standards and specifications issued by various trade associations such as ASHRAE, ASME, AABC, and ADC
3. Air pressure or hydrostatic pressure testing of various systems prior to closing up a wall or ceiling assembly
4. Record drawings
5. Operating and maintenance manuals (O&Ms)
6. Air and water balancing reports
7. System and product identification tags, charts, plaques, and color coding
8. Interim and final inspections by governing authorities and by the design architect/engineer team

Reread the Contract Documents Relating to Testing

Testing and inspection requirements for a construction project can vary considerably from job to job. Don't assume that the same testing and inspection procedures required on that last project apply to the current one. *Read the contract, the specifications, and the general notes on the drawings.*

Contract Requirements

AIA A201, General Conditions of the Contract, contains several sections relating to testing and inspection. Article 13.5 is entitled "Tests and Inspections," and Article 12.1 describes the procedure to be followed if a portion of work is covered without being inspected when the specifications require inspection. In such as case, the architect may require the work to be exposed for inspection and replaced at the contractor's time and expense.

When subcontractors are to provide testing, such as pressure testing on pipe prior to its being enclosed, it behooves them to perform these tests in a manner that does not affect the progress of other trades. If, in fact, that does occur, those affected subcontractors will look to the general contractor for any time/cost impact, and the project superintendent will share some of the responsibility if these delays occurred but could have been avoided with proper notification and scheduling.

Specification Requirements

Testing and inspections can be listed in the specification section to which they refer, that is, concrete in Division 3, structural steel in Division 5, roofing in Division 7, and so forth; or in related sections; or in Division 0 or Division 1 of the specifications book. By scanning the entire specifications book, those provisions dealing with testing and inspections can be highlighted and tabbed. If there is a question regarding testing or inspections, now is the time to resolve these issues. Contact the project manager who can prepare a Request for Information or Request for Clarification and send it on to the architect or engineer. And don't assume that even though testing and inspections are included in specification sections for which subcontracts have been issued, this work may have been excluded by the subcontractor.

Subcontracted Work and the Testing and Inspection Process

When it appears that an inspection or test is required of a subcontractor, notify the subcontractor well in advance by phone call, followed up by a letter or fax, and include that item in the next subcontractor's meeting. If the contract requires testing to be provided by the owner, it is incumbent upon the subcontractor to notify the project superintendent well in advance of the test or inspection, thereby providing the owner with enough time to get the inspector on the job site or to make arrangements for the required test.

Remind the subcontractor that if a request for inspection is not made in a timely manner and the work involved has been closed in by another trade, the cost to uncover or expose the item to be tested, along with the cost to recover or encase the work, will be borne by that subcontractor.

The Preinstallation Conference

The specifications often include a requirement for a preinstallation conference for certain trades, to ensure that the product and installation procedures for the applicable construction component will fully and completely comply with those specification sections. Although product certification may have been obtained via an approved shop drawing, the architect or engineer would like to be assured that the detailed installation and inspection procedures set forth in the related specification section will be followed.

These preinstallation conferences are frequently required whenever a system related to water or air infiltration is involved, that is, roofing, curtain wall, exterior metal siding, masonry wall construction or exterior insulating finish systems (EIFS), windows, flashing, and so forth.

If not specifically required by contract, in many cases, it is a good idea for the superintendent to require subcontractors to prepare for and participate in these types of meetings.

For example, let's look at a structure with a metal stud and exterior-grade gypsum board wall system clad with a brick-veneer surface. There are several areas where less than acceptable workmanship can create a source of future water leaks. The general contractor can incur lots of client wrath and spend lots of time and money to repair work if any leaks occur during the standard 1-year warranty period—or even beyond, if a reliable contractor feels an obligation to correct these defects. A preconstruction conference in this case would involve the following subcontractors:

1. Steel stud and drywall contractor ensures that the joints on the exterior-grade drywall are properly sealed, and that the waterproof integrity of the gypsum board's surface has not been damaged (if so, the subcontractor must repair).
2. Where windows are to be installed in openings in the brick veneer, the window installer should be present to review and discuss flashing details,

blocking for the proper securing of the windows, and acceptance of the correct size masonry opening for the windows.

3. Depending upon which trade will be responsible for supplying and installing all flashing, that subcontractor(s) will be a key participant in this preconstruction conference since the proper installation of flashing is one of the most important aspects of attaining waterproof integrity.

4. And last, but not by degree of importance, the mason subcontractor will be a major factor in this process. Not only will the company be responsible for adhering to its own specification section, but also it must work in concert with other related trades to ensure that the exterior wall assembly, when completed, meets the specifications and "the best practices of the trade."

Time spent reviewing each participant's role in the process under consideration will ensure that these subcontractors fully understand what is expected of them and their intent to comply with all the requirements and instructions discussed at that meeting. The project superintendent should follow up by issuing detailed minutes of this meeting in which all these procedures are reviewed and discussed, and responsibility for implementation is agreed to. Without such documentation, it is possible for any one of the attending subcontractors to misinterpret what was discussed and later dispute or deny responsibility for previously agreed upon actions.

Preconstruction conferences or meetings are particularly useful when dealing with any of the following components of construction:

- Sedimentation control and maintenance
- Cast-in-place concrete work
- Structural steel and metal decks
- Waterproofing—foundations, above-grade walls
- Insulation—in-wall, ceilings, foundations
- Exterior wall assemblies such as stucco, masonry, wall, vinyl, metal siding
- Windows, vents, louvers, and related flashings installed in exterior wall systems
- Roofing, skylights, sheet metal flashings, gutters, downspouts
- Joint sealers—caulking
- Gypsum board and related in-wall blocking
- Kitchen and bath cabinets (in-wall blocking, scribe pieces, etc.)
- Finishes, walls, flooring, signage, window treatments

Prior to the commencement of certain items of work and systems, a re-review of the related plans and corresponding specification sections and approved shop drawings primarily with the subcontractor involved is desirable to ensure that the contract document requirements will be achieved or that some modifications may be required, with the architect/engineer's approval. These print, specification, and shop drawing review meetings are helpful prior to work commencing for

- Elevator installations
- Millwork, cabinetry
- Fire protection systems
- Storm drains—interior and exterior
- HVAC systems—ductwork and equipment
- Electrical systems
- Data and communications systems
- Other low-voltage work, that is, security systems

Sample Panels and Mock-ups

Sample panels are often required when extensive brickwork or other types of decorative exterior masonry unit wall construction are to be built. Some masons use the sample panel process as a way to display their master masonry skills, but they must also take into account the fact that the architect, once the sample panel is erected and approved, will expect to see the same quality in the mass-produced product that follows. Sample panels are often required when metal siding is to be installed as part of a wall system containing other materials. The architect may wish to verify that the flashings or details connecting one material to another are visually correct or may require design modification.

Mock-ups usually include two or more construction products that together will form a wall assembly or a system or a finish schedule. Curtain wall construction often begins by constructing an "in-place" mock-up so that the architect can inspect this sample segment and be assured that a structurally secure, weathertight system will be built. This mock-up process allows all related trades to determine how they will integrate their individual products into the overall system.

To afford the owner a snapshot view of the color scheme as outlined by the finish schedule, an architect may request a sample board be prepared containing a ceiling tile, small gypsum board painted wall, a few floor tiles, or a small piece of carpet.

And it is not solely an architect or engineer's prerogative to request a sample panel or a mock-up. Many general contractors will do so when a particularly complex or complicated exterior wall system is required. All participating subcontractors and vendors are invited to participate in the process to ensure that when the actual in-place work begins, they will have worked out any minor, or sometimes major, details so that the work can continue smoothly and efficiently.

Does the Punch List Qualify for QA/QC?

In a perfect world there would be no need for a punch list. But as we all know, we live in a less-than-perfect world in the construction business. During the final phase of the construction process, one goal is to limit the inevitable punch list by carefully supervising the work as it progresses. However, the project superintendent cannot be everywhere on the site, checking every trade along each step of the operation daily. The project superintendent needs to establish some procedures along the way to assist in striving for a zero-tolerance punch list.

A Clean Site Will Affect Punch List Work

The project superintendent must display an attitude about quality work, and this will filter down throughout the various crews on the project. When a construction site is neat and not littered with debris, where there are sufficient numbers of trash containers placed strategically around the site and in the building, when worker temporary lunchroom areas are policed and kept clean, a clean environment will ultimately lead to higher-quality work. And conversely, a sloppy, trash-strewn site and building sends the message that the project superintendent does not really have too much interest in maintaining order.

So a clean and orderly construction site is the first step in conveying the proper quality message. If cleaning rules and regulations are ignored by subcontractors, depending upon the penalties included in their respective subcontract agreements, strict and prompt enforcement is necessary to bring discipline to the site, which in turn should be reflected in better attention to other work practices.

Matters relating to quality should be addressed at every subcontractor meeting, and the importance of each trade achieving high-quality work ought to be stressed. It must be made perfectly clear that shortcuts will not be tolerated if they shortchange quality. Nonconforming or defective work will not be acceptable, and subcontractor monthly requests for payment may be delayed or partially reduced if defective work becomes a major problem. Each subcontractor crew supervisor will be held responsible for the quality of the respective work teams, and once this message is spread repeatedly, the project superintendent's watchdog activities relating to potential punch list work may be relaxed somewhat.

Reduce Your Punch List—Prepunch the Building

That basic contract document A201, General Conditions, includes a provision placing responsibility on the contractor to inspect the work, in effect a requirement to prepunch the building prior to an inspection by the architect. Document A201 states:

> The contractor shall be responsible for inspection of portions of work already performed under this contract to determine that such portions are in proper condition to receive subsequent work.

This same AIA document continues with a statement relating to punch lists:

> When the Contractor considers that the work, or a portion thereof which the owner agrees to accept separately is substantially complete, the contractor shall prepare and submit to the architect a comprehensive list of items to be completed or corrected.

If you recall from Chap. 6, most subcontractor contracts issued by the general contractor include a provision linking the terms and conditions of the contract between owner and general contractor with those of the subcontractor and general contractor. Therefore these "inspection" requirements, contained in the General Conditions contract included in the general contractor's contract with the owner, will apply to the subcontractor as well. The subcontractor, therefore, has a contractual requirement to inspect the work to "determine that such portions are in proper condition to receive subsequent work," that is, to inspect and prepunch the work.

All trades should be advised that they are *required* to "punch out" their own work before leaving the job. Once they have gone, it is difficult to get tradespeople back to complete or repair minor defective work; they are off on that new project. But before demobilizing, all subcontractors should be made aware of their responsibility to carefully and thoroughly review their work in place, to correct any items that could possibly be picked up on a future punch list. A rapid walk-through with the subcontractor crew supervisor at the beginning of the prepunch list work will show the supervisor work that will be acceptable, and work will show up on the architect/engineer's punch list if it is not corrected at this time.

In fact there is a responsibility to do more than prepunch the work. Subcontractors are obliged to ascertain *that such portions are in proper condition to receive subsequent work.* Does this mean that the painting subcontractor is responsible for inspecting drywall surfaces to ensure that their subsequent painting work will be acceptable, and does it also mean that the flooring subcontractor is to inspect the flooring substrate to determine if it is acceptable for resilient flooring or carpeting? Although it might take the threat of legal action to firmly establish such responsibility, as a project superintendent, it appears that subcontractors ought to have more than a passing responsibility in maintaining levels of quality assurance for the work included in their contract scope.

The subcontractor crew supervisor should willingly participate in this prepunch exercise since she or he will be the recipient of the end result of this process. Failure to correct minor items at this time may result in significant backcharges later on. If a drywall subcontractor fails to correct an unacceptable tape joint now and a semigloss to high-gloss paint is applied over this surface, the drywall subcontractor may be required not only to repair the defective joint, but also to absorb the costs to repaint the entire wall to avoid the appearance of a patch.

Subcontractors must also be reminded that their retainage and final payment are contingent upon completing all closeout requirements, including punch list work. The longer it takes to complete the punch list, the longer it will take for the subcontractor to receive the balance due on the account.

Electronic Aids to the Rescue

Following quickly on the heels of the personal computer and the wireless cellular phone, the *personal digital assistant* (PDA) has become more sophisticated than ever. Although at first the PDA was relegated to the task of keeping a telephone directory and an appointments schedule, much more computing power has been added to the latest editions of PDAs, and software programs including Daily Diary and Punch List Preparation are now on the market. One such firm, Onsyss, headquartered in Rockville, Maryland, developed punch list software that was incorporated into the Meridian Project Systems field management program (Figs. 9-1 and 9-2), used extensively at the Mohegan Sun Casino construction project by Massachusetts-based general contractor Perini, Inc. This fast-tracked project, valued at $870 million and employing 200 tradespeople, generated 21,000 punch list items which were captured on seven handheld PDAs. These lists were regularly transmitted to approximately 30 subcontractors and greatly speeded up the entire punch list process.

Dealing with Subcontractors Who Diligently/ Reluctantly Approach Punch List Work

On every project there are some subcontractors who diligently attack punch list work and complete it promptly. And then there are subcontractors who promise to complete their work quickly, but nibble away, one item at a time, over a period of weeks.

The architect generally will not release retainage until all punch list and closeout work has been completed and accepted. That diligent subcontractor may therefore be penalized and have to wait weeks before receiving funds just because another subcontractor is in the process of correcting its work to satisfy the design team.

If this is the case, suggest a system that will withhold money for incomplete punch list work by one subcontractor while releasing funds for the subcontractor who

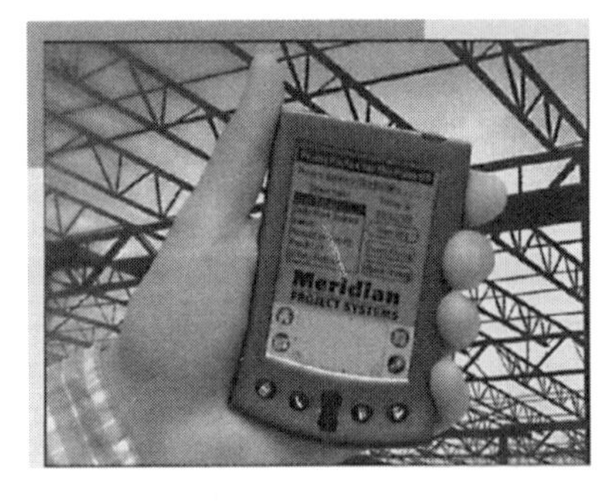

Figure 9-1 Field management software for handheld computers.

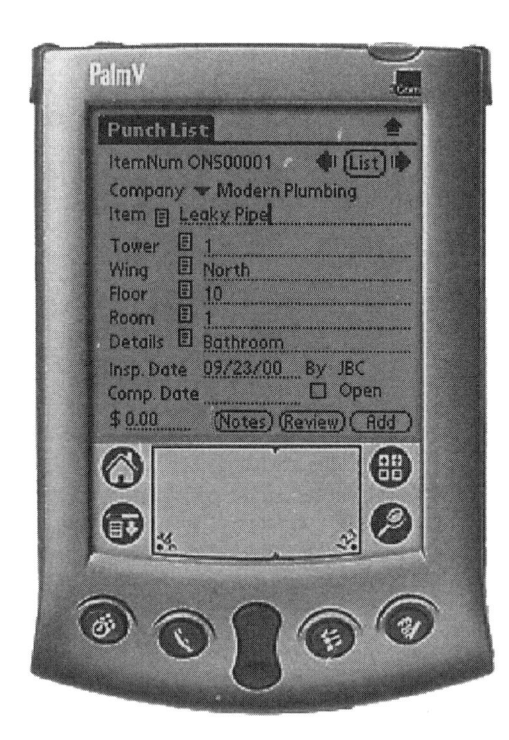

Figure 9-2 Screen shot of hand-held computer with punch list software.

fully cooperated in completing the work. Propose to the architect or engineer that a realistic value be placed on each incomplete item of work; then double or even triple this amount and withhold it from payment. When each item has been completed, the architect will then approve payment.

This system, which works well for everyone, can follow these guidelines:

- The amount retained should be recognized as being sufficient to correct the defective work even if another subcontractor has to be brought on site to do so. In other words, the general contractor and owner have been afforded a degree of protection.
- It allows retainage to be released for all punch list work properly completed and approved, thereby rewarding those subcontractors who performed properly.
- It serves as a recognizable penalty to the offending subcontractor since it represents significantly higher costs than those actually required to complete the work.

To Reduce Your Punch List, Try the Following

1. Inspect for quality on each walk-through of the job, but at a bare minimum inspect for quality once each week.
2. Promptly notify a subcontractor or vendor of any unacceptable work, materials, or equipment, and request immediate replacement or rework.
3. If substandard work or repeated lack of product quality continues, call the subcontractor's owners or manufacturer/vendor representative to the site to inform the person of these quality issues, and obtain the representative's verbal commitment to change and improve. Follow up with a written confirmation of that meeting. If a change in crew supervisor is deemed necessary, so state and make the request.
4. Prior to that particular subcontractor's leaving the site, conduct a walk-through to create a punch list, giving one copy to the subcontractor's crew supervisor and faxing or emailing a copy to the subcontractor's office. Advise both parties that this punch list must be completed, inspected, and signed off before the subcontractor demobilizes. If this is not done, further payments to the subcontractor will be delayed.

5. Advise the subcontractor that *release* by the project superintendent does not relieve the subcontractor from further punch list responsibility but was merely an exercise in reducing the potential for extensive punch list work when the architect/engineers prepare their lists.

The Distinction between Punch List and Warranty Work

During the preparation of the architect/engineer's punch list, certain items on the list may require the contractor to order parts or entire assemblies that could take weeks or months to arrive at the site. When the superintendent is wrapping things up and anxious to leave for that new project as soon as possible, this incomplete work may delay that departure. If the item is included in the punch list and not signed off when the project is substantially or finally completed and occupied, monies will be withheld from the final payment and such items as bond signoff and final releases by the owner may not be forthcoming. But is that last item on the punch list, possibly one that requires replacing a defective motor on a backup water circulating pump, really a punch list item—or is it a warranty item? There is a big difference between the two. *Punch list work* is defined, and accepted by most professionals, as work which has not yet been completed in accordance with the contract scope, or rework required to conform to the project requirements.

Warranty work is "guarantee" work whereby the contractor is responsible to repair or replace any defective materials and equipment, including installation during the warranty period established in the contract. Let's take the circulating pump as an example. If it was furnished and installed in accordance with the contract requirements, but upon start or shortly thereafter it failed, then this is a warranty item. The builder's (and subcontractor's) warranty was issued to cover such situations, and the builder (and subcontractor) has a responsibility to replace, in this case, the defective pump within the warranty period, which is usually 1 year after owner's acceptance of the structure. So a case can be made to the architect that this is not a punch list item and no funds should be withheld until the defective pump is replaced. "Mr. Architect, please delete this item from the punch list, release our final payment, and when the replacement pump arrives, we will install it." The same case can be made for a number of other so-called punch list items—think of a few.

The Preprinted Quality Assurance Checklist Approach

Several general contractors prepare generic quality control type of checklists that also allow for modification to include specific individual project requirements. Although the initial preparation of lists such as these can be very time-consuming, once prepared, they can serve as quick reference documents to check on compliance with contract requirements for various trades and work tasks or to create a punch list.

Figure 9-3 is a concrete checklist that can be used during the entire cast-in-place concrete process, starting with inspection of forms, checking reinforcement steel and embeds, and underground rough-ins. Figure 9-4 on p. 202 is typical checklists for earthwork and doors/frames. (Other quality assurance checklists may be found on the CD that accompanies this book.)

But whichever approach to quality is employed, remember that quality is a state of mind. It begins with a clean and safe working environment and ends with a 10-item punch list.

Concrete Checklist

Building Concrete Work:

Footings:

1) _____ Ensure proper sub grade elevation at building pad.

2) _____ Confirm footing size and location in relation to building stakeout with subcontractor. Contact site engineer if necessary.

3) _____ Check forms for right elevation, width, and depth of footings and piers. Check for the right set backs. Tape all measurements for length and width of building and all interior bearing and non-bearing footings and piers. After step-down footings, landings, and slope footings are formed, set up builder's level and double check elevations before pouring. Make sure forms are well braced, tight, free of debris, and coated with approved compound for ease of removal. Check grading plans for deepened footings and raised stem walls.

4) _____ Check steel to verify proper size and location. Make sure rebar is properly fabricated and installed.

5) _____ Call municipal inspector for footing inspection>

6) _____ Before pouring footings, check for the following:

- A) _____ Check for proper thickness, correct steel and proper concrete mix to ensure that the concrete reaches the required P.S.I.
- B) _____ Get a ticket for every load of concrete poured. Check the type of concrete, cement sack mix, and water ratio. Make sure it is as per contract and soils report. Fill out daily pour sheets. Take cylinder samples and slump tests, as required.
- C) _____ Check anchor bolt, hold-down, post anchor, column base, and post base locations for proper sizes and heights.
- D) _____ Make sure all depths are correct per plans and soils report. Make sure footing trenches and rebar are three inches away from the forms and dirt banks. Make sure trenches are free of debris and dirt clods.
- E) _____ Check for eased edges at stem walls and step downs, as required.
- F) _____ Check installation of special expansion joints at split level, step, or grade break areas.
- G) _____ Make sure concrete is placed properly.

1

Slabs:

7) _____ When grading for the slab area, be sure that all footings, including interior, are exposed, and cleaned of all loose dirt. Keep visqueen off of all footings. Check mil. size of visqueen. Prior to visqueen and sand or gravel fill, the underslab plumbing rough-in should be installed. Check for the following:

- A) _____ Confirm proper type and size of copper line for underslab plumbing rough-in: type "K" or "L."
- B) _____ Make sure there are no splices under slab.
- C) _____ Make sure no copper is exposed to concrete and that all copper installed has approved plastic sleeves to prevent electrolysis.
- D) _____ Recheck forms for correct elevations. Check all buildings for backfill with the proper gravel at raised slab area. Check visqueen - repair any holes or rips. String line all slab areas for four-inch nominal slab. Make sure interior trenches are wheel rolled and filled with good native soil or import sand, whichever is required. Make sure wire mesh is per contract and soils report. Check electrical conduit and floor boxes. Make sure all plumbing and electrical is tied down so that it will not move or float during the pour. Schedule a pre-slab inspection.

8) _____ Before pouring house and garage slabs, check the following:

- A) _____ Check the concrete load tickets for proper type of concrete, cement sack mix, and water ratio.
- B) _____ Make sure the concrete is placed properly and not poured too wet.
- C) _____ Take cylinders and required slump tests.
- D) _____ Make sure mesh is pulled up when pouring and that chairs (if required) are installed. Make sure the "hook-man" does not stand or walk on mesh after it has been centered in the slab. If mesh is hooked and centered, do not allow workmen to stand or walk on it.
- E) _____ Make sure tops of footings are clean. They should be swept or hosed off prior to the pour.
- F) _____ Check for screed pins or grade stakes at proper spacing.
- H) _____ Check for even screed - no highs or lows.
- I) _____ Check for control joints as required (tooled, pre-formed, saw cut).
- J) _____ Check for proper spacing of anchor bolts. Make sure

2

Figure 9-3 Example of checklist prepared for concrete inspections.

Quality Control Check List

Project no.

Section: Earthwork

No.: 02200

Date

1. Soil information report is on job and reviewed. Note elevation of water table.
2. Job survey is reviewed; monuments and stakes are located. Limits of work are establisehd.
3. Job surveyor is on site if required.
4. Observe removal of existing buildings and foundations or other items.
5. Note condition of, or photograph, offsite and onsite improvements to remain, such as paving curbs, gutters and walks before work begins.
6. Existing vegetation to remain is protected.
7. Existing utility lines to remain are located, staked, and protected. Observe conditions of uncovered lines. Verify utility companies have been notified. All lines to be removed or abandoned are properly capped. If unknown lines are encountered, notify proper personnel.
8. Adjacent property is protected. Verify whether adjacent property owner is notified as required by work or code.
9. Shoring and underpinning is provided if required.
10. Extent of grubbing and removal of stumps and matted roots is performed. Depressions are properly filled compacted.
11. Spillage of materials or soil on streets and sidewalks is promptly removed for public safety.
12. Spillage of gas, oil, slurry, etc., is prevented in areas to be planted or near existing vegetation to be retained.
13. Dust control is provided as required.
14. Stripping of site, preservation and depth of removal of topsoil, and location of stockpile are performed and established.
15. Observe that topsoil is not contaminated with subsoil and is free from roots, stones, and other deleterious materials.
16. Check that satisfactory materials are used and unsuitable materials are disposed of in waste areas. Note contamination if it occurs.
17. Observe removal of material, note unusual conditions. Observe subsoil conditions for irregularities such as soft spots, springs, previous debris, etc
18. Excavation is performed in scheduled sequence if required.
19. Excavating does not cause unusual rutting and appears adequate for work to be performed.
20. Observe that over-excavation does not occur.
21. Drainage is provided continuously as excavation progresses: other de-watering methods such as well points are provided, drainage ditches are maintained, and ponding does not occur.
22. Testing, inspection, and compaction are performed during excavation and filling as required.
23. Borrow excavation procedures and materials are adequate.
24. Source and type of imported materials are as approved. Verify samples are tested and approved.
25. Compaction is performed in lifts required.
26. Observe that building layout is properly established, set-backs are observed, and batterboards and elevations are established. See that compacted material extends beyond foundation line as required.
27. Observe foundation excavating for adequacy, bracing, form clearances, type of soil, etc., and that the work is inspected by the contractor.
28. Observe that corrective measures are performed where over-excavation occurs.
29. Observe methods of de-watering foundation excavations and see that footing beds are not disturbed or softened. Methods for surface drainage are provided.
30. Footing drains are installed in manner specified.
31. Backfill materials are from approved source. They are installed in specified layers and adequate compaction equipment is used. Relative density of backfill is checked. Walls are properly cured before backfilling.
32. Waterproof membranes are protected against damage during backfilling operations.

USE REVERSE SIDE FOR ADDITIONAL REMARKS AND COMMENTS

Accepted By ____________________

Quality Control Check List

Project no.

Section: Metal Doors and Frames

No.: 08101

Date

1. Shop drawings and schedule are approved and on site.
2. Doors are as approved; type, design, material, etc.
3. Check panel, lights, louvers, and features. 4. Check defects: dents, buckles, and wraps.
5. Fabrication, construction and workmanship. 6. Smooth edges, joints, finish, and straightness.
7. Additional reinforcement provided for hardware.
8. Observe backing plates during drilling operations.
9. Observe that closure channels are provided.
10. Provisions to receive hardware are adequate.
11. Observe type of factory-installed hardware.
12. Backset is matched to finish hardware.
13. Stile edges, astragals, required for pairs of doors.
14. Fire-rated doors have labels and proper identification.
15. Wire glass is provided.
16. Fusible-link holders provided at louvers.
17. Observe installation and verify proper clearance.
18. Doors are hung straight, level and plumb.
19. Door functions smoothly and easily.
20. Hardware is properly adjusted.
21. Observe glazing operation.
22. Factory prime is retouched.
23. Surfaces are adequate to receive applied finish.
24. Report doors that cannot be properly cleaned.
25. Protection provided to avoid marring.
26. Bumper buttons installed.
27. Fabrication and construction of frame as required.
28. Frames are prebraced if required.
29. Extra reinforcement on frames at head, corners, and hardware.
30. Proper type and number of anchors are provided.
31. Verify adequate anchorage made during installation.
32. Sound-deadening treatment is provided if required.
33. Fire-rated frames have labels and proper identification.
34. Light/sound-proof, lead-lined frames as required.
35. Provide features such as silencer holes if required.
36. Frame is grouted during installation if required.
37. Frame is caulked if required.
38. Frames are installed straight, level and plumb.
39. Frames adequately braced where "built-in".
40. Provide spreaders (masonry walls).
41. Protect threads of hinge plates in buck.

USE REVERSE SIDE FOR ADDITIONAL REMARKS AND COMMENTS

Accepted By ____________________

Figure 9-4 Typical checklists for earthwork and doors/frames.

Chapter

10

The Legal World We Live In: Claims and Disputes

Flip through the Yellow Pages of your phone directory, and you'll find something like this, as the writer did when glancing through the Yellow Pages of the Baltimore city phone directory—6 pages listing contractors and 57 pages for lawyers. An AIA Standard Contract Form for a Stipulated Sum contract, 1939 version, would fit on three 8½ × 11 inch pages. Today's AIA Document A101, Stipulated Sum, requires five pages of paper not including numerous exhibits, addenda, and various other modifications to the standard agreement. There is no need to remind anyone that our world has gotten more complex in the last six decades or so.

At some time or another, chances are that a project superintendent may have to deal with a dispute that escalates to a claim requiring litigation or arbitration proceedings to resolve. The purpose of this chapter is not to create "guard house" lawyers, but to familiarize project superintendents with the more common types of disputes and claims and better prepare them to deal with them, or better yet to avoid them.

Tort versus Criminal Law

Legal matters fall into two broad categories, criminal law and tort law. Criminal law needs little introduction to any TV viewer; it deals with offenses against the public for which the state brings action in the form of criminal prosecution and prison terms for the offender. Tort law may be best described as dealing with a civil, private wrong, one in which the court will hand down a verdict involving monetary penalties rather than jail sentences. Tort law concerns itself with three areas of citizen protection:

1. Protection of personal effects
2. Protection of property
3. Protection from economic loss

The type of conduct defining these personal wrongs can be summed up as follows:

1. *Intentional.* The wrongful person intended to do harm and realized that his or her action would most certainly cause harm.
2. *Negligent.* The wrongful person "failed to live up the standard prescribed by law."
3. *Nonculpable.* The wrongful person committing the wrong did not intend the action to be intentional or negligent.

So it would appear that most legal issues involving construction would fall into the category of tort law, but that doesn't preclude action where violation of the law results in a criminal act, such as a job site fatality where sheer negligence has been displayed.

Although the law is the law, each case presented to a judge and jury must stand on its own merits, and previous decisions in a similar matter can weigh heavily on the decision of the current dispute before the court. For example, in one court case, the judge ruled that a contract stipulating that all earthwork was classified as *unclassified* (in other words, "Mr. Contractor you own all the bad stuff encountered during excavation") voided the contractor's claim for costs associated with unanticipated rock removal. In another similar case involving another contractor, the court ruled in the contractor's favor, stating that the discovery of rock in an unclassified site was cause for reimbursement because the contractor was "misled" by ambiguous language in the bid documents.

How would you rule in this case involving a paving contractor? A paving contractor signed a contract to resurface an existing parking lot based upon bid documents that made no mention of work required on the subbase. When the paving contractor began the work, the weight of the paving equipment caused portions of the subbase to crack. The contractor stopped work and advised the owner that the contractor was not responsible for any damage to the subbase because the bid documents failed to include any mention of subbase responsibility. The owner sued. Can you determine who was right and who was wrong?

The judge sided with the owner stating that the paving contractor failed to use her or his expert experience and knowledge and should have advised the owner that subbase failure may occur during the resurfacing operation. So go figure!

These court decisions cut both ways. Don't assume that the cause is lost because a similar case was not decided in the contractor's favor. And don't assume that a win in a previous case ensures a win in this current dispute. Each claim stands on its own.

What Triggers Claims and Disputes?

Misunderstandings leading to disputes and claims will most likely be triggered by one or more of the following conditions:

1. Plans and specifications that contain errors, omissions, or ambiguities or that lack the proper degree of coordination
2. Incomplete or inaccurate responses or nonresponses to questions or resolutions of problems presented by one party to the contract to another party to the contract
3. Inadequate administration of responsibilities by the owner, architect/engineer, *general contractor,* subcontractors, or vendors
4. Unwillingness or inability to comply with the intent of the contract or to adhere to industry standards in the performance of work
5. Site conditions which differ materially from those described in the contract documents
6. Unforeseen subsurface conditions
7. The uncovering of existing building conditions which differ materially from those reflected in the contract drawings when rehabilitation or renovation work is being undertaken
8. Extra work or change order work
9. Breaches of contract by any party to the contract
10. Disruptions, delays, or acceleration to the work which creates deviation from the initial baseline schedule

11. Inadequate financial strength on the part of the owner, contractor, or subcontractor
12. Inability to meet the performance or quality standards contained in the contract documents

If and when any of these misunderstandings occur, every reasonable effort should be made to resolve them with the normal give-and-take that ought to be expected from all parties to the contract—the essence of negotiation.

Only when efforts of resolution fail is there a need to consult a claims expert and/or the company attorney. If this is done at an early stage, the strengths and weaknesses in both parties' claims may become apparent, and the decision to pursue or dismiss the claim made somewhat more easily.

If the occasion arises and a meeting is scheduled with either a claims consultant or an attorney, the project superintendent must assemble, or have readily available, the following documents as they relate to the dispute or disagreement:

1. Correspondence from and to all parties involved in the dispute—both letters and faxes
2. Daily reports, daily logs, and daily diaries containing entries germane to the issue
3. Appropriate inspection and test reports
4. Any and all requests for payments from all parties and list of payments received
5. Job progress schedules—both baseline and updates
6. Time sheets for all labor expended on the site
7. Shop drawing logs and transmittals attesting to the transfer of these drawings to and from the architect and to and from the subcontractor or vendor
8. Extra work or change order work
9. Interoffice memos, field memos, and telephone conversation memos
10. Estimates, bids, and quotations, either written or confirming telephone bids
11. Change order requests, change order proposals, change order estimates, and all backup data for these proposals
12. Job progress photos
13. Copies of all contracts issued or in progress, and all purchase orders issued or in progress
14. Any as-built drawings, either completed or interim

These are the documents that will be needed if further action is required, either in the process of working with a claims consultant or the company attorney or in preparing for arbitration. If all these documents have been prepared properly as the job progressed, the consultant's or lawyer's job will be much easier and she or he will be able to readily grasp the problem and determine to what extent the proposed claim is creditable. Quite often when the opposition is merely made aware of the depth and completeness of the documentation to be presented to them, resolution of the matter can be achieved quickly.

The Bid Proposal Process

Disputes can often arise even before a contract is awarded. In fact, a dispute can occur before a sealed bid proposal is submitted at a formal bid opening. In private and public construction work, a formal bid procedure is often established. Bids are to be submitted on a preissued bid proposal form; the form is to be completed and signed by an officer of the construction company, sealed in an envelope, and presented to the owner's representative as a predetermined place, time, and date. At times a bid bond is required with the proposal, or a certified check or letter of credit may also be acceptable if presented in the proper amount.

The purpose of the bid bond, certified check, or letter of credit is to ensure the owner that if the contractor's bid is accepted, the owner will be protected if that low bidder is unable or unwilling to accept a contract when offered. Other requirements may also accompany the bid proposal, and in strict accordance with the bidding instructions, any deviations may be cause for rejection of the bid.

In the public sector, strict compliance with all aspects of the bid proposal is standard in order to avoid formal protests that may be lodged by other bidders protesting acceptance of a bid that fails to meet all required criteria.

In the private sector, compliance is not so strict, and the owner may waive any or all requirements in the selection of an acceptable bidder, if it appears to be in the owner's best interest to do so. This option in public bidding is not totally impossible, but it is usually narrowly construed.

Do Late Bids Count?

How many times has the general contractor's representative dashed from the office with an incomplete bid form, several pens, and a cellular phone on the way to the place where sealed bids are to be deposited. The scenario generally is as follows: arrival at the office where the bid will be received, looking for a spot away from the crowd; a call back to the office made for final instructions and late-breaking price adjustments to the bid. With a nearly blank bid form, time is running out, and only a minute or two are left, but many items have to be filled in before that mad dash to the bid opening. And a thought occurs: "What will happen if I make a mistake on this form or enter the wrong number in the wrong place—or worse yet, the battery in my cell phone dies before I'm finished with the office?" The bid form will then be completed as quickly as possible and presented to the bid clerk so it can be date/time-stamped. But the fact that a bid is received after the specified time on a public works project does not mean it will automatically be disqualified.

In private bid situations, the owner is free to waive many prebid qualifications; but in public bid situations, the courts have ruled that the public bidding requirements are there for the public's benefit. If a bid is received a few minutes late and there is no evidence of fraud, collusion, or intent to deceive, not all is lost. If the local authority *refuses* to accept the late bid, an immediate protest must be filed. It can be voiced in the presence of the official refusing to accept the bid and in the presence of witnesses. As quickly as possible, upon return to the office, a written protest should be filed, citing the circumstances involved in the late submission—all other portions of the bid have been met with strict compliance but the bid was delivered at, say, 2:05 p.m. instead of 2:00 p.m. as required.

Whoever delivered this bid, be it a project manager, project superintendent, or administrative assistant, should remain during the entire bid-opening process and keep detailed notes of the other bidders' proposals, noting their competitive bids and any exceptions that they may have taken to the bidding instructions.

If it is deemed in the public interest to accept a bid having minor deviations from the bid documents, the agency may honor the bid. For example, suppose four bidders submitted prices as follows: bidder 1, $500,000; bidder 2, $525,000; bidder 3, $515,000; bidder 4, $485,000. All bidding conditions were met except bidder 4 was

10 minutes late in submitting the bid, or failed to insert a date on page 3. The public agency may waive these two minor discrepancies and award the project to bidder 4.

When One Contract Requirement Contradicts Another

Looking at the General Conditions of the Contract for Construction as outlined in AIA A201, the contractor is directed to refer all questions concerning interpretation of the contract documents to the architect, and that is the correct procedure to follow. In some cases, the contract will stipulate which drawing or which detail takes precedence over another. In some cases the contract language will remain silent in this regard.

Quite often ambiguities in the contract documents are resolved by reasonable parties to the contract taking a reasonable approach. At times reasonableness does not prevail, and one party will take a hard-line approach as far as the interpretation of scope of work is concerned. There are no tried and true procedures for contract interpretation, and each case seems to stand on its own. What documents have priority over others? Do the plans or the specifications take precedence? Will the specifications take precedence over full-scale drawings? Some contract requirements will specifically establish an order of precedence, such as contract requirements are first precedence, schedules are second, drawings are third, and so forth. Typical contract language, when there is a conflict between the plans and specifications, will be

> In the case of conflict between drawings and specifications as to extent of work, or location of materials and/or work, the following order of precedence will govern:
>
> 1. Large-scale drawings
> 2. Small-scale drawings
> 3. Schedules (door, finish, equipment, etc.)
> 4. Technical specifications
>
> In case of conflict as to the type of quality of materials, the specifications will govern.

If the contract does include an order of precedence, this will weigh heavily on the final outcome of the dispute. However, the courts view these questions independently. A *specific* statement will take precedence over a *general* statement. The courts may look for the intended purpose of a specification section relating to the drawing requirements before arriving at a decision. The point is that there are no hard-and-fast rules that apply to interpretation of which document has priority over another, unless, as stated above, a contract order of precedence exists.

Dealing with Inadequate Drawings

A common phrase which architects insert into the contract requires the contractor to advise the architect of a discrepancy, error, or omission before the contractor can submit a cost proposal to correct the discrepancy, error, or omission; and that is, at least, the first step to take when you are dealing with these types of problems.

The construction process is often difficult enough when the project superintendent is presented with a carefully prepared set of plans and specifications. The job becomes frustrating and difficult when the plans lack sufficient detail of key elements of construction or there are omissions of important details, or there is an indication that things just won't fit together. The first order of business is to bring these inadequacies to the attention of the owner and architect/engineer as quickly as possible. A verbal notification can alert them as soon as the problem is discovered, but a detailed written response is required as soon as possible afterward with a response date included.

Although the superintendent and the architect may differ in their opinions of what constitutes a *complete* set of plans and specifications, the courts have made their views known with respect to what is deemed acceptable. In a court case identified as *John McShain v. United States,* 412F.2d 1218 (1969), the contractor stated that the true condition of the drawings was not known at the time of bidding and that, after being awarded the contract for construction, the contractor found that some of the drawings which were illegible at bid time were not replaced with legible ones. The addenda drawings, furthermore, did not correct many of the coordination errors in the bid documents. The general contractor sued to recover damages incurred by the company and its subcontractors because of the inadequate drawings. The U.S. Court of Claims said that although the plans furnished by the owner need not be perfect, *they must be adequate for the purpose for which they were intended.* The court went on to state that the contractor was under no legal or contractual obligation to inspect the drawings to determine their adequacy for construction prior to a contract award. The documents were to be used for estimating purposes only, and it had not been shown that McShain knew or should have known how defective the drawings were.

This court decision amplifies the owner and the architect's responsibility to present the general contractor (GC) with drawings and specifications that are adequate and reasonably accurate. The GC has a right to expect that. If there are considerable problems relating to deficiencies in the documents, the general contractor also has a right to expect compensation for any delays that the substandard or deficient drawings might have caused. The court ruled that if faulty specifications prevent or delay completion of the contract, the contractor is entitled to recover delay damages from the defendant's breach of implied warranty. This breach cannot be cured by the simple expedient of merely extending the time and performance.

Contractor's Guarantee as to Design

When an architect specifies a certain component design and the installation results in poor performance, which party is responsible? In a case brought before the courts in the state of Washington, an architect had modified a curtain wall design. The contract specifications contained a standard clause requiring the general contractor to notify the architect if any materials, methods of construction, or workmanship changes were needed to ensure compliance with the contract documents. The GC did not notify the architect of any changes the GC felt were necessary relating to this curtain wall design.

Once the curtain wall was in place and the building was completed and signed off, a number of leaks appeared in the curtain wall system. The GC refused to correct the leaks, and the owner sued. The court concluded that the leaks were caused by design error. The owner had claimed that the specifications called for the curtain walls to be fabricated and installed by a manufacturer regularly engaged in the manufacture of this type of system, and that the work be first-class and done in a manner so as not to allow any weather infiltration. The court's ruling was that the curtain wall was modified by the architect and was not suited to its use and the leaks were not caused by faulty materials or poor workmanship but were the result of a design defect.

The Spearin Doctrine

A landmark court decision rendered in 1918 still has applicability today. It is the *Spearin case,* sometimes referred to as the *Spearin doctrine.* Spearin, a contractor, bid on a U.S. Navy drydock project which included replacing a 6-foot section of storm sewer pipe, which the contractor did. The replacement sewer proved to be inadequate to carry the volume of water runoff, and it broke due to internal pressure. The Navy held Spearin responsible and told Spearin to replace it. Spearin refused, and the lawsuit went all the way to the Supreme Court.

The resulting "Spearin doctrine" stated that

> If the contractor is bound to build according to plans and specifications prepared by the owner, the contractor will not be responsible for the consequences of the defects in the plans and specifications.

The court continued:

> The responsibility of the owner is not overcome by the usual clauses requiring bidders to visit the site, to check the plans and to inform themselves of the requirements of the work.

Today, the 1997 edition of AIA Document A201, General Conditions, in Article 3 recognizes that the "contractor's review (of the plans and specifications) is made in the contractor's capacity as a contractor and not as a licensed design professional," a further affirmation of the Spearin doctrine.

Problems Relating to Subsurface and Changed and Differing Conditions

Many disputes arise due to problems relating to site work. Since the architect, the engineer, and the contractor do not possess X-ray vision, even with numerous test borings and geotechnical site investigations, conditions uncovered during excavation operations may be at variance with what one perceives to be the contract requirement.

Test borings accurately display the subsurface soil strata in the exact location where they have been taken, but another boring taken just yards away may reveal distinctly different subsurface conditions.

And suppose that in the course of excavating, rock was uncovered where no rock was indicated in the borings on each side of that rock formation. Does the contractor have a legitimate claim for an extra? One answer to that question relates to whether the site is denoted to be *unclassified* or *classified*. An *unclassified site* is one where the contractor "owns" all subsurface conditions necessary to complete the site work, and if unsuitable soils, rock, debris, underground structures, etc., are encountered, the contractor is obliged to remove them and replace them with suitable materials at no cost to the owner. A typical unclassified site section in the specifications will state

> Excavation shall be unclassified and shall comprise and include the satisfactory removal and disposal of all materials encountered regardless of the nature of the materials and shall be understood to include rock, shale, earth, hardpan, fill, foundations, pavements, curbs, piping, and debris.

But is that position by an owner defendable in all cases? The answer is no!

The writer's company entered into a contract with an owner to build two office buildings with a total square footage of approximately 500,000 square feet, and the site was deemed unclassified. Accompanying the plans and specifications were a series of test borings and an isometric drawing showing various soil and rock strata along with a test boring location plan overlaying an outline of the building's footprint (Figs. 10-1 to 10-3). Excavation proceeded in the area of test boring B-23, and rock was discovered at elevation 162, about 8 feet below the surface of the ground.

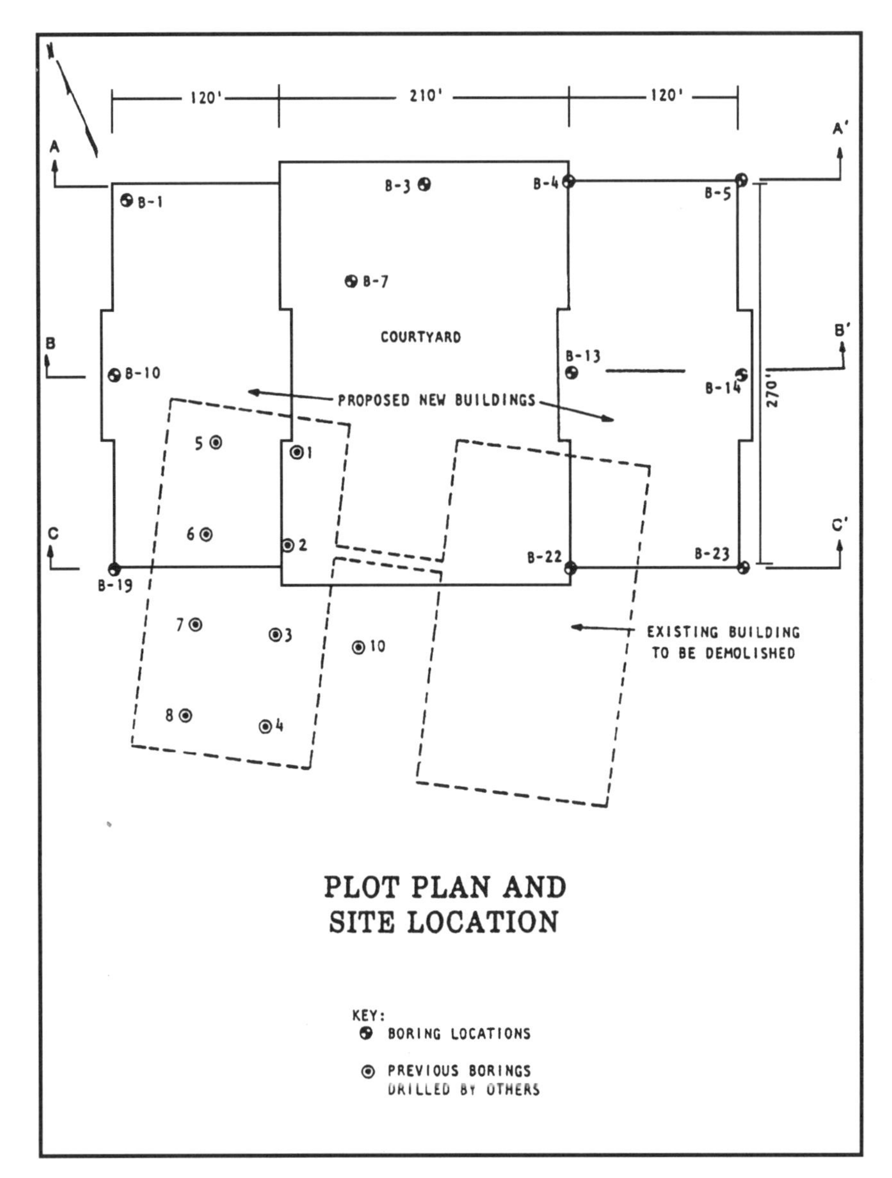

Figure 10-1 Test boring location plan, locating test boring B-23.

Both the isometric and the test boring log clearly indicated the presence of rock at elevation 152, well below the proposed bottom-of-footing elevation. Continued excavation in areas represented by other test borings uncovered rock in areas at lower and higher elevations than those indicated by the borings. The architect and owner were made aware of these disparities, and the contractor continued to excavate, uncovering more rock requiring more blasting. When the site work was complete and the contractor submitted a bill for $288,000, it was dismissed out of hand by the architect and owner since they claimed the site was unclassified and the contractor "owned" all conditions. Based upon the information received during the bid process and the actual conditions uncovered, the contractor felt that he was being treated unfairly.

After weeks and weeks of discussions, many of which were very heated, the architect and owner stood by their interpretation of the contract requirements with respect to the unclassified site and would yield nothing.

The writer made a transparent overlay of the test boring location plan and began to shift it, first to the east and then to the west of the building footprint outline. Suddenly it became evident what had occurred. The geotechnical engineer had

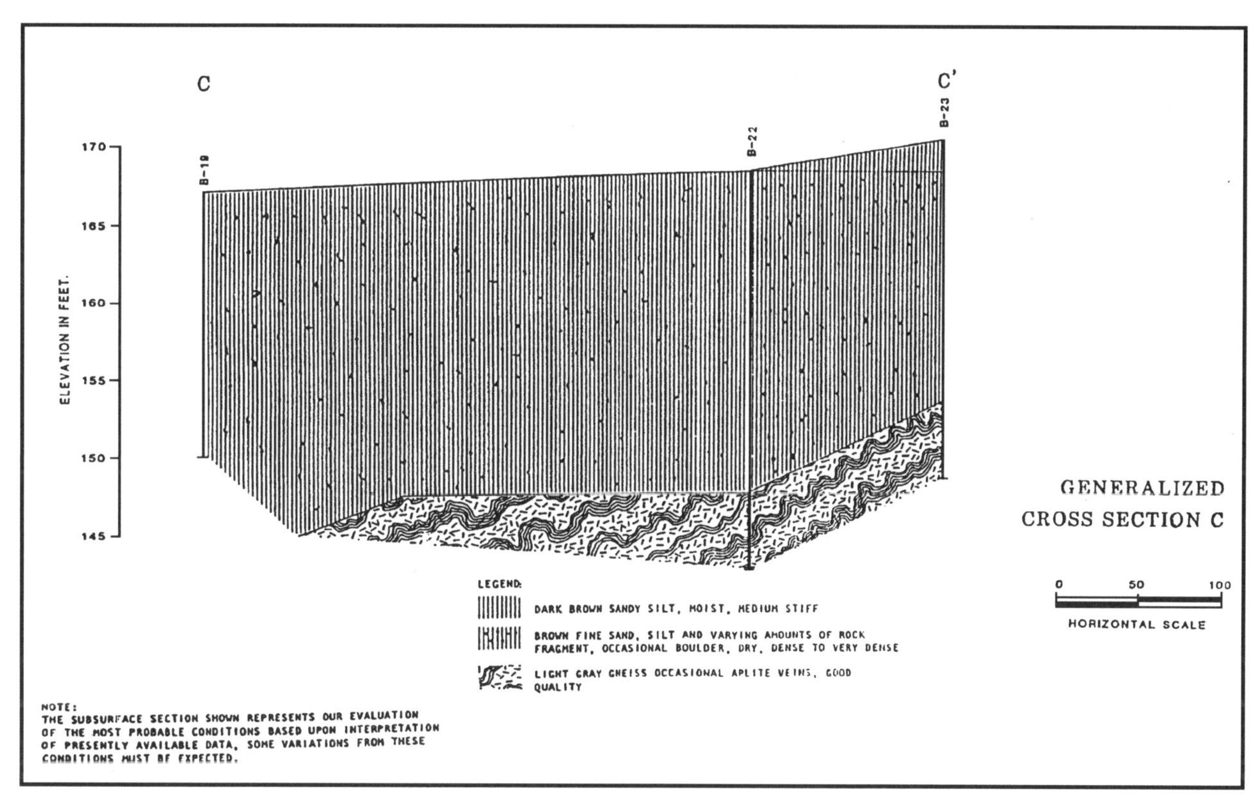

Figure 10-2 Isometric of rock elevations at test boring B-23.

inadvertently shifted the entire test boring location plan approximately 50 feet to the west of where it should have been in relation to the building's footprint. When the overlay was shifted 50 feet to the east, all test boring data matched the conditions that were actually encountered. When these facts were presented to the owner, the owner paid the $288,000.

On another occasion the writer, while administering another contract with an unclassified site, encountered considerable unsuitable soils below the *payline* (the bottom of footing/foundation level as indicated on the contract drawings). A proposed change order was presented to the owner, who dismissed it, claiming that the contractor was responsible for all unsuitable soils, no matter where uncovered, because, there again, the site was unclassified. The writer argued that a contractor "owns" all soils *above* the payline in an unclassified site, but this does not mean that the contractor "owns" *all* unsuitable soils *under* the payline. This interpretation fell on deaf ears, and the claim was again denied.

At the next meeting to review this issue, the writer drew a round circle meant to represent the earth and placed the project in question on top, drawing a deep V meant to represent excavation to the bottom of the circle (China) and said, "So you're telling me that we are obligated to dig all the way to China if necessary?" The owner's representative said, "Well, no, that's unreasonable." So the concept of degree of "ownership" of responsibility was established, and the claim was settled by agreeing on a limit to the amount of soil the contractor was contractually responsible to remove at no cost—somewhere between the payline and China.

The point is that the project superintendent should not give up the claim for compensation for unforeseen subsurface conditions, *even in the face of restrictive language*

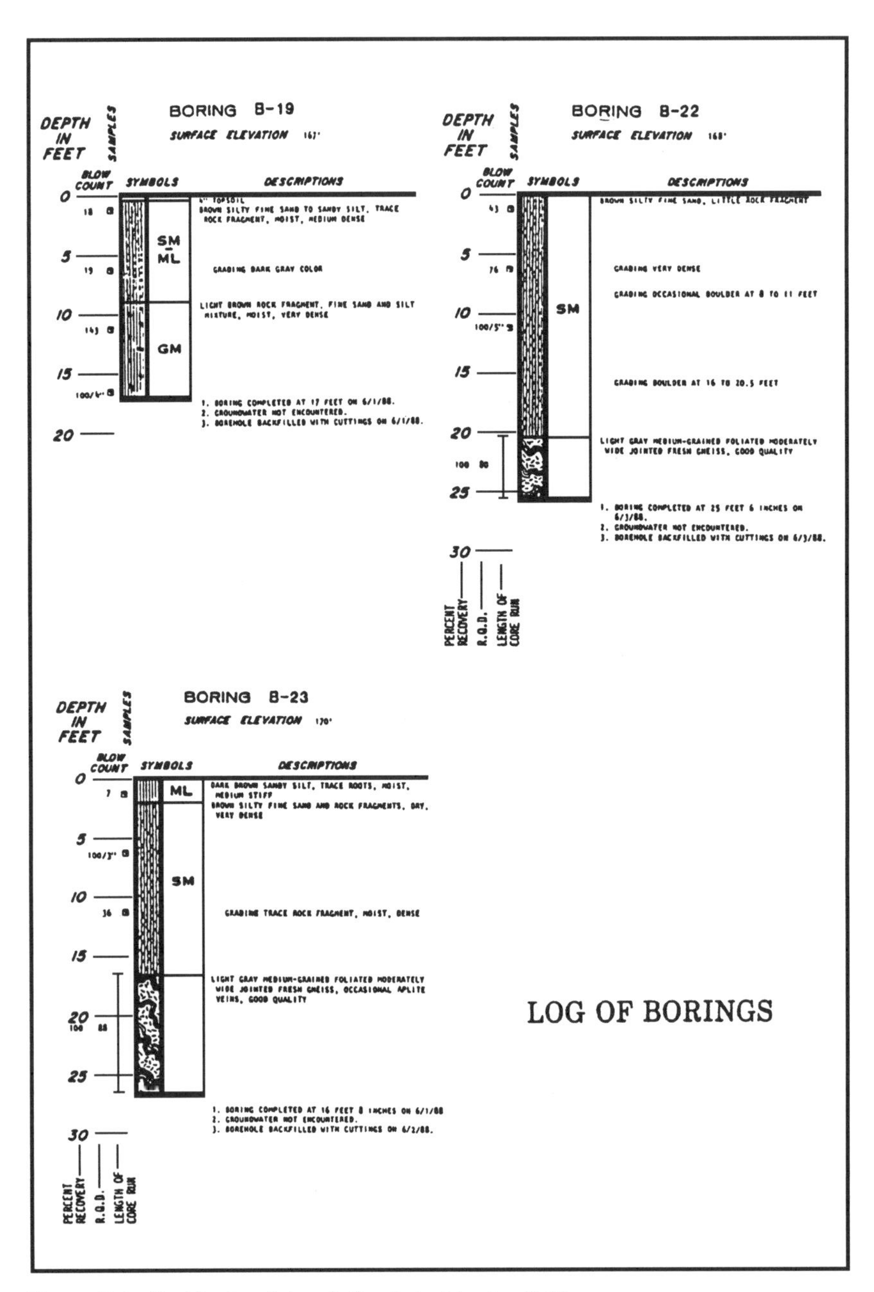

Figure 10-3 Test boring data relating to test boring B-23.

in the contract. If something appears to be unfair, it usually is, and by continuing to investigate the situation the contractor may be able to develop enough information and documentation to refute the unfairness of the architect/owner's ruling.

Dealing with Exculpatory Statements in the Contract

Most contracts include statements inserted to defuse any claims the contractor may present when encountering unforeseen site conditions. They can take one of several forms:

1. The contractor may be required to visit the site prior to submitting the bid, and any condition visually observed or reasonably assumed by this pre-bid visit will be interpreted by the owner as being *disclosed.*

If that is the case, a close and careful inspection of a site is necessary to document not only that all existing conditions were observed, but also the extent to which other hidden conditions could not be observed and therefore not anticipated.

2. The "no damages for delay" clause will prevent the contractor from claiming additional costs beyond direct costs if unforeseen conditions are encountered and delays occur.

If this clause cannot be stricken from the contract, it must be passed through to all subcontractors, thereby eliminating any potential for subcontractor delay claims if the project is delayed when unforeseen subsurface conditions are encountered.

3. A clause that requires the contractor to examine the contract documents and report, in writing, any obvious errors, omissions, and ambiguities within a specific time period, or else a claim will not be considered. Although it may be difficult to uncover all the "errors and omissions" as soon as the project commences, every effort ought to be made to review the plans and specifications promptly, to search out obvious problem areas. Ask all subcontractors to quickly review their particular plan and specification sections with an eye to uncovering deficiencies.

4. Geotechnical reports included in the bid documents that state that they "are not be to relied on" since the contractor must draw his or her own conclusions as to the representation of the information contained in these geotechnical reports.

The general contractor has to rely on the information presented by the geotechnical engineer in order to prepare the site work estimate, which will include a contingency of some sort for unknown or unanticipated conditions. If the contractor can show that the information in the bid documents was misleading or insufficient and therefore affected the amount of a *reasonably assumed* contingency, the contractor may be able to overcome this exculpatory contract statement.

5. Disclaimers in the specifications or on the drawings such as "groundwater may vary" or "subsurface information is not to be relied on" or "test borings are for information only."

The contractor must be able to show that he or she prepared an estimate that, based upon the contractor's experience in site work and the information presented in the bid documents along with a site visit, included these reasonable assumptions.

Differing Conditions—Preparing to Document

There are two distinct different types of *differing conditions* claims, type I and type II, as defined by the federal government in the Federal Acquisitions Regulations (FAR). A type I claim requires the following:

1. Representation of the subsurface conditions is reflected in the contract documents' plans and specifications.
2. The contractor made a reasonable interpretation of these contract representations.
3. The contractor had every right to rely on these contract representations.
4. The subsurface or latent physical conditions that were encountered were materially different from those represented by the contract documents.
5. The subsurface or latent physical conditions encountered were unforeseeable.
6. The additional costs incurred by the contractor were due solely to these materially differing conditions.

The contractor must be able to show that the contract documents related to this differing-conditions problem were either inaccurate or misleading and that the bid was based upon the information contained in these inaccurate or misleading documents. Obvious conditions that could have been observed during a prebid site visit will void any claim. (Remember Article 1 of AIA Document A201?)

The type II claim is slightly different from type I in that the owner will not have represented the subsurface conditions in the bid documents. Type II claims require the following:

1. The subsurface or latent physical condition was unknown.
2. The subsurface or latent physical condition was unusual and could not have been anticipated from a review of the bid documents or was not evident after a site inspection.
3. The materials or conditions encountered were materially different from those ordinarily encountered and/or generally encountered in the type of work being undertaken.

To prepare a viable change order for additional costs utilizing differing conditions as the basis for that claim, a contractor must show that

1. The contract documents reflect certain conditions that formed the basis for the contractor's estimate.
2. The contractor's interpretation of the documents was reasonable and based upon previous experience in such matters.
3. The subsurface conditions actually encountered differed materially from those represented in the documents. Some public works contracts stipulate that if conditions encountered exceed 20 percent of those "reasonably" anticipated, the definition of differing conditions will have been met. This standard is often embraced by the private sector as well.
4. The actual conditions encountered could not have been *reasonably* anticipated.
5. The costs claimed must be solely attributable to the materially differing conditions.

The contractor should include in her or his narrative

- A clear statement as to the *usual* conditions a contractor would have expected to encounter on the site
- The conditions actually encountered
- How these conditions differed materially from the known and usual
- Costs and delays incurred because of the encountered conditions

Claims due to Scheduling Problems

AIA Document A201, General Conditions, as it relates to the contractor's schedule has changed over the years. The 1970 edition required the contractor to submit the schedule for the architect's *approval*; the 1976 version required submission for the architect's *information*; and the 1987 edition of A210 requires the contractor to prepare and submit the schedule *promptly*. The current 1997 version of A201 requires the contractor to submit a schedule that complies with the contract completion date.

Schedules are dynamic and change quite frequently. The latest architect requirement per AIA A201 provides the general contractor with greater flexibility in modifying the schedule without being held accountable for changing sequences or time frames for various work tasks, as long as the completion date is not extended.

The use of the *critical path method* (*CPM*) schedule permits the representation of the relationship of one work task to another. These CPM schedules, easily created by many software programs, play a key role in any delay claim presented by the contractor. The baseline schedule—the initial schedule prepared at the beginning of the project—ought to have been assembled with a great deal of input from subcontractors and vendors with the project manager and project superintendent acting as coordinators for this critical task.

Once published, the baseline schedule becomes the "official" road map, and any changes to the sequence or time allotment for selected activities will affect the baseline and either retain the contract completion schedule or extend it. Depending upon the reasons for the extension of a project completion date, a request for a time extension or, in the case of a nonexcusable delay, a recovery plan ought to be prepared to ensure a timely completion.

When a contractor prepares a delay claim and *consequential damages* are mentioned, the term *Eichleay formula* is often heard. Although AIA Document A201 does not permit the contractor to submit a claim for consequential damages (Article 4), there may come a time when this document is not a part of the contract and consequential damages are allowed. *Consequential damages* are damages (costs) beyond the easily recognizable bricks and mortar costs. The project superintendent should be at least aware of this term and the concept of Eichleay.

The Eichleay Formula

Back in 1954, Eichleay Corporation, a general contracting firm, filed a claim against the federal government that included a significant delay in completing the project. Eichleay said that during the year in which this delayed project was being constructed, the company had estimated its corporate overhead costs for that year based upon completing that project, as scheduled, during that year. When completion of the project was delayed and it was not completed until the following year, Eichleay claimed that the company did not have enough sales volume to absorb all the corporate overhead, and therefore the company requested the government to reimburse it for this "underabsorbed" overhead. The company won its case and collected these costs. Eichleay created a formula to show the government how this delayed completion affected its corporate overhead. For decades after, the Eichleay formula was used by hundreds of contractors in putting together delay claims in both the public and private sectors.

The CPM schedule becomes a two-edged sword. Once published, it will be accepted by both owner and subcontractor as the time frame for the completion of individual tasks within the overall framework of the entire project's schedule for completion. Delays created by one subcontractor may affect other subcontractors as well as the overall project completion date. The other affected subcontractors will certainly look to the general contractor for assistance and possibly additional compensation to complete their work on time in spite of the delays caused by others.

If a delay claim to the owner is being considered, a well-documented series of CPM schedules graphically showing one delay or a series of delays and the impact on the overall completion date will be strong evidence to support the claim.

One mistake many project superintendents make is to assume that somehow a minor delay will continue to remain minor and therefore no request for a time extension is necessary.

A good rule of thumb to follow is this: *Whenever a delay occurs, document either in the daily log or in a memo to file the reasons for the delay, the circumstances surrounding the delay, the person or persons responsible for the delay, any actions taken to make up for lost time, and person(s) notified of the delay, either verbally or in writing.*

Claims against Professionals

Architects and engineers, when presented with claims for design deficiencies, turn to their standard professional liability for malpractice insurance that refers to *errors of commission* and *errors of omission*. The development of a set of error-free plans and specifications for a major project is nearly an impossible task. It is certainly understandable that the complexity involved in creating the construction documents may result in a lack of a few minor details, dimensions, or less than perfect coordination.

Most of the problems that involve architects, engineers, and contractors seem to be due to the following shortfalls:

- Drawings are not coordinated properly among mechanical, electrical, fire protection, structural, and architectural trades.
- There are conflicts between small and large details and between written specifications and graphic drawings.
- Boilerplate requirements on the drawings or in the specifications are often lifted from other projects and therefore don't apply to the project at hand.
- There is a lack of communication between the various design consultants, so that a change made by one is not transmitted to the others to determine whether it has any impact on their design.
- Consultation with owners is insufficient to afford them the opportunity to participate in decisions that ultimately affect the way in which their program is being developed.
- There is insufficient time for proper and thorough review of all contract documents because the owner has demanded an unrealistically compressed time frame for drawing production and submission for bidding purposes. (Only once did the writer hear an engineer tell the owner, "You can have accuracy or you can have expediency, but you can't have both—which do you want?" The owner, backing down on the demand for drawings as soon as possible, replied, "Accuracy," thereby giving the design engineers the time required to prepare, and check, a proper set of drawings.)

Incidentally, this list was paraphrased from one prepared by a design professional who was expressing concern over the growing cost of malpractice insurance. The professional was making the point that problems are not being created by what the professionals are *doing*; they are being created by what professionals are *not doing*.

Preparing a claim against a design professional is extremely difficult and certainly limited as far as most general contractors are concerned. First, the GC's contract is with the *owner* and not the architect, so there is no contractual relationship to be breached between architect and contractor. But such a lawsuit can be instituted in some cases and would be known as a third-party claim. One such lawsuit that comes to mind originated when the architect failed to process shop drawings within a reasonable period of time.

In *Peter Kiewits Sons Co. v. Iowa Utility Co.,* 355 F. Supp. 376, 392, S.S.Iowa (1973), a claim was made to collect damages because the contractor suffered losses due to an unjustified delay in the architect's processing of shop drawings.

In another case a general contractor sued an architect, claiming that the architect prepared erroneous design drawings and erroneous design documents, knowing that the owner would furnish them to the successful bidder and fully aware of the injury that would be caused to that contractor. The appellate court held that the allegations were sufficient to establish a special relationship between the parties, even though there was no direct contract between the contractor and architect.

Acceleration—What It Is and How It Is Used

The legal term *acceleration* should be a part of every project superintendent's vocabulary. It occurs, in the legal sense, when an owner recognizes that there have been delays to a construction project, thereby extending its completion date, but the project superintendent or project manager is directed to maintain the original project completion schedule. These instructions are known as a *demand for acceleration.*

There are two types of acceleration: actual and constructive. *Actual acceleration* occurs when the owner directs the contractor to complete the project ahead of the completion date contained in the baseline (accepted) schedule. *Constructive acceleration* occurs when the contractor is delayed by some owner/architect's action or inaction. This can occur when you are dealing with change order work, or when an owner's or architect's delay in reaching a decision on a question posed by the contractor has both cost and time implications.

If the contractor has requested a project completion extension and is not granted one and the owner subsequently demands that the contractor complete the project according to the initial completion date, a condition of constructive acceleration has been created. By claiming the condition of constructive acceleration, a contractor can attempt to receive some monetary relief. The legal elements required to establish acceleration in a claim are as follows:

1. There is an *excusable delay* which entitles the contractor to a time extension.
2. The contractor submits a written request to the owner for the time extension.
3. The request for the time extension is refused, but the contractor proceeds with the work.
4. The owner issues a directive to accelerate performance to complete the project within the original time frame.
5. The contractor proceeds with the work at an accelerated pace and documents all costs associated with this speed-up process.
6. The contractor then notifies the owner of the intent to file a claim to recover costs incurred in this accelerated effort, if the owner rejects the contractor's change order for the work.

These situations will arise, and the project superintendent should be aware of the steps outlined above in order to recover costs caused by these undue demands by the owner.

What Is a Mechanic's Lien?

A *mechanic's lien* is a charge against the property that effectively notifies the property owner that some portion of the labor and/or materials and equipment costs installed in the building have not been paid. A mechanic's lien can be filed only against work in connection with contracts in the private sector; the filing of a mechanic's lien against unpaid goods and services in a public works project is not allowed by law.

The Miller and Little Miller Acts

The Miller Act, passed by the federal government years ago, does not permit liens to be placed against property of the U.S. government. Similar acts passed by most state governments, known as "little Miller Acts," prevent liens from being placed against state and local government property. The remedy for claims against unpaid labor and materials costs by subcontractors and vendors in government projects rests with the obligation of the general contractor to provide payment and performance bonds for those types of projects. Claimants for monies owed but unpaid by the general contractor can file a claim with the appropriate federal, state, or local agency and *call the bond.*

Mechanic's Liens in the Private Sector

When a mechanic's lien is filed, the title to the property is "clouded" and the property can't be sold until that lien is "satisfied." The unpaid sum creating the lien can be either paid or *bonded,* in effect requiring the bonding or insurance company to guarantee payment if the claim remains unresolved.

The normal expiration date of a lien may vary from state to state; but if the lien remains unsatisfied, the issuer of the lien can institute foreclosure proceedings. Lien proceedings are started when the company attorney files a claim stating that a valid lien has been placed upon the property and that certain sum is due their client. Proof of the amount owed is necessary, as is proof that the lien was properly filed against the correct owner of the property, the property was correctly identified, and the lien was filed within the time frame established by law.

In theory, when foreclosure takes place, the property is sold and the lien holder receives payment from the proceeds of the sale. In practice, the sale of the property rarely occurs, and the lien is eventually satisfied by being paid in full or bonded.

The rules and regulations governing the filing of liens also vary with the state, but generally a lien must be filed within 90 days of the last date on which work was performed on the project in question. When the filing of a lien is contemplated, there are certain pitfalls to look out for and to avoid:

1. *Make certain the lien is filed within the filing limit time.* If the filing time is within 90 days after the last date that work was performed on the job, make certain that filing is done before the deadline. If it is not, the right to file a lien is lost. The court is aware of some tricks that can be played to comply with the 90-day requirement. The date of last work must be the last date of *meaningful work.* For example, a contractor realizing that the lien rights will expire, say, tomorrow, cannot send a mechanic back to the site to replace a filter in HVAC equipment or adjust a door or replace a broken light fixture lens. The courts will interpret this as a means to circumvent the *intent* of the lien rights, and the lien will probably be declared invalid.
2. *Make certain the lien is filed against the correct property.* Although this might seem rather simple, when urban property is involved, it can get complicated, particularly in subdivisions or projects composed of a number of different parcels. If the wrong property is described, the lien will be invalid.

   The writer was involved with a large senior living community several years ago, and the concrete subcontractor had not paid its ready-mix concrete supplier (although the subcontractor signed monthly lien waivers indicating that the supplier had been paid!). The irate owner of the property notified the writer's company that the current requisition would not be honored until "the enclosed lien is satisfied and proof of removal of the lien is furnished." The only problem was that the lien was filed improperly against *another* property also owned by that owner for which the writer's company was not involved. Payment of the requisition could not be denied.
3. *Is the lien filed against the proper owner?* The proper owner can sometimes be found on the land records in the tax assessor's office, but quite often defining legal ownership is difficult. When a project involves a limited liability corporation (LLC), a joint venture between corporations or individuals, a syndication or a "shell" corporation, it may be difficult to identify the proper owner of record.

A Word about Lien Waivers Submitted by Subcontractors

As mentioned above, not all subcontractors will faithfully and honestly complete their lien waivers. On more than one occasion a subcontractor has been known to falsify a lien waiver. Although the subcontractor may sign the lien waiver, indicating payment for all labor and materials placed in the building during the period covered by the waiver, either knowingly or unknowingly the information may be false. Some subcontractors may not be aware that their subcontractors have not paid *their* bills, for example, the mechanical subcontractor who engages a pipe insulator and never requests a lien waiver from that company or never questions whether the company has paid its suppliers. There are other subcontractors who just plain lie because they need the money to pay for other materials and equipment, and more than likely, its materials and equipment for other projects.

So don't assume that the submission of a lien waiver from subcontractors ensures that no liens will be filed.

Most subcontract agreements include a clause stipulating that the subcontractor cannot hire a second-tier subcontractor without approval from the GC. It may be difficult to determine if all lien waivers have been submitted if the identification of all second- and third-tier subcontractors is unknown.

The project superintendent can help in these matters by maintaining a list of all subcontractors working on the project, so the project manager can be aware of the need to obtain lien waivers from those companies. Last, when a false lien waiver is discovered, along with a threat of legal action, that subcontractor should be advised that from that time forth, joint checks may be issued to all the subcontractor's suppliers and subcontractors.

Arbitration and Mediation

Look once again at that key contract document, AIA A201, General Conditions. Article 4.6 requires mediation and arbitration as the initial and secondary methods to resolve disputes. This clause is actually helpful to the general contractor since it eliminates the costly process of litigation as a means of instituting and resolving claims of less than monumental proportions.

If a contractor, for example, has had a $10,000 claim denied by the architect and owner, the time and costs to litigate may be such that, economically speaking, this claim may not be pursued through legal action. With the lower costs associated with the mediation and arbitration process, pursuit of such a claim would be economically feasible.

Article 4.6 of AIA Document A201 sets forth the procedure for commencement of the dispute resolution process and requires mediation (Article 4.6.1) as the first step in that process.

Mediation

Mediation is *nonbinding*. This means that if it is employed as a first step in resolving a dispute, either party to the process, at any time, may decide to withdraw from the proceedings. When the mediation sessions have been concluded, the parties are under no legal obligation to accept and abide by the conclusions. However, many of the facts "discovered" during the process could be used against that party in either arbitration or litigation proceedings at a later date.

This process involves engaging a professional mediator to review the facts surrounding the dispute and to attempt to get each party to give a little, or sometimes more than a little, to resolve a dispute.

Typically the mediator will start the proceeding by announcing to both parties the steps he or she plans to take to bring about resolution. The mediator discusses the strong and weak points of each party's claim, and if there is a genuine desire by both parties to negotiate a settlement, the mediator will act as the go-between to do so.

The mediator will separate each party, assigning one group to one room and the other group to a second room. By shuttling back and forth and presenting the mediator's opinion of the strong and weak points of each party, the mediator will attempt to negotiate a settlement.

As of the old saying goes, a successful negotiation session is one in which each party feels he or she did not win. But resolving a dispute usually means giving up something, and the mediator's job is to attempt to make each party reduce its initial demand and resolve the dispute so they can get on with their business as usual.

Mediators can be located by thumbing through the Yellow Pages, or the American Arbitration Association (AAA) can also be contacted to provide a list of mediators. If mediation fails, then each party can ratchet the dispute up one further notch and request arbitration.

There again, the contract documents will set forth the notification required to demand arbitration, and the American Arbitration Association can provide all the details and fees regarding the arbitration process.

Although the arbitration process was initially established to reduce or eliminate participation by lawyers, nowadays many law firms have developed specific departments that specialize in arbitration hearings. In most cases, attorneys not only will assist in the development of the facts and documentation required for the arbitration process, but also will attend all hearings.

Partnering—A 1990's Buzzword?

In 1988, the U.S. Army Corp of Engineers sought to find a better way to settle claims quickly and more economically and, in the process, attempt to develop a less adversarial environment when working with contractors. The partnering process that evolved out of this effort was based upon the realization that when all parties better understood the goals and aspirations of the other parties to the contract, they would find that they all share common goals, albeit in different ways.

By defining these common goals before the project starts, developing a means of attacking problems *as they occur,* and resolving them *quickly and equitably,* successful completion of the project would be achievable.

Since the introduction of the formal process known as *partnering,* any number of public agencies and private firms have embraced the process, and many have included the process of partnering as a contract requirement.

Common to all such projects is the use of a *facilitator,* a trained professional whose responsibility it is to guide all participants through the process.

How Partnering Works

All parties to the construction process attend one or more partnering sessions—the owner, the architect/engineer, general contractor, subcontractors, and vendors will be invited and urged to attend these sessions. When the facilitator assembles all these seemingly disparate groups and requests that they express their project goals and aspirations in writing, it becomes clear, when all these statements are combined, that each party has the same basic goals.

- Make a fair profit
- Receive prompt payment for work performed
- Reduce or eliminate change orders which often create problems
- Produce a high-quality project
- Achieve rapid construction and complete the project on time or ahead of time
- Eliminate or substantially reduce any disputes

One of the more important procedures created during the initial partnering sessions is the mechanism to resolve disputes quickly. The process that usually emerges includes

- A resolve to settle simple problems at the lowest management level possible. This generally means the general contractor's project superintendent and the owner's representative on the site.
- Preparation of a structured process for escalating an unresolved issue to the next-highest management level, which might mean involvement on the first level between the project superintendent and the owner's on-site representative, escalating to the next-higher level of the contractor's project manager and the owner's project manager. A still higher level may be at the project executive level or divisional manager and the level above that the general manager and possibly the construction company's owner.
- Agreement to adhere to a rigid time frame for the resolution of any disagreement or dispute. For example, at the field supervisory level, problems *must* be resolved within 24 hours; problems escalated to the project management level for resolution must be concluded within 2 working days; problems at the executive level *must* be resolved within one week.

This dispute-solving procedure may be one of the most important contributions of the partnering process. After all, who wants the boss to see that employees are incapable of resolving disputes without relying on their superiors to do so?

Summary

Just remember that the construction industry is an industry of contracts. To deal with the inevitable conflicts that occur through misunderstandings of one's contract obligations, first and foremost, a thorough understanding of one's rights and obligations under the contract is essential. So, project superintendent, carefully read the contract and the specifications, and review the drawings in detail.

Second, it is important to pay prompt attention to an impending dispute, most of which does not disappear if not resolved. Third, not all disagreements can be resolved quickly, but attempts to do so are greatly enhanced when complete and accurate documentation relating to the dispute has been prepared and assembled along the way. And, most importantly, many disputes can be resolved by viewing them from the other party's perspective and approaching resolution with an open mind and an attitude of reasonableness.

Chapter

11

Effective Letter Writing for Project Superintendents

Prompt and accurate communication is an essential task in the construction business, and the project superintendent may be called upon, from time to time, to compose an effective letter when certain situations require this form of documentation. To be effective, a letter should include certain basic elements, but the real challenge is to state your case without wasting a word.

Effective letter writing does not come naturally to many people. We all have to work on it—not to win any writing awards, but to be able to express ourselves clearly and precisely when the occasion arises.

There are all kinds of guidelines to follow when composing a letter, but the two that seem to make the most sense are the three C's and the five W's.

The Three C's of Writing—Clarity, Correctness, and Courtesy

Clarity

Decide what you want to say and say it. What points am I trying to make? What ground do I want to cover in this letter? Think through the content of your letter before you sit down at the computer or with pencil and paper in hand.

Correctness

Use the correct words and the correct spelling. Certain words are most expressive and have greater impact than others. Your job is to search out those correct words. Long, complicated words are not necessarily the best approach to impress your intended reader. Sometimes short, concise words are more effective. Most of the word processing software today alerts the writer to misspelled words by highlighting these words in red. If you are not using a computer with one of these spell check programs, are writing a letter by hand, and are unsure of the spelling of a word, go to the dictionary. Nothing turns off a reader faster than finding misspelled words in your letter or words that are used incorrectly. If you are not sure that the word you are using is correct, select one that you can use with assurance.

Courtesy

Although your emotions may be running high and you want to let the reader know that you are really angry or upset about the situation in hand, be careful of how you express your anger and the words you use to express those feelings. Ask yourself, How would I respond if I received such a letter? Control your thoughts and present your case, objectively—without those four-letter words.

The Five W's—Who, What, Why, Where, and When

The five W's method of letter writing may have had its origin in the criminal investigation field or certainly in newspaper reporting. To compose a letter or a report, certain questions need to be answered, and that is the basis of the five W's approach:

Who

Who is the intended recipient of this letter, not only the primary recipient, but also others who either may receive a copy directly or may have an interest in your letter. When you are writing to an architect, explaining the reasons behind a recently disputed item of work involving, say, the electrical contractor, do you want to send a copy to the electrical engineer and/or the electrical subcontractor, or do you not want anyone else to be aware of what you are doing?

What

What do you want to convey? What is to be accomplished by this letter? Do you want to ask a question? If so, then you will want to include a statement that you will expect an answer—don't assume that the reader will respond. Request a written response, and if the issue is time-sensitive, include a reasonable date by which you would like to receive a response. Are you stating a position on a specific matter or issue? Are you responding to a previous letter you received? That's the essence of the "what."

Why

Somewhat similar to the W above, this also asks the question, *Why* am I writing this letter? Do you want to document a verbal direction received or one that you initiated? Do you want to explain your position on a certain matter, or do you want to defend your position?

Where

Where did the events included in your proposed letter take place, or where are the events to take place? Were they the result of a field discussion or a project meeting or a meeting on the site?

When

When did the event in your proposed letter take place, or when do you expect the event to take place?

Use the Draft Approach to Letter Writing

It is often difficult to produce the final letter at first writing. Professional writers often wrestle for hours or days over the right word to use in a particular sentence or two, or whether they are clearly expressing all the thoughts and facts they intended to express. The draft approach to letter writing is one way of coping with *writer's block*—the inability to pick the words that describe the situation you wish to present.

By scribbling your first thoughts in that first draft, you will gain a starting point. Someone once said that a thought was not a thought until it was expressed verbally or in writing. Although you may have thought that you had the letter composed in your head, when you try to put it down on paper, you may find these thoughts don't come easily or clearly.

In many cases, it is more important that the content of your letter be correct and complete than that the letter be transmitted quickly. However, a letter written too late after an event loses much of its impact. "If it was so important, why did he or she wait 10 days before sending me this letter?" So there is the balance between accuracy and completeness and promptness.

It is also helpful to put the draft away for a few hours or even overnight, so that a fresh look later on will either reinforce what you have already written or change its content. Don't be discouraged if you find yourself preparing three or four drafts before you create the final one. When that letter is finally typed, you will gain some satisfaction in knowing that it contains the correct words and phrases and conveys the exact thoughts that you set out to convey.

Some Basic Letter-Writing Principles

1. Before you write that letter, establish a clear purpose for the letter. What do I want to accomplish by writing this letter? Do I want to notify someone of an impending problem? Do I want to clarify my company's position on a certain matter? Do I just want to document some important verbal directive or confirm, in writing, a position I have stated orally? Do I want to put someone on notice that a contract violation is occurring or will occur unless some action is taken?

2. Learn to express your thoughts on paper in the same way as you would when talking. This will allow you to write in a more conversational manner. But you shouldn't use the word *ain't* or drop the letter g from the end of *beginning* or *assisting*. Don't try to insert big words to impress. Don't use "expletive deleted" words (even if you'd really like to). Pretend you are having a conversation with the person to whom you are writing. Would you use the phrase *set forth herein,* or would you use the word *attached* or *enclosed?* The answer is probably the latter and not the former, so try to avoid these stilted expressions.

Many words or expressions ought to be avoided in your letters:

Instead of:	Use:
Hold in abeyance	Wait
Pursuant to your request	As you requested
Prior to	Before
Subsequent to	After
Conceptualize	Think of
Due to the fact that or Inasmuch as	Because
At an early date	Soon
Taken into consideration	Consider
Ascertain the data	Get the facts
As this point in time	Now

Can you think of any others? If so, add them to your own "use instead of" list.

3. Be concise. Try to use as few words as possible to express yourself. Searching for that right word might be difficult, but once you have found it, it

will be worth the effort. Often the right word can take the place of three or four less effective words. As an example, suppose you are writing a letter to an architect or an owner, advising the person that an addendum or contract modification or change order has not been received on the date promised. You could have written the following:

> As we discussed last week, you indicated that we would receive a (copy of Addendum No. 2) or (a copy of a proposed change) or (Change Order No. 4) on June 5, 2003; however, we did not receive this (document) on that date as promised.

This sentence consists of 24 words, exclusive of dates and items in parentheses. You could also state the same thing by using only 7 words (exclusive of the item in parentheses and the date).

> We did not receive (document) on June 5, 2003, as promised.

Consider the following expressions and how fewer words express the same thought better.

If you want to say:	Shorten it to:
Engage in conducting an inspection	Conduct an inspection
Long or lengthy period of time	Long time
Enclosed herewith	Enclosed
Decide at a meeting to be held on Monday	Decide at Monday's meeting
During the period of 2003	During 2003
There is one point that is clear and that is	One point is clear
The cost was higher than I expected it to be	The cost was higher
That is (was) the situation at that time	That is (was) the situation
During the course of the meeting	During the meeting
The trouble with the compressor was that it was too noisy	The compressor was too noisy

4. Research before you begin to write. Ask yourself, What documents will I need to refer to as I write this letter? You may need to refer to

- A specific contract requirement
- Specific plans or details, sections
- The specification book
- Inspections, architect/engineer reports
- Previous correspondence
- Shop drawings

If you need to refer to any of these documents, then assemble them, make copies, or have the originals on hand as you prepare the letter. Those "sticky notes" can be used to flag pages in the specifications or drawings. If any of these documents are to be attached to the letter, get them beforehand so that you can refer to them as needed; and after all, you will need them when the letter is completed. Avoid long interruptions during the letter-writing process due to the need to locate a referenced document, so that your train of thought will not be broken.

5. Organize the letter and the topics to be included in the letter. Sometimes it is easier to create an outline to establish the order in which each part of a lengthy or complex letter is to be prepared. This "skeleton" can then be fleshed out with the facts once the outline has established the orderly presentation of events or facts.

6. Finally, check the spelling of the person's name to whom the letter is addressed. The person's title and company address are important. You don't

want to misspell the recipient's name, nor do you want to list an incorrect title. Nothing turns off a reader more quickly than having her or his name misspelled or title or position in the company listed inaccurately.

Writing effective letters may be difficult to many people, but it occupies an important role in every manager's business and personal life. With patience and practice and a little advance preparation, everyone can produce concise, clear, and effective letters.

Appendix

Earthwork and Compaction*

*This appendix is made up of pp. 13 and 24 to 51 from Sidney M. Levy, *Construction Site Work, Site Utilities, and Substructures Databook*, McGraw-Hill, 2001; reprinted by permission of the source organization.

1.6.0 Soil Compaction

Soil compaction has been practiced by man for thousands of years.

The first attempts at earthen dams and irrigation ditches demonstrated the value of compaction by adding strength and some measure of protection against moisture damage. Early tamped earth buildings depended on thorough compaction for stability.

However, it was not until road building became a highly developed art, during the period of the Roman Empire, that the importance of soil compaction was fully appreciated. Roman roads, which are still in use, were built with careful attention to subsoil conditions and with thorough compaction of gravel and clay base. (Figure 1)

The Roman road builders knew that their cut stone road surfaces were only as good as the foundation on which they rested.

Some years ago, most compaction was done only on large construction sites, such as roads and airports. Very large, heavy machinery was used.

Only in the last few decades has the importance of **confined area** compaction been recognized.

With the introduction of self-contained portable rammers and vibratory plates, confined area compaction became practical.

Now soil compaction is commonly specified for building foundations, trench backfills, curbs and gutters, bridge supports, slab work, driveways, sidewalks, cemeteries and other confined area work.

Properly done, soil compaction adds many years to the useful life of any structure by increasing foundation strength and greatly improving overall stability.

Figure 1 - Roman Road

1.6.1 What Is Soil?

Soil is any natural material found on the surface of the earth except for embedded rock, organic plant and animal material.

Soil may be divided into four major groups according to particle size. As shown in Figure 2, the major groups are:

Clay — with particle sizes 0.00024" (.006 mm) and smaller;

Silt — with particle sizes ranging from 0.00024" (.006 mm) — 0.003" (.076 mm);

Sand — with particle sizes ranging from 0.003" (.076 mm) — 0.08" (2.03 mm);

Gravel — with particle sizes ranging from 0.08" (2.03 mm) — 3" (76.2 mm).

Figure 1.6.1 *(By permission: Wacker Corporation, Menomonee Falls, Wisconsin)*

1.6.2 Soil Classification According to Size

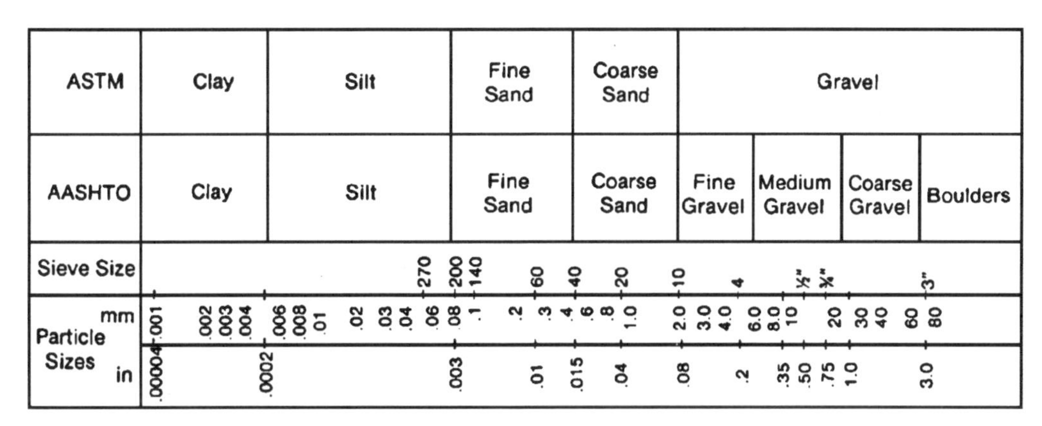

Figure 2 - Soil Classification According to Size

1.6.3 AASHTO Group Classification

The American Society for Testing and Materials (ASTM) and the American Association of State Highway and Transportation Officials (AASHTO) classify soil as **granular** or **cohesive** on the basis of a sieve analysis of the soil. See Figure 3.

Granular soil consists mainly of sands and gravels. **Cohesive soil** consists mainly of silts and clays.

In granular soil, the particles are held in position due to the frictional force that exists at the contact surfaces. In the dry state, granular soil particles can be easily separated and identified. In a moist state, a granular material, such as sand, may be formed to desired shapes but will crumble easily as soon as it is disturbed.

In cohesive soil, the molecular attraction between soil particles is the force which holds the soil in place. As these particles are very small in size, high in number, and densely arranged, the cohesive force within the soil is very high. Cohesive soils are very hard in the dry state. When moist, they are plastic and can be molded or rolled to almost any shape.

General Classification	Granular Material							Cohesive Material More than 35% of Total Sample Passing No. 200			
Group Classification	A-1				A-2						A-7
	A-1-a	A-1-b	A-3	A-2-4	A-2-5	A-2-6	A-2-7	A-4	A-5	A-6	A-7-5 A-7-6
Sieve Analysis % Passing											
No: 40 ____	30 max	50 max	51 min								
No. 200 ____	15 max	25 max	10 max	35 max	35 max	35 max	35 max	36 min	36 min	36 min	36 min

Figure 3 - AASHTO Group Classification

1.6.4 How to Test for the Correct Amount of Moisture

The moisture content of a soil, either granular or cohesive, is a critical factor in the workability of that soil.

Moisture acts as a lubricant between soil particles. Too little moisture will not allow soil particles to move into a dense arrangement. Too much moisture will saturate a soil taking up space which would normally be filled by soil particles.

Figure 1.6.4 *(By permission: Wacker Corporation, Menomonee Falls, Wisconsin)*

Figure 4 - Simple Soil Moisture Test

A simple method of testing whether the soil contains the right amount of moisture for compaction, is to take a handful of the soil to be compacted and squeeze it into the size and shape of a tennis ball and drop it on the ground from about 1 foot. See Figure 4. Uniform gravel or mostly sandy soil do not react well to this test.

At optimum soil moisture — The ball breaks apart into a small number of fairly uniform fragments.

If too dry — The soil does not form into a ball at all, and moisture must be added to the soil.

If too moist — The soil does not break apart (unless the soil is very sandy), and soil should be allowed to dry if possible.

1.6.5 Grain Size Distribution

Since a soil may contain different particle sizes, it is useful to know the amount of each size present in the soil. To do this, a sample of soil is dried, crumbled to separate the particles and then run through a series of standard sieves of different sizes. The amount of soil retained on each sieve is noted and calculated as a percentage of the total sample weight. The percentages obtained are plotted against sieve sizes to give a **Grain Distribution Curve** for the soil under investigation as shown in Figure 5.

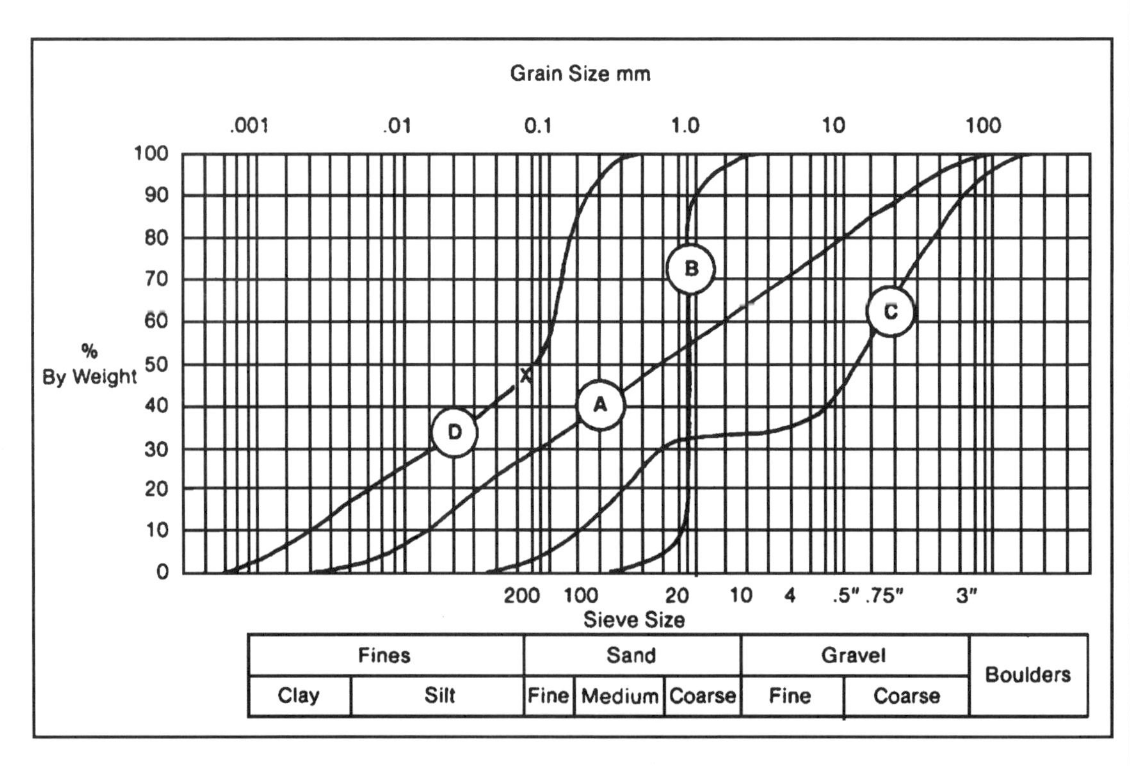

Figure 5 - Grain Distribution Curves

Figure 1.6.5 *(By permission: Wacker Corporation, Menomonee Falls, Wisconsin)*

The shape of curve so obtained gives an indication of the **gradation** of the soil. A "well graded" soil is defined as a soil which contains a broad range of grain sizes. A well graded soil is distinguished by a curve with a fairly uniform incline similar to Curve A in Figure 5.

A "poorly graded" or "uniform" soil is a soil that contains limited range of grain sizes. A steep curve is characteristic of this soil as shown in Curve B in Figure 5.

A soil that is missing certain particle sizes will have a curve with a horizontal portion as indicated in Curve C in Figure 5. Such soil is termed as "gap graded".

Point X on Curve D in Figure 5 shows that 48% by weight of that soil is finer than #200 sieve, meaning a very cohesive soil.

A well graded soil compacts to higher density than a poorly graded soil, and, therefore, has a higher load bearing capacity. This is because the finer grains can be vibrated or compacted into the cavities between the larger particles. If the fine grains were not present, those cavities would stay unfilled, resulting in air voids and lower load bearing capacity of the soil.

Response to Moisture

The response of soil to moisture is important, since the soil has to carry the load year round, rain or shine. Rain, for example, may transform soil into a plastic state or even into a liquid form. In these forms, the soil has little or no load bearing capacity.

Soil Classification Systems

Various soil classification systems exist to indicate the quality of soil as construction material. These classification systems take into consideration particle sizes, grain size distribution, and the effect of moisture on the soil. One of the soil classification systems is the Unified Soil Classification System (USC). A summary of USC is indicated in Figure 6. The meaning of the System Code Letters is also indicated in Figure 6 to make the Chart simple and easy to understand.

Group Symbol	Brief Description	Suitable as Construction Material
GW	Well graded gravels	Excellent
GP	Poorly graded gravels	Excellent to Good
GM	Silty gravels	Good
GC	Clayey gravels	Good
SW	Well graded sands	Excelllent
SP	Poorly graded sands	Good
SM	Silty sands	Fair
SC	Clayey sands	Good
ML	Inorganic silts of low plasticity	Fair
CL	Inorganic clays of low plasticity	Good to Fair
OL	Organic silts of low plasticity	Fair
MH	Inorganic silts of high plasticity	Poor
CH	Inorganic clays of high plasticity	Poor
OH	Organic clays of high plasticity	Poor
PT	Peat, mulch and high organic soils	Not Suitable
Code:	G = Gravel	W = Well Graded
	S = Sand	P = Poorly Graded
	M = Silt	L = Low Liquid Limit
	C = Clay	H = High Liquid Limit
	O = Organic	PT = Peat

Figure 6 - Unified Soil Classification

Figure 1.6.5 *(cont.)*

1.6.6 What Is Soil Compaction?

Soil compaction is the process of applying energy to loose soil to consolidate it and remove any voids, thereby increasing the density and consequently its load-bearing capacity.

Why is soil compaction necessary?

Nearly all man-made structures are ultimately supported by soil of one type or another. During the construction of a structure, the soil is often disturbed from its natural position by excavating, grading or trenching. Whenever this occurs, air infiltrates the soil mass and the soil increases in volume. Before this soil can support a structure over it or alongside it, these voids must be removed in order to be a solid mass of high strength soil.

Residential construction as well as commercial construction can benefit from Soil Compaction.

Soil Compaction provides the following benefits:

Increases Load Bearing Capacity - Air voids in the soil cause weakness and inability to carry heavy loads. With all soil particles squeezed together, larger loads can be carried by the soil because the soil particles support each other better.

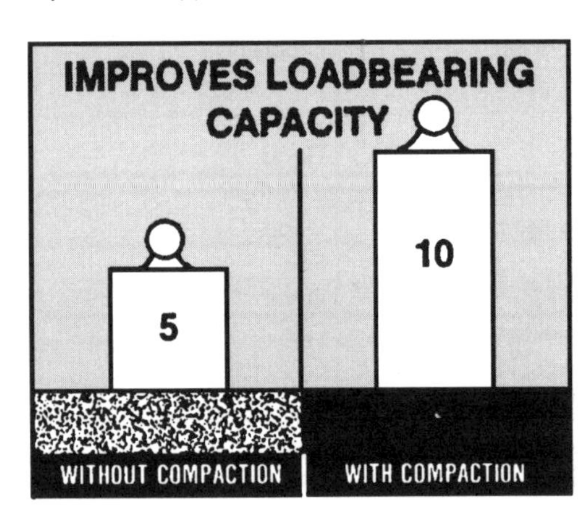

Prevents Soil Settlement - If a structure is built on uncompacted or on unevenly compacted soil, settlement of soil occurs causing the structure to deform. If the settlement is more pronounced at one side or corner, cracks or complete failure can result.

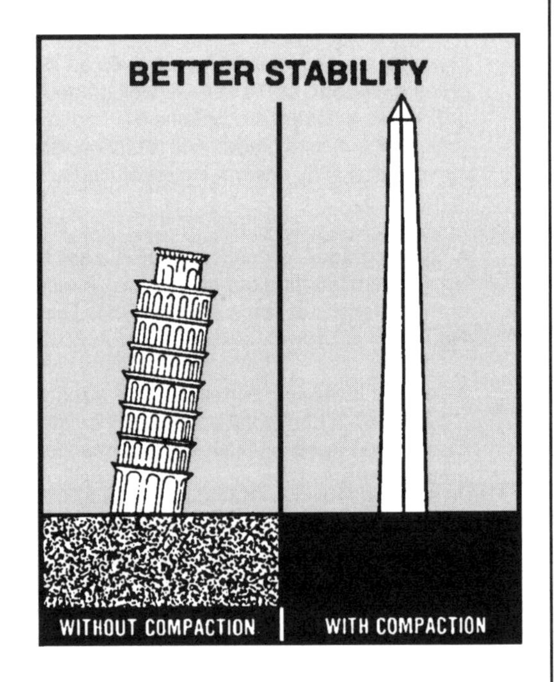

Reduces Water Seepage - Compacted soil reduces water penetration. Water flow and drainage can then be brought under control.

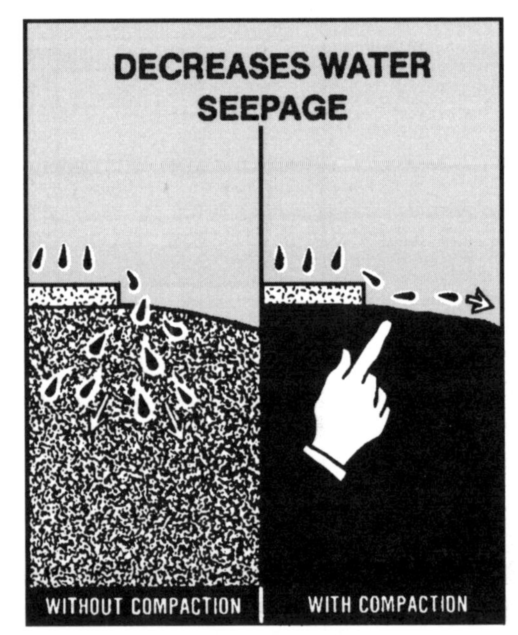

Figure 1.6.6 *(By permission: Wacker Corporation, Menomonee Falls, Wisconsin)*

Reduces swelling and contraction of soil - If air voids are present, water may penetrate the soil to fill the air voids. The result will be a swelling action of the soil during the wet period and a contraction action during the dry season.

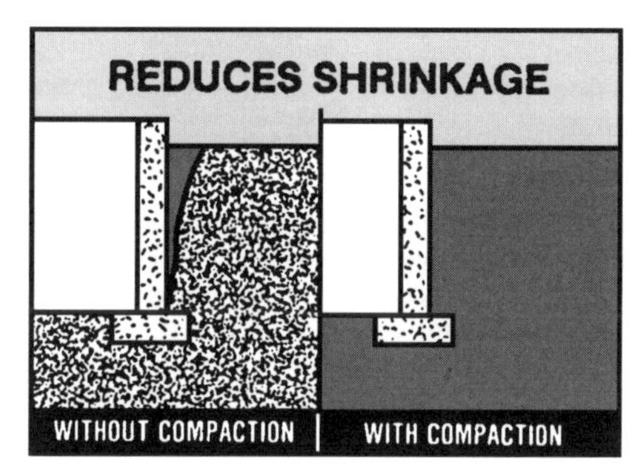

Prevents frost damage - Water expands and increases its volume upon freezing. This action often causes pavement heaving and cracking of walls and floor slabs. Compaction reduces these water pockets in the soil.

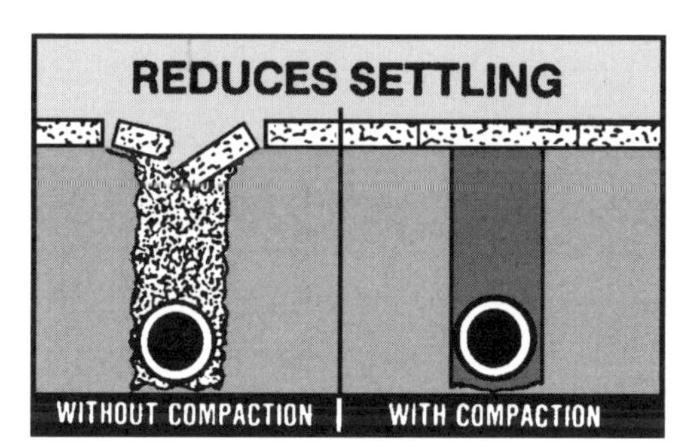

In summary, compaction should be used every time the soil is disturbed. Effective compaction means densely packed soil without any voids.

1.6.7 Methods to Compact Soil

Three major methods are used to compact soil.

1.) **Static force** - Compaction is achieved using a heavy machine whose weight squeezes soil particles together without the presence of vibratory motion. **Example**: A static roller. See Figure 7.

Figure 7 - Static Roller

2.) **Impact force** - Compaction comes from a ramming shoe alternately striking and leaving the ground at a high speed, literally "kneading" the ground to increase its density. **Example**: A rammer. See Figure 8.

Figure 8 - Wacker Rammer

3.) **Vibration** - Compaction is achieved by applying a high frequency vibration to the soil. **Example**: A vibratory plate. See Figure 9.

Figure 9 - Wacker Vibratory Plate

Figure 1.6.7 *(By permission: Wacker Corporation, Menomonee Falls, Wisconsin)*

1.6.8 Choosing the Correct Method

Granular soils are best compacted by **vibration**. This is because the vibration action reduces the frictional forces at the contact surfaces, thus allowing the particles to fall freely under their own weight. At the same time, as soil particles are set in vibration, they become momentarily separated from each other, allowing them to turn and twist until they can assume a position that limits their movements. This settling action and repositioning of particles is compaction. All the air voids that were previously present in the soil mass are now replaced by solidly packed soil.

Cohesive soils are best compacted by **impact force**. Cohesive soils do not settle under vibration, due to natural binding forces between the tiny soil particles. These soils tend to lump, forming continuous laminations with air pockets in between.

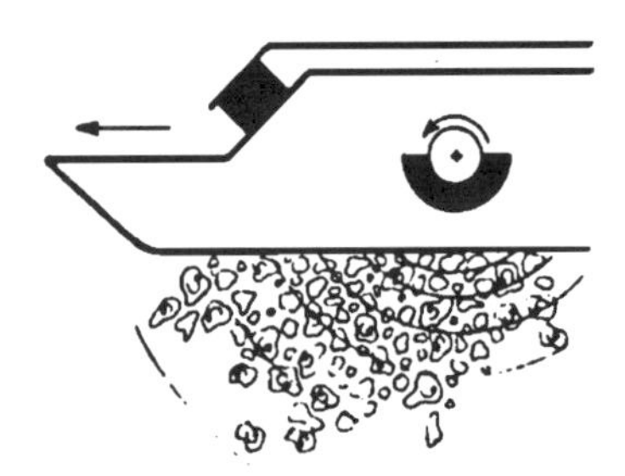

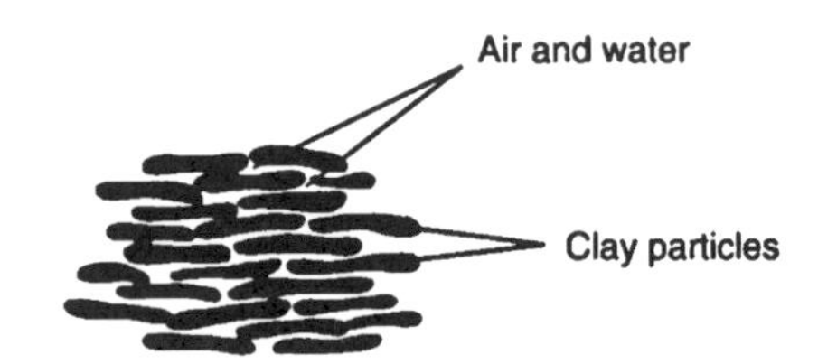

Clay particles especially present a problem because of their extremely light weight which causes the clay to become very fluid when excess moisture is present. Also, clay particles have a flat, pancake shape appearance which prevents them from dropping into voids under vibration. Therefore, cohesive soils, such as silt and clay, are effectively compacted using **impact force** which produces a shearing effect that squeezes the air pockets and excess water to the surface and moves the particles closer together.

Combinations of impact force and vibration are also used. For example, large vibratory plates and vibratory rollers combine static weight with vibration to achieve compaction.

1.6.9 Soil Testing

Compacted Soil is measured in terms of **Density** in pounds per cubic foot (lb./cu. ft.).

For Example:
Loose soil may weigh 100 pounds per cubic foot. After compaction, the same soil may have a density of 120 pounds per cubic foot. This means that by compaction, the density of the soil is increased by 20 pounds per cubic foot (PCF).

Laboratory Testing

To determine the density value of a soil from a given job site, a sample of the soil is taken to a soil test lab and a **Proctor Test** is performed. See Figure 10.

The purpose of a Proctor Test is twofold. The Proctor Test (1) measures and expresses the density attainable for any given soil as a standard; and (2) determines the effect of moisture on soil density.

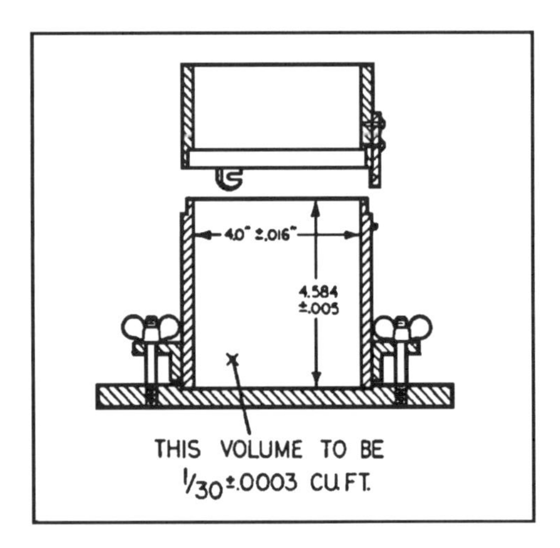

Figure 10 - Cylindrical Mold for Performing Proctor Soil Tests

1.6.10 Standard and Modified Proctor Tests

In this test, a sample of soil is compacted in a standard container 4" dia x 4.59" high which has 1/30 cubic foot capacity. The container is filled in 3 layers. Each soil layer is compacted using a 5.5 lb. weight which is lifted through a distance of 12" and dropped 25 times evenly over each soil layer, yielding a soil sample which has received a total of 12,375 ft. lb. of energy per cubic foot, determined as follows:

1 ft. x 5.5 lb. x 25 drops x 3 layers x 30 =
12,375 ft. lb./cu. ft.

After striking off the surface of the container, the soil sample is weighed immediately after the test (wet weight) and then weighed again after drying the soil in an oven (dry weight). The difference between the wet and dry weights represents the weight of water that was contained in the soil. The density of the dry soil can now be expressed in terms of lb. per cubic foot. The amount of water or moisture may also be expressed **as a percentage** of the dry weight.

The procedure is repeated, adding different amounts of water to the soil with each repetition and the soil weights as well as the percentages of moisture are recorded as described previously.

Example: For a given 1/30 cu. ft. soil sample:
Wet weight = 4.6 lb.
Dry weight = 4.0 lb.
Weight of water lost = 0.6 lb.

We then calculate:
Soil Dry Density = 4.0 lb. / 1/30 cu. ft. = 120 lb./cu. ft.
% Moisture = 0.6 lb. / 4.0 lb. x 100 = 15%

Plotting the data on a graph, a curve similar to the one shown in Figure 11 is obtained.

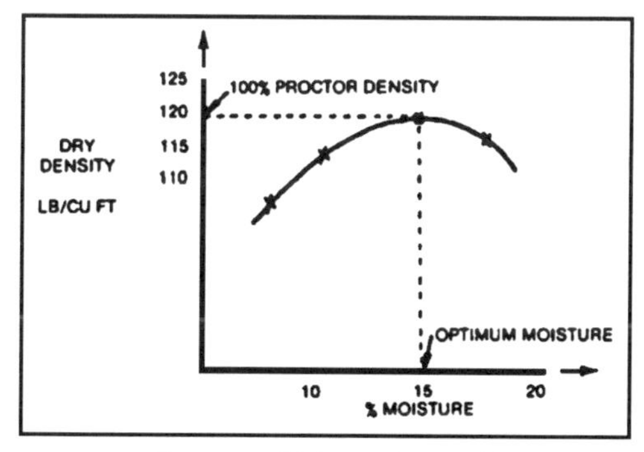

Figure 11 - Moisture Density Curve

The curve is referred to as the **Moisture-Density Curve** or **Control Curve.**

Conclusions

1. At a certain moisture, the soil reaches a maximum density when a specific amount of compaction energy is applied.
2. The maximum density reached under these conditions is called **100% Proctor density.**
3. The moisture value at which maximum density is reached is called **optimum moisture.**
4. When compacting soil at above or below optimum moisture, and using the same compacting effort, the density of the soil is less than when compacted at optimum moisture.
5. The 100% Proctor value thus obtained in a laboratory test is used as a basis **for comparing the degree of compaction of the same type of soil on the job site.**

Example:

For the soil under investigation, 100% Proctor represents a density of 120 lb. per cubic foot. (Refer to Figure 11). Assuming the same soil is compacted to a dry density of 115 lb. per cubic foot. The degree of compaction for the soil is then expressed as:

$$\frac{115}{120} \times 100 = 96\%$$

In other words, the soil is being compacted to 96% Proctor.

The above lab test for soil density was developed by R.R. Proctor, a Field Engineer for the City of Los Angeles, California, back in the early 1930's. It is now universally accepted throughout the construction industry and is known as the **Standard Proctor Test.**

The trend to construct heavy structures, such as Nuclear Power Plants and jet runways, has increased the demand for tougher compaction specifications. For those structures, a **Modified Proctor Test** was developed. The principles and procedures for both tests are very similar.

Figure 1.6.10 *(By permission: Wacker Corporation, Menomonee Falls, Wisconsin)*

1.6.11 Requirements for Standard and Modified Proctor Density Tests

Specifications	Standard Proctor	Modified Proctor
Weight of the Hammer	5.5 lb.	10 lb.
Distance of Drop	12 inches	18 inches
Number of Soil Layers	3	5
Number of Drops on Each Layer	25	25
Volume of Test Container	1/30 cu ft.	1/30 cu ft.
Energy Imparted to Soil	12,375 ft. lb. per cu. ft.	56,250 ft. lb. per cu. ft.

Each soil behaves differently with respect to maximum density and optimum moisture. Therefore, each type of soil will have its own unique control curve. See Figure 13.

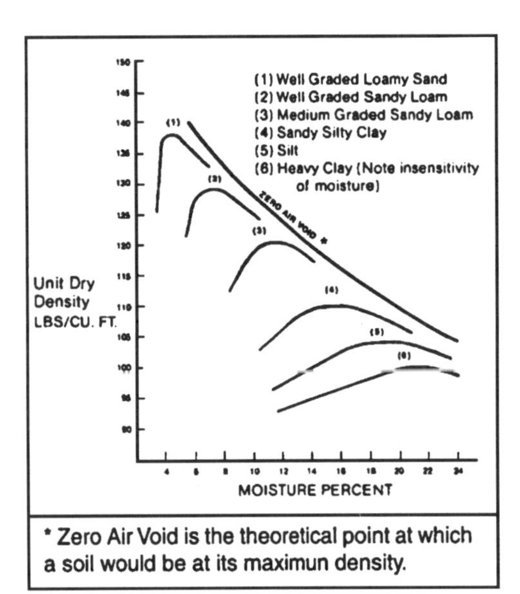

* Zero Air Void is the theoretical point at which a soil would be at its maximun density.

Figure 13 - Control Curves of Various Soils

The Proctor Test is usually conducted in the laboratory, not on job sites.

It is quite possible that a soil may be compacted to more than 100% Proctor, for example, to 104%. This is because the 100% Proctor value is obtained by using a specific amount of energy during compaction. If more energy is put into the soil, higher densities are to be expected.

The Standard Proctor Test has been adopted as AASHTO, Standard T-99 and ASTM, Standard D 698. Modified Proctor has been adopted as AASHTO, Standard T-180 and ASTM, Standard D 1557, respectively.

On establishing the Proctor curve for the soil and determining its 100% density, the architect/engineer is now in a position to specify the percent Proctor to which the soil must be compacted. Actual compaction then takes place in the field.

Field Testing Methods

There are two major methods used for field compaction testing today.

1.6.12 The Sand Cone Method

A hole about 6" wide and 6" deep is dug into the compacted soil. The soil removed from the hole is weighed, then completely dried and weighed again. The amount of water lost, divided by the dry weight gives the % moisture in the soil. A cone and jar apparatus containing special fine-grained uniform sand is placed over the hole, and the hole is filled with sand. See Figure 14. The jar is weighed before and after filling the hole, and in this way, the volume of sand required to fill the hole is determined.

Dividing the dry weight of soil removed by the volume of sand required to fill the hole gives the density of the compacted soil in lb./cu. ft.. The density so obtained is compared to the maximum density from a Proctor Test and the relative Proctor Density is obtained. The sand cone method has been in use for a long time and is well known and accepted.

Figure 1.6.12 *(By permission: Wacker Corporation, Menomonee Falls, Wisconsin)*

However, several common mistakes are sometimes made by field testers, that can render the sand cone test inaccurate.

The use of uniform sand assumes that the sand is not compactible. However, since sand particles are not completely round, job site vibration can compact the sand and the test will show a lower density than actual.

Several types of sand are used in sand cone testing, but each sand cone device is calibrated to use only one type of sand. Errors can result if the wrong type of sand is used in a particular test device.

Since a sand cone test takes several hours to perform, testing after each compaction pass is unfeasible, and too much or too little compaction is a possibility.

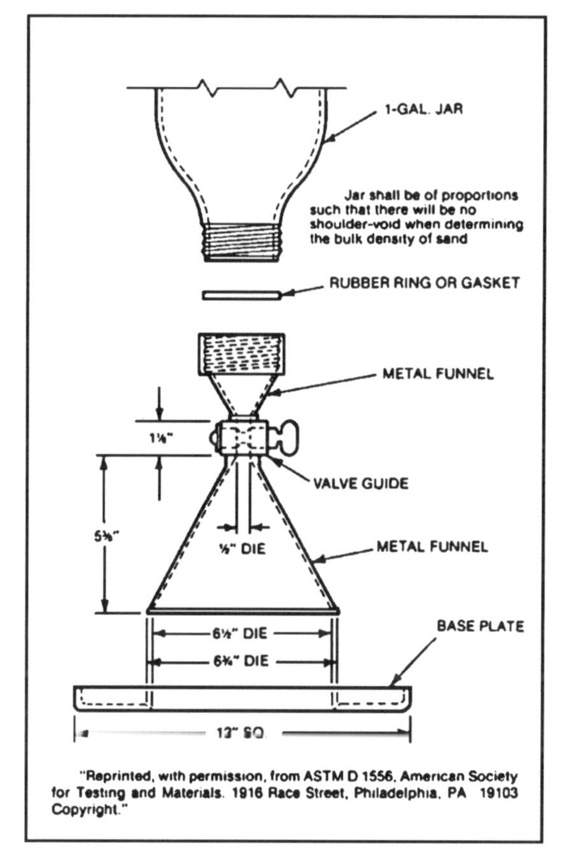

Figure 14 - Sand Cone Testing Device

1.6.13 The Nuclear Method

The Nuclear Density/Moisture Meter operates on the principle that dense soil absorbs more radiation than loose soil. The Nuclear Meter is placed directly on the soil to be tested and is turned on. See Figure 15. Gamma rays from a radioactive source penetrate the soil, and, depending on the number of air voids present, a number of the rays reflect back to the surface. These reflected rays are registered on a counter; and the counter reading visually registers the soil density in lb. per cu. ft.

This density is compared to the maximum density from a Proctor Test and the relative Proctor Density is obtained as before.

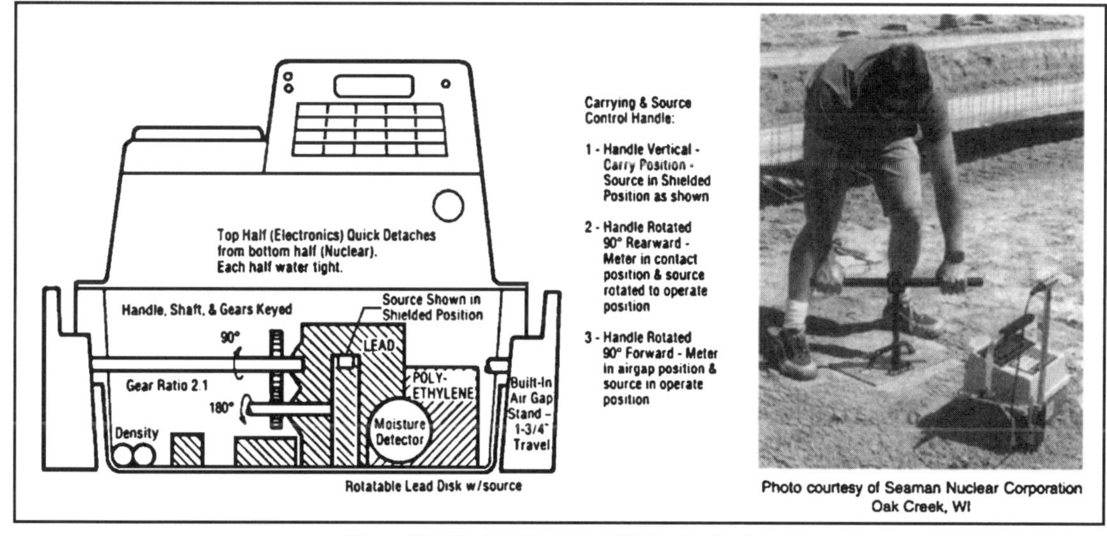

Figure 15 - Nuclear Meter and Meter Application

Figure 1.6.13 *(By permission: Wacker Corporation, Menomonee Falls, Wisconsin)*

The Nuclear Method has gained in popularity because it is accurate and fast — test results are obtained in 3 minutes — and the soil is not disturbed. Newer Nuclear Meters incorporate "quick indicator modes" for instantly checking the density after each pass the equipment makes. In addition, the Nuclear Meter method helps to quickly establish optimum compactor usage by eliminating over compaction, equipment wear and abuse, and wasted operator time. The initial cost of a Nuclear Meter can be upward of several thousand dollars; however, the time saving per test is considerable when compared to other methods.

The latest development in soil and asphalt density testing is called Density On The Run. With Density On The Run, the nuclear density meter is actually attached to the roller and a display screen is positioned near the operator for a continuous, visual read-out of the density measure. So the operator can precisely check the density of soils or hot asphalt without stopping to take a reading or rolling for miles with poor compaction results.

This compactor/density meter combination reduces roller hours, prevents over compaction, and encourages more efficient and productive man hours.

Figure 16 - Nuclear Density Meter on Soil and Asphalt
Photos courtesy of Seaman Nuclear Corporation Oak Creek, WI

Figure 1.6.13 (*cont.*)

1.6.14 Equipment Types and Selection

How does one choose the right compactor for the job? The answer is not always straightforward or simple, because a number of factors must be considered, mainly, soil type, physical conditions of the job site, compaction and specifications to be met.

The above factors must be evaluated with two purposes in mind. First, to determine which machines are able to do the job, and secondly, to recommend the one which will do the job most economically.

Let's discuss each factor separately.

Soil Type

As stated earlier, soil may be granular or cohesive in nature. See Figure 17.

Figure 17 - Granular and Cohesive Soils

Granular Soils

For **granular** soils, compaction by vibration is most effective and economical. Vibration decreases friction between soil particles allowing them to rearrange themselves downward into a tightly packed configuration, eliminating all air voids. The effect of vibration penetrates deep into the soil, meaning that large layers of soil may be compacted, which contributes to the economy of the compaction process.

Vibratory plates are the machines commonly specified for use on granular soils because they are dependable, relatively inexpensive and very productive.

Vibratory rollers are used where even higher production rates are necessary.

The various granular soils have different **Natural Resonant Frequencies**, defined as that frequency which causes the greatest soil particle motion.

The **smaller** the soil particle, the **higher** the natural frequency; the **larger** the particle, the **lower** the natural frequency. See Figure 18.

That is why a lightweight vibratory plate, of 183 lb., with a high frequency of 6250 vibrations per minute and low amplitude, is the best compactor for fine and medium sands. Other vibratory compactors with lower frequencies and higher amplitude are necessary for coarse sands, gravels, and mixes containing more cohesive particles.

For optimum compaction, a plate with a frequency approximately equal to the natural frequency of the soil particle mix being compacted should be used.

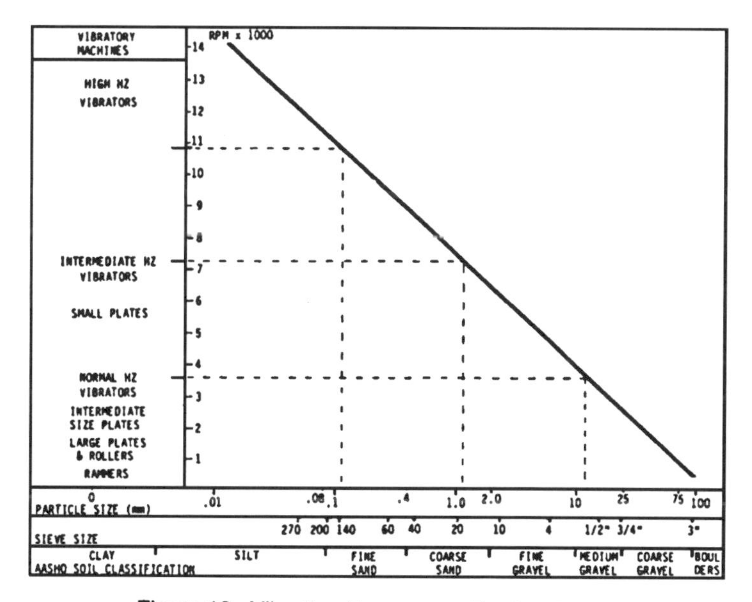

Figure 18 - Vibration Frequency - Particle Size

Figure 1.6.14 Choosing compaction equipment based on soil conditions. (*By permission: Wacker Corporation, Menomonee Falls, Wisconsin*)

1.6.15 Pea Gravel

A Word About Pea Gravel

One common misconception about Pea Gravel, a granular soil, is that it is not compactable and, therefore, does not require compaction.

Pea Gravel **is** compactable because the stones are not all perfectly round, making them subject to settling if compaction is not performed.

See Figure 19: Whether the soil settles 1/2 inch or 4 inches does not matter, as there is no support for the structure above it either way. So, be sure that Pea Gravel is always compacted.

Slurry Mixes also tend to settle in the same manner, and these mixes should NOT be specified **without** compaction.

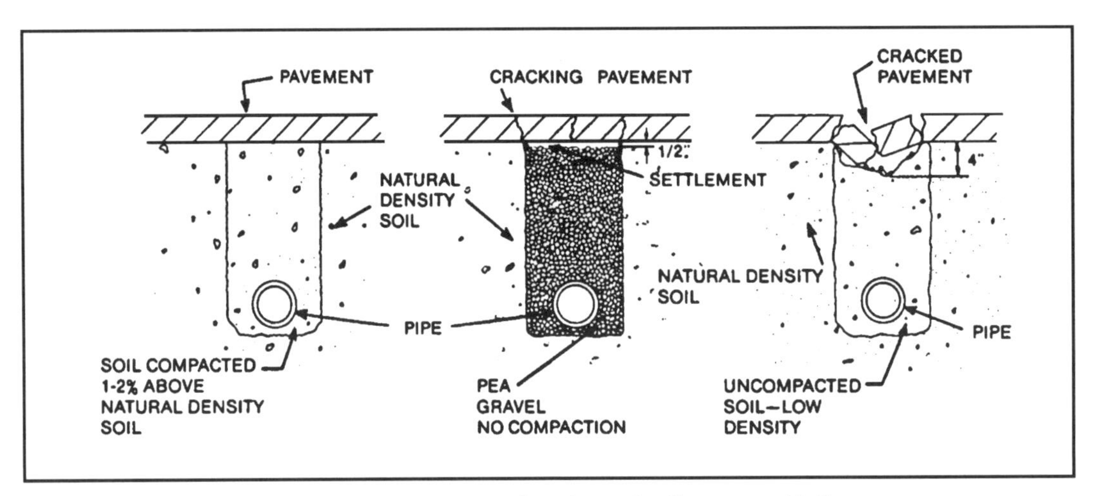

Figure 19 - Uncompacted Pea Gravel as well as Uncompacted Soil will Settle and Cause Damage to Support Structures

1.6.16 Cohesive Soils

For **cohesive** soils, impact type machines must be used. The impact force produces a shearing effect in the soil, which binds the pancake shaped particles together, squeezing air pockets out to the surface.

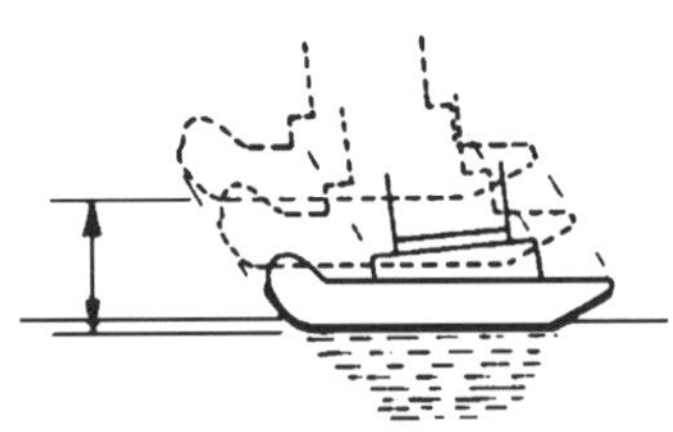

A high shoe lift of the ramming machine is very desirable in order to provide a high impact energy and to make the forward advance possible. A high rammer speed, in the range of 500–700 impacts per minute, also creates a vibratory action that is desirable with granular as well as with cohesive soils. Vibratory trench rollers with special cleated drums also perform well on cohesive soils because of their shearing action.

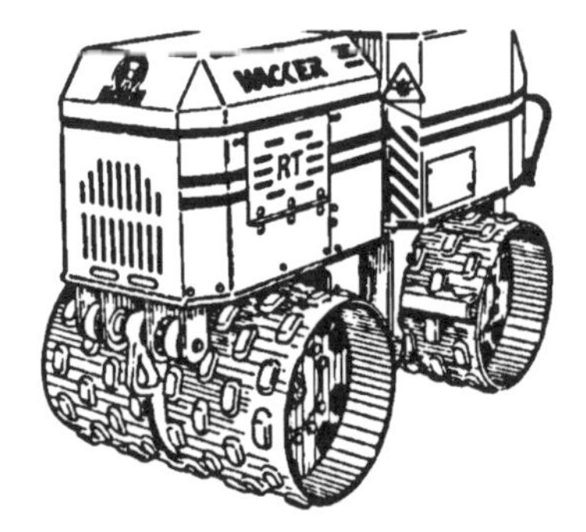

Thus, as a general rule:

For **granular soils**, the first choice should be a vibratory plate or smooth drum vibratory roller. See Figure 20. Your second choice could be a rammer for narrow trench applications. See Figure 21.

For **cohesive soils**, a rammer or vibratory trench roller should be used.

Figure 1.6.16 *(By permission: Wacker Corporation, Menomonee Falls, Wisconsin)*

1.6.17 Matching the Machine to the Job

WP 1550A Vibratory Plate

RSS 800A Vibratory Roller

Figure 20

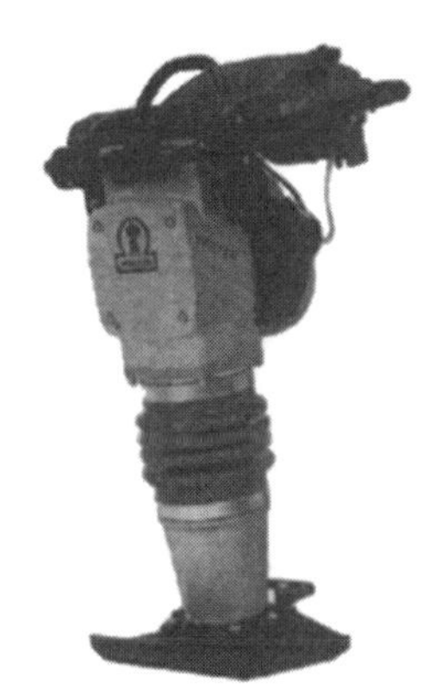

BS 600 Vibratory Rammer

DS 720 Diesel Rammer

Figure 21

Physical Conditions of the Job Site

In a trench or next to a foundation wall, the space available often determines the machine model.

A 6" wide utility trench necessitates using a rammer with a shoe size not exceeding 6" wide.

A 24" wide trench with granular fill may be compacted using a rammer or vibratory plate, with the vibratory plate being the faster machine. If there is no room at the end of the trench to turn a unidirectional plate around, then a reversible vibratory plate should be used.

On the other hand, when compacting the granular base for a large warehouse or driveway, only a large gasoline or diesel vibratory plate or vibratory roller will provide the area capacity needed to do the job in a reasonable time.

Figure 1.6.17 *(By permission: Wacker Corporation, Menomonee Falls, Wisconsin)*

1.6.18 Specifications to Be Met

Many compaction specifications are still written as the "method" type which specify the type of machine to use, the soil depth or lift, and the number of passes. Machine selection in this case, is dictated by the specifications.

Many method type specifications indicate ignorance as to the "state of the art" of modern vibratory compaction equipment. For instance, it is completely unrealistic to specify a vibratory plate with 8000 lb. of centrifugal force to run over lifts of 4", and specify a minimum of 20 passes.

These type of specs not only waste many man-hours, but drastically increase the maintenance cost of compactors and, most of all, do not achieve compaction.

Most specifications, however, are the "end result" type, allowing the use of any equipment which will achieve the specified Proctor Density.

Soil is compacted in layers which are called "lifts", and most manufacturers rate their equipment as to the maximum lift each machine can compact under ideal conditions.

In the case of "end result" specs, recommend a machine according to the soil type and job physical dimensions, and be sure it has a lift rating greater than the depth of soil layer to be compacted.

However, for modern compaction machines, the lift to be compacted should not be less than 1/3 the maximum ratod lift. If thin layers must be specified, a lighter machine must be used to achieve proper compaction.

During any compaction process, it is very important that the soil be at or as close to **optimum moisture** as possible as this will insure achieving the density required and expending the least amount of energy and making the minimum number of passes of the equipment.

Performing compaction on soil that is so dry that the compaction effort raises a cloud of dust is a waste of time and money. Such dry soil (See Figure 22) will not accept compaction energy, and even many passes will not compact the soil to an acceptable density.

All Wacker compaction equipment is designed and rated to provide 95% or better Standard Proctor Density with three to four passes, when the soil moisture is near optimum.

Figure 22 - Soil Is Too Dry For Good Compaction Results

Soil Should not be Overcompacted

As soon as the specified density is reached, compaction should be stopped. If the machine is continued to be run over a compacted area, soil particles will start to move sideways, under the effect of continued compactor pressure; thus breaking up a stable soil which results in a decrease in density.

If possible, soil testing should be started immediately after the first and each pass thereafter, continuous monitoring will eliminate the possibility of damaging the machine.

Summary

When recommending equipment for a compaction job, the soil types have to be taken into consideration.

Soil Type:

Cohesive Soils	— Use a rammer or cleated vibratory trench roller.
Granular Soils	— Use a vibratory plate or roller. — Rammers can also be used.
Mixed Soils	— Use any rammer or trench roller. Some faster vibratory plates and rollers can handle mixed materials.

Physical Dimensions and Restrictions of the Site

Match the size of the machine to the job.

Wacker offers optional narrow shoes for rammers and special narrow plates when compacting in extremely narrow areas.

Reversible vibratory plate are available for use in trenches and open areas where turning is impossible or inconvenient.

Specifications: Density requirements, site size, optimum moisture, and number of passes — Match the equipment to these special requirements.

Figure 1.6.18 *(By permission: Wacker Corporation, Menomonee Falls, Wisconsin)*

1.6.19 Vibratory Rammers

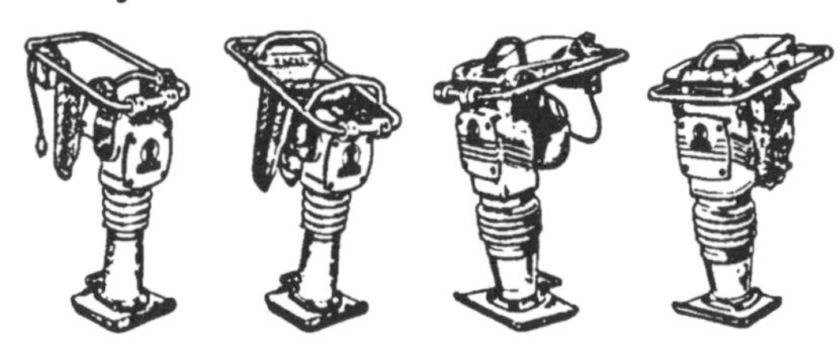

Figure 23 - Vibratory Rammers

Rammers produce an **impact force** which is necessary for the compaction of **cohesive soils**. See Figure 23.

Wacker rammers are classified as **vibratory impact rammers**. That is due to their high number of blows per minute, ranging from 450 to 800. At this high impact rate, vibration is induced in the soil. Because of their vibratory action, coupled with impact, vibratory impact rammers can also be used on **granular and mixed soils**.

An efficient rammer should provide:

1. **High impact power** (ramming shoe must come off the ground 2"–3")
2. **Good balance** (easy to guide and good shock isolation to reduce operator fatigue)
3. **Durability** (to withstand the high stresses created in the rammer)
4. **Easy maintenance**

Certain points should be considered when purchasing or comparing rammers. See Figure 24.

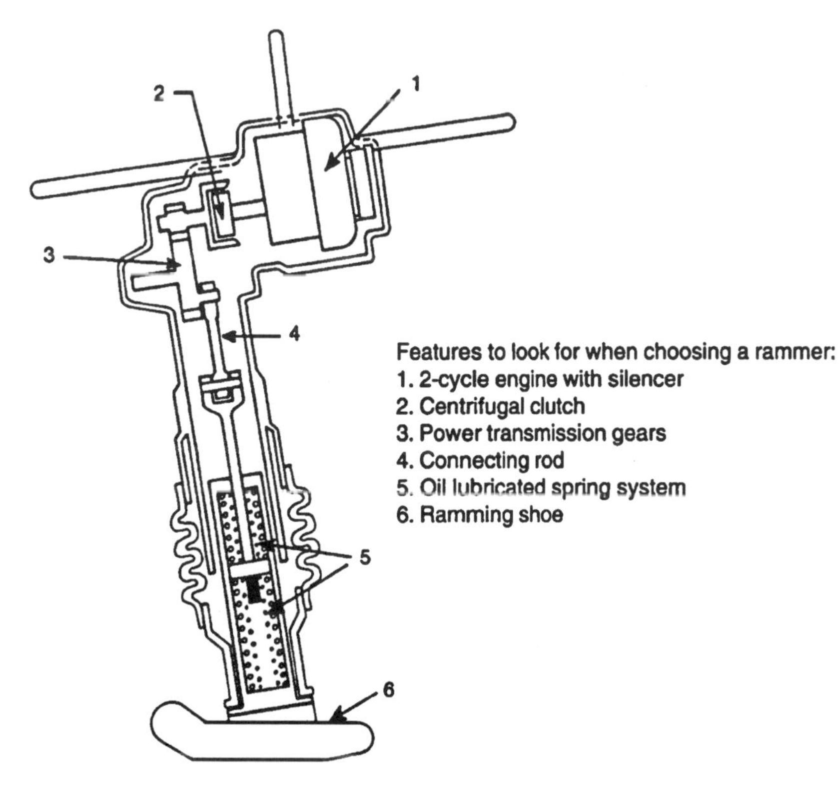

Features to look for when choosing a rammer:

1. 2-cycle engine with silencer
2. Centrifugal clutch
3. Power transmission gears
4. Connecting rod
5. Oil lubricated spring system
6. Ramming shoe

Figure 24 - Rammer Cutaway

Figure 1.6.19 *(By permission: Wacker Corporation, Menomonee Falls, Wisconsin)*

Two-Cycle Engine

Lubrication of the moving parts is provided by oil mixed with the fuel. A 4-cycle engine with the lubricating oil placed in the sump does not allow for hard impact jumps and side tilting of the machine because when tilted, the oil runs towards the cylinder and splashes against the piston causing heating, foaming and lubrication problems. Therefore, a 2-cycle engine is necessary to allow for the up and down jumping action, tilting when operating on slopes and transportation in the horizontal position. The 2-cycle engine has no valves and is light in weight.

Centrifugal Clutch

A centrifugal clutch allows easy engine starting with the ramming system disengaged, and also allows the engine to idle during short work stoppages.

Power Transmission Gears

The power which is developed by the engine must be transferred to the ramming system to produce 450 to 800 impact blows per minute. This is the function of the power transmission gears.

In all Wacker rammers, hardened gears machined from forgings do the job. Belt power transmissions are used in some rammers to reduce the cost of manufacturing; however, transmission via belts can prove to be quite costly to the end user in terms of increased maintenance and down time.

Ramming System

Also called spring system. It has two functions. One is to store the energy developed by the engine and then release it to the shoe during the downward stroke. The second is to work as an elastic buffer between the oscillating ramming shoe motion and the circular rotation of the upper mechanical components, that is the gears, clutch and engine.

Wacker spring systems are comprised of two sets of springs, with a piston in between. Each set of springs consists of two or three separate coil springs placed inside each other. Therefore, the total number of springs in any Wacker rammer is four or six springs depending on machine size. Lesser number of springs, or no springs at all, reduce the efficiency and impact power.

Oil Bath Lubrication

High quality rammers incorporate a sealed oil bath lubrication system. The oil is splashed throughout the machine, providing reliable and continuous lubrication for all internal parts. A periodic oil change is all the attention that an oil lubricated machine needs. Daily or weekly greasing is eliminated. Grease lubrication is only used in rammers which were designed decades ago.

Silencer

Equipment owners and operators as well as government agencies do not tolerate noisy equipment; therefore, most rammers are equipped with engine silencers as standard equipment.

High Impact Force

A **high impact** machine is desirable so that deeper soil lifts may be compacted. High impact force is possible only with a well designed spring system and a long shoe stroke.

A short shoe stroke reduces the impact force and prevents machine advance on slopes and unleveled soil.

Figure 1.6.19 *(cont.)*

1.6.20 Vibratory Plates

Vibratory plates apply high frequency, low amplitude vibrations to the ground, and are used mainly for compacting granular soils such as sand and gravel; mixes of granular and cohesive soil; and asphalt mixes, both hot and cold. See Figure 25.

Figure 25 - Wacker Vibratory Plate

Usually powered by small gasoline or diesel engines up to about 10 hp., some vibratory plates are available with electric motors for use where noise or fumes must be minimized.

A vibratory plate consists essentially of two masses, upper and lower. The upper mass includes the engine, centrifugal clutch and engine console. The lower mass, which includes the base plate with the vibration producing exciter unit rigidly bolted onto it.

Vibratory plates are generally high production machines, in terms of volume of soil compacted (cubic yards per hour), because of their fast forward speed, deep effective lifts and large plate contact areas. See Figure 26.

Figure 26 - Two Wacker Gasoline Vibratory Plates

Rubber shockmounts isolate the upper mass from the vibrating lower mass. The power transmission from the engine to the exciter is achieved using V-belts. The guiding handle can be mounted on the upper or lower mass and is usually rubber shockmounted to reduce operator fatigue. Figure 27 shows the main components of a vibratory plate.

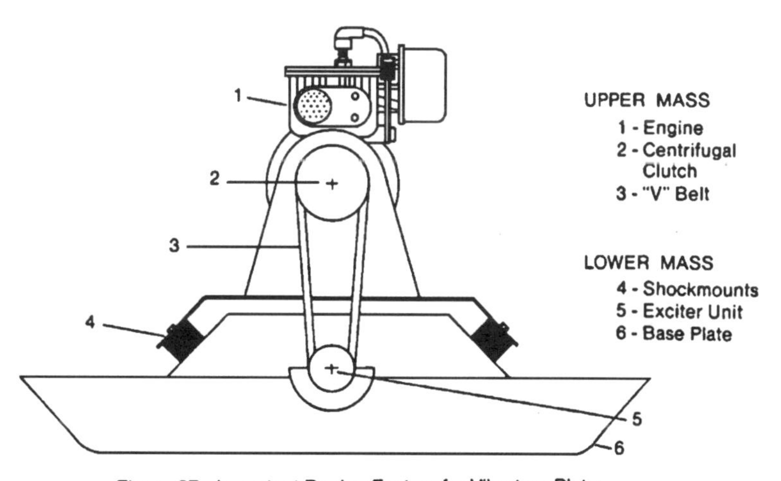

Figure 27 - Important Design Factors for Vibratory Plates

Figure 1.6.20 *(By permission: Wacker Corporation, Menomonee Falls, Wisconsin)*

Important Design Factors for Vibratory Plates

The total static weight, the exciter design, the exciter frequency, and the positioning of engine/exciter all play an important part in the efficiency and performance of the vibratory plate.

Static Weight

The static weight of a small vibratory plate (150-300 lb. weight class) is usually negligible compared to the centrifugal force that is generated in the exciter. Here, the vibratory force is the dominant force which acts on soil particles during the compaction process. For larger vibratory plates (above 300 lb.), the vibration action, as well as the static weight have a combined effect on soil particles. The total effect is to vibrate and squeeze soil particles together to achieve compaction.

Exciter Design

The exciter unit of any vibratory plate can be thought of as the "heart" of the machine. See Figure 28.

Exciter units operate on the principle of turning an unbalanced "eccentric" weight at high speed to produce centrifugal force. This centrifugal force causes the machine to vibrate, move forward and compact the soil.

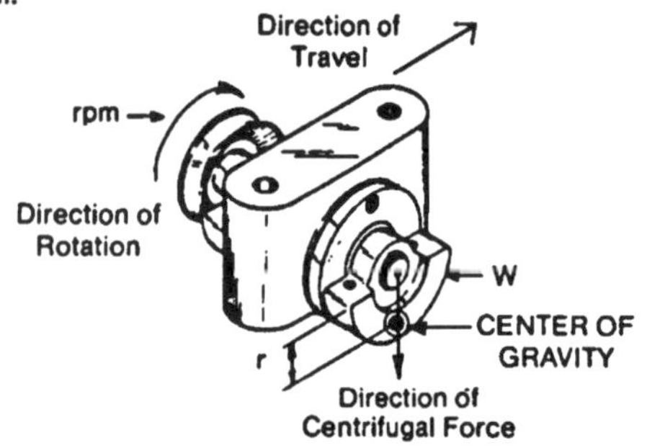

Figure 28 - Exciter Unit of a Wacker Vibratory Plate

The formula showing the centrifugal force produced by an exciter is:

Centrifugal Force $= w \times r \times rpm^2 \times K$

Where w = weight of the eccentric

r = radius from center of shaft to center of gravity of the eccentric weight

rpm^2 = speed of the exciter unit square

K = A constant factor $= \frac{2\pi}{60}$

The fact that centrifugal force produced varies with the square power of the exciter speed is of practical importance in the operation of a vibratory plate for two reasons.

If the engine is overspeeded just a little bit, centrifugal force increases a lot; which will overload the exciter bearings.

If the engine runs underspeed just a little bit, centrifugal force will be much too low, causing poor performance, slow forward speed and low compaction effort.

Therefore, it is extremely important that any vibratory plate engine be set to the manufacturer's recommended speed with a tachometer.

Exciter units can be either oil or grease lubricated. On an oil lubricated exciter, it is important that the exact amount of oil specified be contained in the unit. Too little oil will cause the bearings to burn up; too much oil will allow the exciter weights to churn the oil, causing foaming, overheating and poor performance.

Exciter Frequency

Each soil particle size responds differently to the various exciter frequencies. Laws of physics state that a small mass responds favorably to rapid vibration and that a larger mass responds favorably to slower vibration. Therefore, an attempt should be made to match the frequency of the exciter to the dominant particle size of a soil mix.

As the exciter frequency approaches the resonant frequency of particles, sympathetic vibration occurs and soil particles vibrate with maximum amplitudes.

Engine/Exciter Layout

The relative position of the exciter mounting to the engine is also an important design factor.

A centrally mounted exciter (Figure 29) is one placed in the direct center of the base plate, directly under the engine. This provides uniform amplitudes at the front and rear of the plate.

Figure 1.6.20 *(cont.)*

A front mounted exciter (Figure 29) is placed at the front of the base plate, and the engine is mounted in the rear. The amplitude at the front of the base plate is larger than that at the back. The result is faster forward speed and the ability to compact soil with a certain amount of cohesive material content. This design also allows for lower overall center of gravity which contributes to the stability of the machine.

ENGINE
EXCITER
Amplitude Run

a. Amplitude of Machine with centrally mounted exciter.

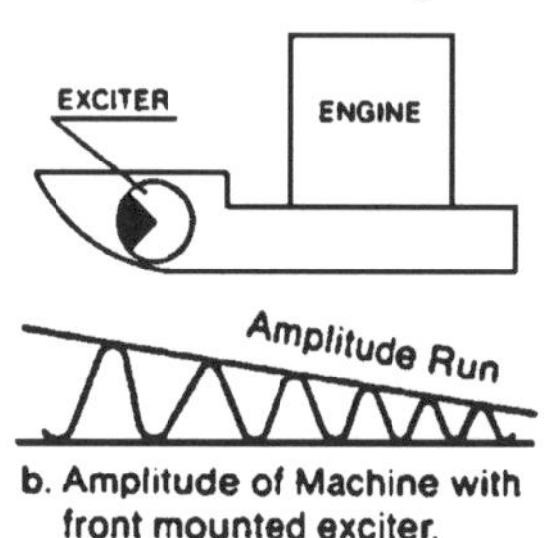

b. Amplitude of Machine with front mounted exciter.

Figure 29 - Types of Vibratory Plates

Reversibility

Some larger vibratory plates are unique due to the fact that they are **reversible**. The exciter system of a reversible vibratory plate has two eccentric weights that revolve in opposite directions. (See Figure 30).

These eccentric weights are arranged in a way that will move the plate in the opposite direction every time the relative position of one eccentric is changed 180° with respect to the other. This is done by a special spring and cam changing device that insures 180°° change in relative position with each shift without changing the direction of rotation of the two eccentric weights.

The change of direction of travel of a reversing plate occurs instantaneously at full shaft speed, without having to bring the plate to a neutral position or stop.

Modern designers use hydraulic power to change the eccentrics in infinite increments from full forward to full reverse.

This design allows a reversible plate to do spot compaction, i.e., no forward/reverse motion but the full centrifugal force of all eccentrics is used for compaction only. No energy is wasted to propel the machine.

The latest vibratory plate design features 2 pairs of reversible eccentrics in the exciter housing. These plates are fully steerable at the touch of a finger. And for additional operator safety -- some of these steerable plates are remote controlled. The operator stands on top of a trench and the unit works down in the trench. This safety feature may save costly trench shoring in many places.

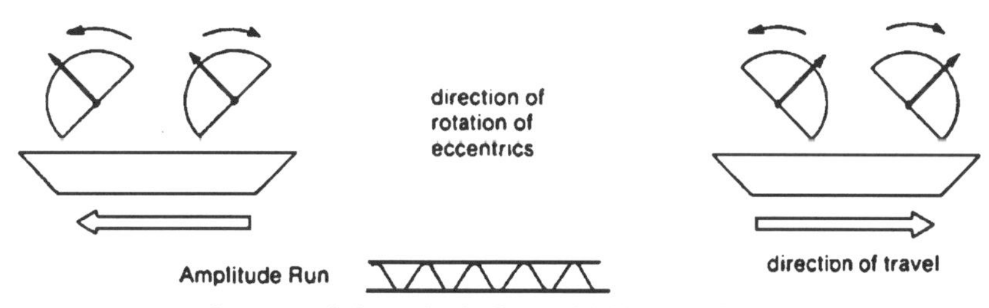

Figure 30 - Schematic of a Reversible Vibratory Plate

Diesel Power

Diesel powered vibratory plates offer several advantages over similar gasoline models. One is economy of operation in terms of both lower overall fuel consumption and lower cost per gallon of diesel fuel versus gasoline. Secondly, maintenance costs are also lower because diesels have no spark plugs, carburetor, or electrical ignition components to service. Third, one can expect longer engine life due to the sturdier construction of diesels.

A diesel powered plate has higher initial cost, although its lifetime cost per hour of operation is lower than that of a gasoline engine.

Figure 1.6.20 *(cont.)*

1.6.21 Rollers

In general, rollers may be classified as static, vibratory, sheepsfoot, or pneumatic tire.

Static rollers (Figure 31) rely on their intrinsic weight to achieve compaction. Their use for soil compaction is steadily diminishing with the introduction of vibratory rollers because the static roller must have high intrinsic weight in order to handle even moderate soil lifts. The heavy static weight means higher component costs and increased size which make handling and transportation difficult.

Static rollers, however, are still used for asphalt rolling as they provide the desirable smooth surface.

Figure 31 - Static Roller

Vibratory rollers (Figure 32) have exciter weights in one or more drums and provide vibration action (dynamic force) in addition to the static weight, vibratory rollers produce superior compaction, particularly on granular soil because the vibratory impulses break up the frictional force between soil particles, thus allowing deeper layers of soil to vibrate and settle. The vibratory action permits the use of larger lifts and provides quick and effective particle rearrangement i.e., compaction.

Figure 32 - Wacker Vibratory Roller
For Use on Soil or Asphalt

Sheepsfoot rollers, static or vibratory, (Figure 33) have drums with many protruding studs, each similar in shape to a sheepsfoot, that provide a kneading action on the soil. The total force becomes concentrated on the small protruding sheepsfoot.

These machines can effectively compact cohesive soils as they break hard soil lumps and homogenize the soil into a dense layer. Sheepsfoot rollers are sometimes used for drying areas saturated with water because they create multiple indentations in the soil, increasing the exposed surface area, thus speed up drying.

Figure 33 - Wacker Sheepsfoot Roller

On **Pneumatic tired rollers,** (Figure 34) the combination of gross static weight, number of wheels, tire size, inflation pressure and travel speed, all have a bearing on the compaction performance. The pneumatic wheels are usually arranged so that the rear ones will run in the spaces between the front ones; theoretically leaving no ruts. Pneumatic tired rollers are mainly used for compacting granular soil bases as well as for compacting asphalt surfaces.

Because of their static roller design, Pneumatic tired rollers are primarily surface compactors with effective compaction depths up to 6"

Figure 34 - Pneumatic Tired Roller

1.6.22 Other Compacting Rollers

There are other specialized rollers, such as the segmented pad and sanitary landfill type compactors; however, these are outside the scope of this booklet.

For confined area soil compaction, the double-drum vibratory rollers dominate the roller field.
See Figure 35.

Figure 35 - Wacker Double Drum Walk-behind Roller

In a roller design such as the Wacker Model W74, both drums vibrate. While one drum is in its upper motion, the other drum is moving downwards and hitting the ground; as shown in Phases A & C in Figure 36. This means that there is always one drum in contact with the ground, and that the impact force is always directed downwards. The horizontal forces in each drum (Phases B & D) are equal and opposite in direction, and, therefore, they cancel each other out.
See Figure 36.

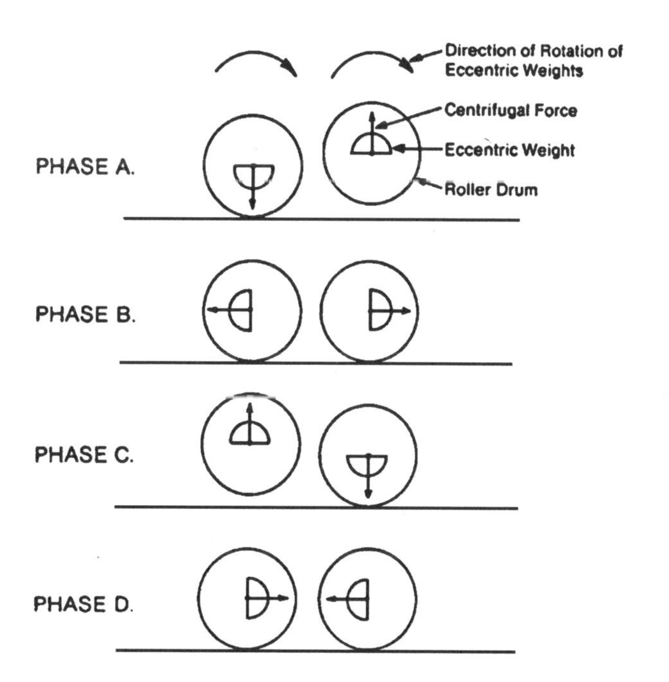

Figure 36 - Direction of Centifugal Forces from a Wacker Model W74 Vibratory Roller

Figure 1.6.22 *(By permission: Wacker Corporation, Menomonee Falls, Wisconsin)*

Because of the absence of a horizontal force, the forward motion of the machine has to be maintained through a direct gear drive, which propels both drums in both a forward and a reverse mode.

When selecting a walk-behind double drum vibratory roller for confined area soil compaction, the following specifications and parameters should be considered.

- Drum width for suitability to the application.
- Static weight for ease of handling and transport.
- Centrifugal force output for deep soil compaction.
- Dynamic linear force which is the centrifugal force per inch of the drum width.

$$\text{Dynamic Linear Force} = \frac{\text{Centrifugal force}}{\text{Number of drums} \times \text{Drum width}}$$

For the Wacker roller, Model W74, the Dynamic Linear Force is:

$$\frac{9000 \text{ lb.}}{2 \times 29.5 \text{ in.}} = 152.5 \text{ lb./in.}$$

- Static linear force which is the static force per inch of the drum width.

$$\text{Static Linear Force} = \frac{\text{Static Weight}}{\text{Number of drums} \times \text{Drum width}}$$

For the Wacker Roller, Model W74, static linear force is:

$$\frac{1863 \text{ lb.}}{2 \times 29.5 \text{ in.}} = 31.6 \text{ lb./in.}$$

Using Wacker roller, Model W74, as an example, the dynamic linear force is shown to be 5 times the static linear force. For this reason, a small vibratory roller can achieve higher compaction results than a static roller 5 times its size.

Other considerations in the selection of a roller compactor are adequate engine horsepower for reserve power in handling tough applications and grades; the sealing of all moving components from dust and moisture and the availability of a water sprinkling system for blacktop application (water is used to lubricate and clean the drum surfaces). Also keep in mind; the frequency of maintenance, accessibility, of components for service and maintenance, and the ready availability of parts and service.

The advantages of using vibratory rollers are: they are used for larger area compaction where width and speed are beneficial, they are used on inclines where vibratory plates do not have enough gradability, in trenches and around footings where the roller can propel itself without requiring lifting equipment to place it as well as on rough and uneven ground where the positive drive makes for easier travel. See Figures 37 and 38.

The designs of roller compactors are many and varied. Often the application dictates the design used. It is well to remember that it is not always the largest design that produces the most work.

Figure 37 - Double Drum Walk-behind Compacting Soil

Figure 38 - Double Drum Walk-behind Compacting Asphalt

Figure 1.6.22 *(cont.)*

1.6.23 Equipment Maintenance

Poor maintenance is the main cause of premature construction equipment failure. Because soil compaction equipment operates exclusively under conditions of heavy dust and vibration, proper machine maintenance is vital for long life.

The following are maintenance and repair tips which apply to different types of compaction equipment.

Rammers

Dust is rated as enemy number one for every engine. Under normal operating conditions, the elements will not require cleaning and should not be removed from the machine. If the elements do become plugged with dirt, the engine will begin to lose power. In this case, the air cleaner elements can be removed and cleaned as described below. Replace an element if it becomes so plugged with dirt it can no longer be cleaned.

Erratic operation or drop in rammer performance is usually caused by clogged outlet ports at engine exhaust. If this happens, clean ports and inspect the muffler. If the clogging problem occurs frequently, check fuel/oil mix. The correct ratio is 50:1. Too much oil accelerates carbon buildup and promotes spark plug fouling. To little oil causes excessive wear because of lack of lubrication.

In-tank fuel filters should be periodically cleaned with a solvent and the tank flushed to remove any sediment.

Lubrication of the rammer's spring system (lower unit) should be checked. This is done from the outside through an oil sight glass. Change system oil, first after 50 operating hours, then after each 300 operating hours.

Damaged ramming shoes and broken mufflers are often caused by loose bolts. Always tighten plow bolts on the shoe and inspect all nuts holding the muffler on engine.

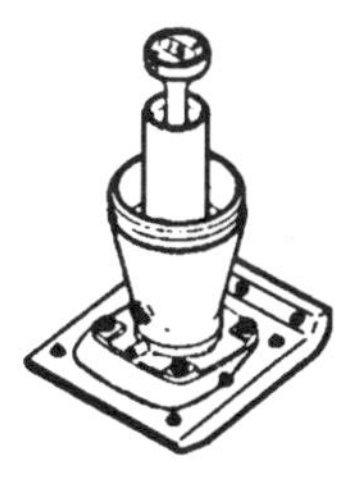

The operator must know how to operate the rammers so that the shoe hits parallel to the soil surface. Not on its toe, nor on its heel. This prolongs the service life of the shoe.

Avoid overthrottling the rammer. This damages the shoe and contributes to erratic operation and difficult handling. Best compaction is achieved when the rammer produces smooth, rhythmic tamping action.

Care should be taken when loading and unloading rammers; springs may break, if machine is dropped from a high distance on hard ground.

The 2-cycle, air-cooled rammer engine always takes a minute or two to warm up on initial start. It is during this period that carburetors are often mis-adjusted because the machines do not appear to be running properly. This practice should be avoided. However, the Wacker rammer engine has a tinker-proof carburetor and adjustments are eliminated.

Vibratory Plates

Vibratory plates are relatively simple machines and usually require little maintenance.

As with rammers, the air filter must be kept functional at all times. Inspect and clean the air filter every 8 operating hours or more often as necessary. Replace the air filter cartridge on gasoline engine driven units, as recommended in the Service Manual. Change oil in the oil bath type air filter, for diesel engines, every 8 operating hours or more frequently when operating under dusty conditions. **Never run an engine without an air filter!**

Check and change engine oil as recommended for each model. Use good quality detergent oil of the recommended viscosity to suit different seasons.

Check and change exciter oil. Do not overfill exciter with oil. Overfilled exciter causes excessive heating and foaming of oil which may result in power loss (as most of engine energy would be expended in moving and splashing the oil unnecessarily); belt overheating and failure; loss of lubricating effect of exciter oil as a result of excessive heating or bearing failure.

Figure 1.6.23 *(By permission: Wacker Corporation, Menomonee Falls, Wisconsin)*

Follow instructions in the service manual on recommended filter changes and hydraulic fluid types.

If a gasoline engine loses power, and if all other normal components such as spark plugs, points, high-tension cable, etc, check all right, then the engine combustion chamber must be cleaned of combustion deposits. Note here that unleaded gasoline may be used in vibratory plate engines with some added advantages such as a cleaner spark plug and valves, and less contamination of lubricating oil.

Inspect rubber shockmounts under engine console or at handle and replace those that are defective, i.e., cracked or hardened.

Check belt tension to prevent slipping (slack belts) and to prevent overloading of bearings and shockmounts (tight belts).

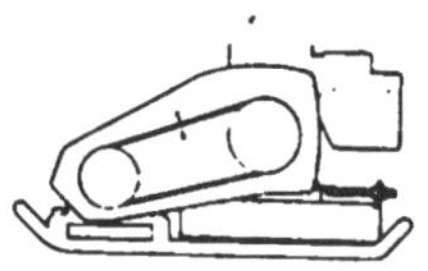

All large plates are equipped with lifting eyes for crane handling. Do not lift machines at other points to avoid breakage of machine components.

Keep base plate clean and free of soil accumulation. Dirt accumulation only increases the load unnecessarily, resulting in exciter oil overheating and a drop in performance.

Rollers

Wacker rollers are hydraulically driven.

Inspect and change engine oil regularly. Replace fuel filter at recommended intervals. This is a standard procedure for any internal combustion engine which should not be neglected.

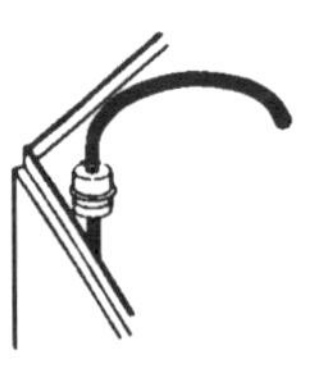

The other type of drive systems used on Wacker rollers are hydraulic or hydrostatic transmission. A hydraulic system makes use of a fixed displacement pump supplying hydraulic fluid flow to steering cylinders, drum drive motors, and exciter drive motors. Since the output of the pump is continuous, control valves are required allowing operator control of the various machine functions.

By contrast, the hydrostatic system has an infinitely variable drive. It depends on a variable displacement hydraulic pump to provide totally smooth speed control from full forward, through neutral, to full reverse.

A hydraulic system is fairly easy to maintain. Cleanliness is No. 1 when it comes to servicing hydraulic systems. **Keep dirt and other contaminants out of the system.** Key maintenance problems are usually: not enough oil in system, incorrect oil, clogged or dirty filters, and loose lines.

Figure 1.6.23 *(cont.)*

1.6.24 Compaction Costs

As cost is a major factor in any construction project, the cost of compaction must be taken into consideration.

The examples below are an indicator of how cost of compaction can be calculated and how machine performance greatly affects these costs. Because machine prices are subject to change, the example is a comparison between two vibratory Plates.

The price information is hypothetical but the performance figures and formulae are **actual** and may be used by job estimators to forecast costs of compaction.

Descrpition	Unit of Measure	Machine "A"	Machine "B"
Weight	lb.	183	267
Plate width	in.	21	21
Centrifugal force	lb.	2850	4700
Lift	in.	10	13
Speed	ft./min	66	87
Fuel consumtion	qt/h	1.6	1.7
Production rate	cu yd/h	200	366
Hypothetical list price	$	1300	1800

Let us assume that we want to compact the granular backfill around the outside foundation walls of a high rise building. The building is rectangular, 400 x 300 ft. The width of trench around the building is 10 ft. The depth of trench is 10 ft. (See Figure 39).

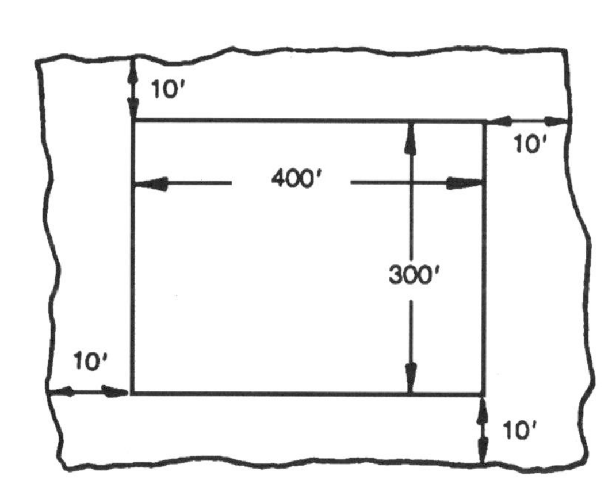

Figure 39 - Jobsite Layout

Volume of soil to be compacted:

= 2 [(300 x 10 x 10) + (420 x 10 x 10)]
= 2 x (30,000 + 42,000)
= 2 x 72,000
= 144,000 cu. ft.
or
$= \frac{144{,}000 \text{ cu. ft.}}{27}$
= 5333 cu. yd.

By using the maximum effective lift for each machine, 95-100% Proctor Density is expected to be reached after 3 passes provided that moisture is close to optimum.

To determine the total cost of owning and operating the two vibratory plates for this job, assume that:

1. Machines have a life expectancy of 3000 hours. (4 hours per day, 250 working days per year, 3 years).
2. Depreciation is determined by pro-rating the price of the machine over their 3000 hour life.
3. Maintenance cost is 80% of the depreciation cost.
4. Fuel cost is $1.20 per gallon.*
5. Labor cost is $17.00 per hour.*

*These costs will vary.

Figure 1.6.24 *(By permission: Wacker Corporation, Menomonee Falls, Wisconsin)*

Using the above assumptions, the depreciation cost is:

Machine A: $\frac{\$1300}{3000 \text{ hr.}}$ = $0.43 per hour

Machine B: $\frac{\$1800}{3000 \text{ hr.}}$ = $0.60 per hour

The maintenance cost is:

Machine A: $0.43 x 80% = $0.34 per hour

Machine B: $0.60 x 80% = $0.48 per hour

Fuel costs are:

Machine A: $\frac{\text{1.6 quarts per hour}}{\text{4 quarts per gallon}}$ x $1.20 per gallon
= $.48 per hour

Machine B: $\frac{\text{1.7 quarts per hour}}{\text{4 quarts per gallon}}$ x $1.20 per gallon
= $.51 per hour.

Thus the total for machine depreciation, maintenance and fuel is:

Machine A: $0.43 per hour
0.34 per hour
0.48 per hour
Total: $1.25 per hour

Machine B: $0.60 per hour
0.48 per hour
0.51 per hour
Total : $1.59 per hour

Using Vibratory Plate Machine "A"

Production rate per hour 200 yd³/h

Production rate per 3 passes $\frac{200}{3}$ = 66.7 yd³/h

Time needed for compacting 5333 yd³ $\frac{5333}{66.7}$ = 80 hours

Based on the equipment costs of — $1.25 per hour; and labor costs of — $17.00 per hour;
Total cost of using Machine "A" for this job: ($17.00 + $1.25) x 80 hours = $1460.

Using Vibratory Plate Machine "B"

Following the same steps as before:
Production rate per hour 366 yd³/h

Production rate per 3 passes $\frac{366}{3}$ = 122 yd³/h

Time needed for compacting 5333 yd³ $\frac{5333}{122}$ = 43.7 hours

Based on equipment cost of $1.59 per hour and labor cost of $17.00 per hour:
The total cost of using Machine "B" for this job: ($17.00 + $1.59) x 43.7 hours = $812.38
Summarizing preceding calculations:

	Machine "A"	Machine "B"
Equipment costs (Depreciation, maintenance, and fuel) per hour:	$1.25	$1.59
Labor costs per hour:	+ $17.00	+ $17.00
Total costs per hour:	$18.25	$18.59
Time required to finish job (hours)	x 80 h	x 43.70 h
Total cost for this job:	$1460.00	$812.38
Savings by using Machine "B":		$647.62

Figure 1.6.24 *(cont.)*

Note that:

1. The hourly cost of depreciation, maintenance and fuel for Machine "B" is 27.2% greater than that of Machine "A" ($1.59 vs. $1.25).
2. Machine "B" required a little more than half the time needed by Machine "A" to finish the job (43.7 hours vs. 80 hours).

The above should not lead us to believe that Machine "B" is costly to own and operate. Its high rate of production offsets the higher cost of initial investment.

Since labor constitutes the major expense of any Confined Area Compaction operation, it is, therefore, always advisable to recommend the most productive machine, rather than the least expensive one.

The smaller machine costs less in terms of ownership but more in labor. As one combines both equipment and labor costs, the total cost to finish the job is $1460 for the small machine as compared to $812.38 for the larger machine. That is to say, the smaller machine, "A", costs almost twice as much to finish this job as the larger machine, "B". This final cost is what determines the profit or loss of the compaction process.

In general, for each compaction situation, one should look at the cost of equipment, as well as the cost of labor involved, and then try to strike a balance between equipment and labor costs.

Conclusion

This booklet has reviewed soil types, soil compaction tests and the factors that influence the selection of compaction equipment. Also, the general principles and applications of rammers, vibratory plates and rollers were covered.

With this information, one should now have a better understanding of soil compaction and the equipment connected with it and should also be in a position to be able to recommend the right equipment for the job.

If additional confined area compaction application assistance is needed, please consult with the Wacker Sales Engineering Department for expert technical advice; it is free and your inquiries are always welcome.

Figure 1.6.24 *(cont.)*

1.4.0 U.S.A./Metric Sieve Sizes

This chart shows the various sieve-size openings and their metric conversions.

U.S.A. Sieve Series and Equivalents—A.S.T.M. E-11-87

Sieve Designation		Sieve Opening		Nominal Wire Diameter	
Standard (a)	Alternative	mm	in (approx. equivalents.)	mm	in (approx. equivalents.)
125 mm	5"	125	5.00"	8.00	.3150"
106 mm	4.24"	106	4.24"	6.40	.2520"
100 mm	4"(b)	100	4.00"	6.30	.2480"
90 mm	3.5"	90	3.50"	6.08	.2394"
75 mm	3"	75	3.00"	5.80	.2283"
63 mm	2.5"	63	2.50"	5.50	.2165"
53 mm	2.12"	53	2.12"	5.15	.2028"
50 mm	2"(b)	50	2.00"	5.05	.1988"
45 mm	1.75"	45	1.75"	4.85	.1909"
37.5 mm	1.5"	37.5	1.50"	4.59	.1807"
31.5 mm	1.25"	31.5	1.25"	4.23	.1665"
26.5 mm	1.06"	26.5	1.06"	3.90	.1535"
25.0 mm	1"(b)	25.0	1.00"	3.80	.1496"
22.4 mm	7/8"	22.4	0.875"	3.50	.1378"
19.0 mm	3/4"	19.0	0.750"	3.30	.1299"
16.0 mm	5/8"	16.0	0.625"	3.00	.1181"
13.2 mm	.530"	13.2	0.530"	2.75	.1083"
12.5 mm	1/2"(b)	12.5	0.500"	2.67	.1051"
11.2 mm	7/16"	11.2	0.438"	2.45	.0965"
9.5 mm	3/8"	9.5	0.375"	2.27	.0894"
8.0 mm	5/16"	8.0	0.312"	2.07	.0815"
6.7 mm	.265"	6.7	0.265"	1.87	.0736"
6.3 mm	1/4"(b)	6.3	0.250"	1.82	.0717"
5.6 mm	No. 3-1/2(c)	5.6	0.223"	1.68	.0661"
4.75 mm	No. 4	4.75	0.187"	1.54	.0606"
4.00 mm	No. 5	4.00	0.157"	1.37	.0539"
3.35 mm	No. 6	3.35	0.132"	1.23	.0484"
2.80 mm	No. 7	2.80	0.11"	1.10	.0430"
2.36 mm	No. 8	2.36	0.0937"	1.00	.0394"
2.00 mm	No. 10	2.00	0.0787"	.900	.0345"
1.70 mm	No. 12	1.70	0.0661"	.810	.0319"
1.40 mm	No. 14	1.40	0.0555"	.725	.0285"
1.18 mm	No. 16	1.18	0.0469"	.650	.0256"
1.00 mm	No. 18	1.00	0.0394"	.580	.0228"
850 μm	No. 20	0.850	0.0331"	.510	.0201"
710 μm	No. 25	0.710	0.0278"	.450	.0177"
660 μm	No. 30	0.600	0.0234"	.390	.0154"
500 μm	No. 35	0.500	0.0197"	.340	.0134"
425 μm	No. 40	0.425	0.0165"	.290	.0114"
355 μm	No. 45	0.355	0.0139"	.247	.0097"
300 μm	No. 50	0.300	0.0117"	.215	.0085"
250 μm	No. 60	0.250	0.0098"	.180	.0071"
212 μm	No. 70	0.212	0.0083"	.152	.0060"
180 μm	No. 80	0.180	0.0070"	.131	.0052"
150 μm	No. 100	0.150	0.0059"	.110	.0043"
125 μm	No. 120	0.125	0.0049"	.091	.0036"
106 μm	No. 140	0.106	0.0041"	.076	.0030"
90 μm	No. 170	0.090	0.0035"	.064	.0025"
75 μm	No. 200	0.075	0.0029"	.053	.0021"
63 μm	No. 230	0.063	0.0025"	.044	.0017"
53 μm	No. 270	0.053	0.0021"	.037	.0015"
45 μm	No. 325	0.045	0.0017"	.030	.0012"
38 μm	No. 400	0.038	0.0015"	.025	.0010"
32 μm	No. 450		0.00126"	.0011	
25 μm	No. 500		0.00098"	.001	
20 μm	No. 635		0.00079"	.0008	

(a) These standard designations correspond to the values for test sieve apertures recommended by the International Standards Organization Geneva, Switzerland.
(b) These sieves are not in the fourth root of 2 Series, but they have been included because they are in common usage.
(c) These numbers (3-1/2 to 400) are the approximate number of openings per linear inch but it is preferred that the sieve be identified by the standard designation in millimeters or microns (1000 microns = 1 mm.)

Appendix

B

Concrete Facts

Concrete is a mixture of 5 to 7 percent portland cement, 14 to 21 percent water, and 60 to 75 percent aggregate. The way in which it is mixed, worked, consolidated while being placed, and allowed to harden can create a durable, long-lasting material of construction or one that fails the test of time and fails to meet specifications. The consistency and hardening properties of ready-mix concrete can be changed by the addition of *admixtures,* which are chemicals that can increase workability, accelerate or delay curing, entrain air for better weathering capability, or adjust other properties in the mix.

Different Types of Portland Cement for Different Purposes

Type I is a general-purpose cement used for pavements, floors, reinforced concrete buildings, and precast concrete products.

Type II is used where moderate sulfate attack may take place and in structures of considerable mass—large piers and heavy abutments. This cement generates less heat during hydration and at a slower rate than if type I is used.

Type III is known as *high early* because it provides higher strength at an earlier stage of curing.

Type IV is used in concrete mixes where the heat of hydration must be minimized. As a result, this cement type develops strength over a longer period of time. This cement is used in massive concrete pours such as dams.

Type V is used in concrete mixtures that will be subjected to severe sulfate action, such as placement in soils or areas where groundwater contains a high sulfate content.

White portland cement is similar in content to type I or type III cement, but the manufacturing process is controlled so that the finished product will be white rather than the standard gray color associated with cement.

Blended Hydraulic Cements

Other by-product materials added to and blended with portland cement have been developed with an eye to energy conservation. They include such blends as these:

Type IS is granulated blast furnace slag added to portland cement.

Type IP is created by adding pozzolan with portland cement clinker or portland blast furnace slag cement to pozzolan.

Type I (PM) is pozzolan-modified portland cement.

Type S is a blend of ground granulated blast furnace slag and portland cement or combination cement, slag, and hydrated lime.

Note: The term *pozzolan* refers to a siliceous or aluminous material having no cementitious value but that can chemically react with other properties of cement to form concrete.

Mixing

Freshly mixed concrete should be plastic—pliable and capable of being molded. When it is properly mixed, each grain of sand, cement, and aggregate is encased and held in suspension. During shipment from the batch plant to the construction site, the constituents of the concrete mixture should not separate and should remain in a plastic state while being unloaded and placed at the site. The degree of plasticity can be controlled, not by the addition of water, but by the addition of admixtures known as *superplasticizers.*

Water and Its Impact on the Quality of Concrete

Reducing water in the concrete mix has many advantages:

- It increases compressive and flexural strength.
- It increases weatherability.
- There is less volume change.
- Shrinkage crack tendencies are reduced.
- There is an increased bond between successive layers and increased bond with reinforcing steel or wire mesh.
- It lowers permeability to water penetration.

Workability

The ease with which concrete is placed and finished is called its *workability. Slump* is a term used to measure the consistency of concrete, but some supervisors consider slump and workability to be one and the same. Although a low slump (thick consistency) may make for difficult placement, increased workability achieved by adding water to increase slump is not the answer. Increased water may lead to *bleeding,* where water migrates to the top surface of the freshly placed concrete as cement, sand, and aggregate settle to the bottom of the mix. One result of excessive bleeding is the creation of a weak top layer that lacks durability. Entrained air provides some increase in workability and reduces the tendency of the mix to bleed.

Why Vibration Is Used

When ready-mix concrete is placed, vibration reduces the friction between its particles and makes the mixture more fluid. Stiff mixtures benefit the most from vibrating, but when concrete is workable enough (having a much wetter consistency) to consolidate by hand rodding, vibration may not be required.

Heat of Hydration and the Hardening Process

Portland cement is not a simple compound; 90 percent of its weight is composed of tricalcium silicate, dicalcium silicate, tricalcium aluminate, and tetracalcium aluminoferrite. The chemical action between this compound and water is called *hydration.* The most important chemical reaction between cement and water occurs when a calcium silicate hydrate gel is formed and creates the setting and hardening strength and dimensional stability of the concrete mixture.

When concrete sets, its gross volume stays almost the same, but the number of pores filled with water and air does not contribute to the strength of the mix; only the paste and aggregate create the strength. Usually more water than necessary for complete hydration is added to the mixture, but reduction in water content to minimum levels is essential in creating a strong product.

Heat created as cement progresses through the hydration process can be helpful in certain instances and a problem in others. During cold weather pours, the heat of hydration helps in preventing freezing of the mix; but when massive pours occur, heat of hydration may cause undue stress during the hardening-cooling process.

Curing of Concrete

Concrete gains strength as long as any unhydrated cement remains in the mix, provided that the concrete remains moist. When relative humidity within the concrete mixture falls below 80 percent or when the temperature of the pour drops below freezing, hydration and therefore strength gain cease. That is the primary rationale behind the moist curing techniques specified by architects and engineers. Concrete must retain enough moisture during the curing period to allow all the cement to hydrate.

Concrete Admixtures

Air-Entraining Admixtures and Their Functions

The intentional entrapment of millions of tiny (1-millimeter or larger) bubbles in concrete improves resistance to freezing during exposure to water and deicing chemicals. This resistance to freeze-thaw cycles makes it a standard mixture in concrete exposed to the environment. As water freezes, it expands about 9 percent, and with no place for this expansion to be accommodated, hydraulic pressure will cause the cured concrete to flake or peel. The millions of microscopic bubbles provide areas of expansion as these small voids become filled with ice upon freezing, thereby theoretically relieving hydraulic pressure on the surrounding concrete.

The effects of entrained air on concrete properties are listed below.

Property	Effect
Abrasion	Little effect but increased strength does increase abrasion resistance somewhat
Absorption	Little effect
Bleeding	Significant reduction in bleeding
Bond to steel	Decreased
Compressive strength	Reduced 2 to 5 percent per percentage point increase in air
Freeze-thaw	Significantly improves resistance
Heat of hydration	No significant effect
Scaling	Greatly reduced
Shrinkage during drying	Little effect
Slump	Increases approximately 1 inch per ½ to 1 percent of air
Unit weight	Decreases with increased air
Workability	Increases with air

The total amounts of air content for concrete mixes are listed below.

Nominal max. aggregate size, inches	Exposure levels		
	Severe	Moderate	Mild
3/8	7½	6	4½
½	7	5½	4
¾	6	5	3½
1	6	4½	3
1½	5½	4½	2½
2	5	4	2
3	4½	3½	1½

Water-Reducing Admixtures

Recall that the quality of concrete is improved when the addition of water to the mix is reduced. Water-reducing admixtures can be used either to reduce the quantity of water required to produce a certain slump or to increase slump while reducing water/cement ratios. While regular water-reducing admixtures reduce the water content by 5 to 10 percent, there are high-range admixtures referred to as *superplasticizers* that can reduce water content by 12 to 30 percent and still provide workability. Water-reducing admixtures can often create increased cracking during the drying process, and some water-reducing admixtures can retard the concrete's setting time.

Retarding Admixtures

The introduction of a retarding admixture into the concrete mix retards the setting time. These types of admixtures are frequently used to offset the accelerating effect of the concrete pour during very hot weather, or when an exposed aggregate finish is required and delaying of the finishing process is desirable. An alternative to using retarders during hot weather pours can be achieved by requesting the ready-mix company to chill either the mixing water or the aggregate, or both, at the batch plant.

Reduction in strength during early stages of curing is another factor to be considered when you request the addition of a water-reducing admixture.

Accelerating Admixtures

These types of admixtures are used when it is desirable to accelerate the strength of the concrete mix at an early stage after the pour. Calcium chloride is one type of admixture often used for this purpose, and as any superintendent knows, chlorides attack metal over the long haul. The advantages of this admixture are outweighed by its tendency to cause increases in drying shrinkage, potential discoloration of concrete, and corrosion of embedded steel reinforcement.

Other accelerating admixtures contain chemicals such triethanolamine, sodium thiocyanate, calcium formate, and calcium nitrite or nitrate.

Superplasticizers

A form of water-reducing admixture, superplasticizers when added to low-slump concrete create the same workability as high-slump concrete. Similar to water-reducing admixtures, superplasticizers are more effective, but more expensive. A superplasticizer added to 3-inch slump concrete can produce the same effect as a 9-inch slump. However, the increased flowability characteristic is short-lived, often no more than 30 to 60 minutes; and this admixture, as opposed to others which are added in the batch plant, is generally added to the concrete at the job site. The flowability characteristic of this admixture is especially helpful

1. When the pour is relegated to thin section placements
2. Where there are closely spaced areas of reinforcing steel and the concrete must flow around these congestions
3. When the Tremie pipe delivery system is used in underwater concrete placement
4. When concrete is being pumped a long distance and reduced pump pressure is required

Concrete Strength

Tests for compressive strength are the tests most familiar to project superintendents, generally expressed in terms of pounds per square inch (abbreviated psi) when the test cylinders have been properly aged for 28 days. Generally test cylinders are made so that they can be broken in 7 days, 14 days, or 28 days; and sometimes extra cylinders are created in case a 56-day break is required. Reports of these tests are quickly prepared and sent to the field in order to properly monitor the quality of all future pours. But when are test results of early cylinder breaks in the acceptable range?

According to the Building Code Requirement for Reinforced Concrete as set forth in American Concrete Institute (ACI) Reference 318, if the average of all sets of three consecutive strength tests equal or exceed the specified 28-day strength requirement, and if no individual cylinder's test result is more than 500 pounds per square inch below design strength, then the compressive strength of the concrete can be considered satisfactory.

Recommended Slumps for Various Types of Concrete Construction

Unreinforced footings and substructure walls	3
Reinforced foundation walls and footings	3
Slabs	3
Pavements	3
Beams and reinforced walls	4
Columns (building)	4

Allowable Tolerances in Cast-in-Place Concrete

The American Concrete Institute in its *Manual of Concrete Practice* states that no structures are exactly level, plumb, or precisely straight and true, but some variation, or tolerance, is required to provide both the designer and the contractor with acceptable levels of performance.

With respect to cast-in-place concrete:

Plumb: In lines and surfaces of columns, piers, and walls

In any 10 feet	¼ inch
Maximum for total height of structure when structure does not exceed 100 feet	1 inch

For exposed corner columns and other conspicuous lines

In any 20 feet	¼ inch
Maximum for total height of structure when total height does not exceed 100 feet	½ inch

Level: From grades and elevations specified

Slab soffits, ceilings, beam soffits, measured before removal of supporting shores

In any 10 feet	±¼ inch
In any bay or in any 20 feet	±⅜ inch
Maximum for total length of structure	±¾ inch

Exposed sills, lintels, parapet, horizontal grooves

In any bay or in 20 feet	±¼ inch
Maximum for total length of structure	±½ inch

Cross-sectional dimensions of columns, beams, walls, and slab thicknesses:

Up to 12 inches	+⅜ inch
	−¼ inch
More than 12 inches	+½ inch
	−⅜ inch

Footings:

Horizontal dimensions (formed)	+2 inches
	−½ inch
Horizontal dimensions (unformed)	+3 inches
To receive masonry construction	
Alignment in 10 feet	±⅓ inch
Maximum for entire length—50 feet	±½ inch
Level in 10 feet	±¼ inch
Maximum for entire length—50 feet	±½ inch

Slabs based upon class of tolerance:

Class AA. Depressions in slabs between high spots shall be not greater than ⅛ inch below a 10-foot straightedge.

Class AX. Depressions in slabs between high spots shall be not greater than 3⁄16 inch below a 10-foot straightedge.

Class BX. Depressions in slabs between high spots shall not be greater than 5⁄16 inch below a 10-foot straightedge.

Class CX. Depressions in slabs between high spots shall not be greater than ½ inch below a 10-foot straightedge.

What 1 Cubic Yard of Concrete Will Place

Thickness, inches	Square feet
2	162
2½	130
3	108
3½	93

4	81
4½	72
5	65
5½	59
6	54
6½	50
7	46
7½	43
8	40
8½	38
9	36
9½	34
10	32.5
10½	31
11	29.5
11½	28
12	27
15	21.5
18	18
24	13.5

And remember that adding 1 gallon of water to 1 yard of 3000-psi concrete

Increases slump by about 1 inch

Reduces compressive strength by as much as 200 pounds per square inch

Wastes the effect of ¼ bag of cement

Increases potential for shrinkage by about 10 percent

Decreases freeze-thaw resistance by 20 percent

Decreases resistance to attack by deicing salts

Appendix

C

Hot-Weather and Cold-Weather Concreting, Nondestructive Methods for Testing Strength of Hardened Concrete, and Control of Cracking*

*This appendix is made up of pp. 10, 130 to 150, and 172 to 175 from S. H. Kosmatka and W. C. Panarese, *Design and Control of Concrete Mixtures*, 13th ed., Portland Cement Association, 1988; reprinted by permission of Portland Cement Association, Skokie, Ill., www.portcement.org.

CHAPTER 11

Hot-Weather Concreting

Weather conditions at a jobsite—hot or cold, windy or calm, dry or humid—may be vastly different from the optimum conditions assumed at the time a concrete mix is specified, designed, or selected. Hot weather can create difficulties in fresh concrete, such as

—increased water demand
—accelerated slump loss
—increased rate of setting
—increased tendency for plastic cracking
—difficulties in controlling entrained air
—critical need for prompt early curing

Adding water to the concrete at the jobsite can adversely affect properties and serviceability of the hardened concrete, resulting in

—decreased strength
—decreased durability and watertightness
—nonuniform surface appearance
—increased tendency for drying shrinkage

Only by taking precautions to alleviate these difficulties in anticipation of hot-weather conditions can concrete work proceed smoothly.

WHEN TO TAKE PRECAUTIONS

The most favorable temperature for freshly mixed concrete is lower during hot weather than can usually be obtained without artificial cooling. A concrete temperature of 50°F to 60°F is desirable but not always practical. Many specifications require only that concrete when placed should have a temperature of less than 85°F or 90°F.

For most work it is impractical to limit the maximum temperature of concrete as placed because circumstances vary widely. A limit that would serve successfully at one jobsite could be highly restrictive at another. For example, flatwork done under a roof with exterior walls in place could be completed at a concrete temperature that would cause difficulty were the same concrete placed outdoors on the same day.

The effects of a high concrete temperature should be anticipated and the concrete placed at a temperature limit that will allow best results in hot-weather conditions, probably somewhere between 75°F and 100°F. The limit should be established for conditions at the jobsite based on trial-batch tests at the limiting temperature rather than at ideal temperatures.

EFFECTS OF HIGH CONCRETE TEMPERATURES

As concrete temperature increases there is a loss in slump that is often unadvisedly compensated for by adding more water at the jobsite. At higher temperatures a greater amount of water is required to hold slump constant than is needed at lower temperatures. Adding water *without* adding cement results in a higher water-cement ratio, thereby lowering the strength at all ages and adversely affecting other desirable properties of the hardened concrete. This is in addition to the adverse effect on strength at later ages due to the higher temperature even without the addition of water.

As shown in Fig. 11-1, if the temperature of freshly mixed concrete is increased from 50°F to 100°F, about 33 lb of additional water is needed per cubic yard of concrete to maintain the same 3-in. slump. This addi-

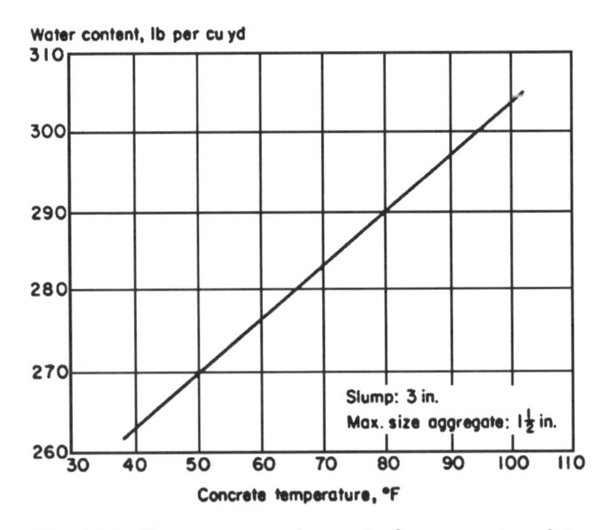

Fig. 11-1. The water requirement of a concrete mixture increases with an increase in concrete temperature.

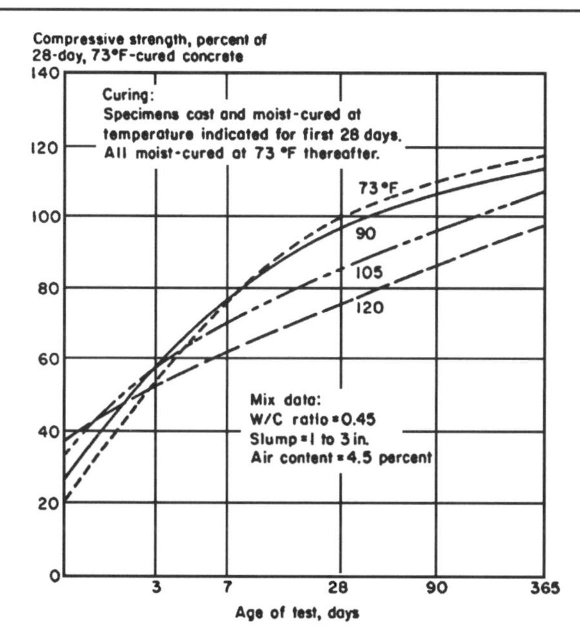

Fig. 11-2. Effect of high concrete temperatures on compressive strength at various ages.

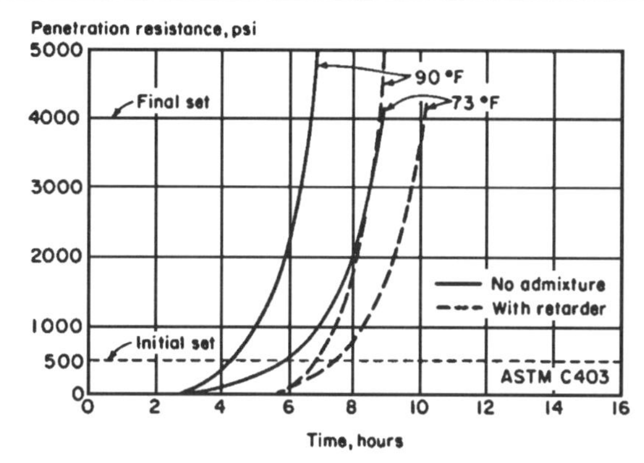

Fig. 11-3. Effect of concrete temperature and retarder on setting time.

tional water could reduce strength by 12% to 15% and produce a compressive strength cylinder test break of less than the specified compressive strength (f'_c).

Fig. 11-2 shows the effect of high initial concrete temperatures on compressive strength. The concrete temperatures at time of mixing, casting, and curing were 73°F, 90°F, 105°F, and 120°F. After 28 days, the specimens were all moist-cured at 73°F until the 90-day and one-year test ages. The tests, using identical concretes of the same water-cement ratio, show that while higher concrete temperatures give higher early strength than concrete at 73°F, at later ages concrete strengths are lower. If the water content had been increased to maintain the same slump (without increasing cement content), the reduction in strength would have been even greater than shown.

Besides reducing strength and increasing the mixing-water requirement, high temperatures of freshly mixed concrete have other harmful effects. Setting time is significantly reduced—high temperatures increase the rate of concrete hardening and shorten the length of time within which the concrete can be transported, placed, and finished. Setting time can be reduced by 2 or more hours with a 20°F increase in concrete temperature (Fig. 11-3). Concrete should remain plastic sufficiently long so that each layer can be placed without development of cold joints or discontinuities in the concrete. Retarding admixtures, ASTM C 494 Type B, can be beneficial in offsetting the accelerating effects of high temperature.

In hot weather, the tendency for cracks to form is increased both before and after hardening. Rapid evaporation of water from freshly placed concrete can cause plastic-shrinkage cracks before the surface has hardened (discussed in more detail later in this chapter). Cracks may also develop in the hardened concrete because of increased drying shrinkage due to a higher water content or thermal volume changes at the surface due to cooling.

Air entrainment is also affected in hot weather. At elevated temperatures, an increase in the amount of air-entraining admixture is required to produce a given air content.

Because of the detrimental effects of high concrete temperatures, operations in hot weather should be directed toward keeping the concrete as cool as is practicable.

COOLING CONCRETE MATERIALS

The usual method of cooling concrete is to lower the temperature of the concrete materials before mixing. One or more of the ingredients should be cooled. In hot weather the aggregates and water should be kept as cool as practicable, as these materials have a greater influence on temperature after mixing than other components.

The contribution of each material in a concrete mixture to the initial temperature of concrete is related to the temperature, specific heat, and quantity of each material. Fig. 11-4 shows graphically the effect of temperature of materials on the temperature of fresh concrete. It is evident that although concrete temperature is primarily dependent upon the aggregate temperature, it can be effectively lowered by cooling the mixing water.

The approximate temperature of concrete can be calculated from the temperatures of its ingredients by using the following equation:

$$T = \frac{0.22(T_a W_a + T_c W_c) + T_w W_w + T_{wa} W_{wa}}{0.22(W_a + W_c) + W_w + W_{wa}}$$

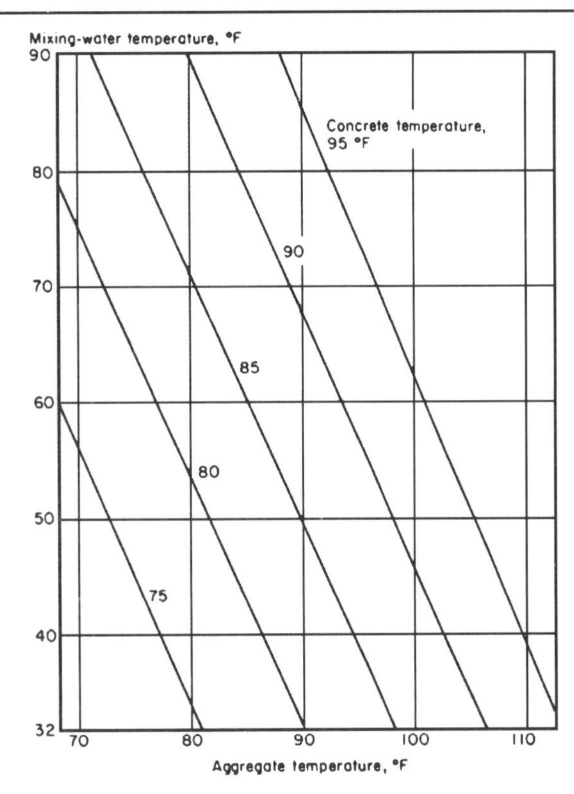

Fig. 11-4. Temperature of freshly mixed concrete as affected by temperature of its ingredients. Although the chart is based on the following mixture, it is reasonably accurate for other typical mixtures:

Aggregate	**3000 lb**
Moisture in aggregate	**60 lb**
Added mixing water	**240 lb**
Cement at 150°F	**564 lb**

where

T = temperature of the freshly mixed concrete

T_a, T_c, T_w, and T_{wa} = temperature of aggregates, cement, added mixing water, and free water on aggregates, respectively

W_a, W_c, W_w, and W_{wa} = weight of aggregates, cement, added mixing water, and free water on aggregates, respectively

Example calculations for initial concrete temperature are shown in Table 11-1A.

Of the materials in concrete, water is the easiest to cool. Even though it is used in smaller quantities than the other ingredients, cold water will effect a moderate reduction in the concrete temperature. Mixing water from a cool source should be used. It should be stored in tanks that are not exposed to the direct rays of the sun. Tanks and pipelines carrying the mixing water should be buried, insulated, shaded, or painted white to keep water at the lowest practical temperature.

Fig. 11-5. Liquid nitrogen water-cooling installation at a ready mixed concrete plant.

Water can be cooled by refrigeration, liquid nitrogen, or ice. Fig. 11-5 shows a liquid nitrogen water-cooling installation at a ready mixed concrete plant. Liquid nitrogen can also be added directly into a central mixer drum or the drum of a truck mixer to lower concrete temperature. Ice can be used as part of the mixing water provided it is completely melted by the time mixing is completed. When using crushed ice, care must be taken to store it at a temperature that will prevent the formation of lumps.

When ice is added as part of the mixing water, the effect of the heat of fusion of the ice must be considered, so the equation for temperature of fresh concrete is modified as follows:

$$T = \frac{0.22(T_a W_a + T_c W_c) + T_w W_w + T_{wa} W_{wa} - 112 W_i}{0.22(W_a + W_c) + W_w + W_{wa} + W_i}$$

where W_i is the weight in pounds of ice.

The heat of fusion of ice in British thermal units is 144 Btu per pound. Calculations in Table 11-1B show the effect of 75 lb of ice in reducing the temperature of concrete. Crushed or flaked ice is more effective than chilled water in reducing concrete temperature. The

Table 11-1.

A. Effect of Temperature of Materials on Initial Concrete Temperatures

Material	Weight, W, lb/cu yd	Specific heat	Btu to vary temperature 1°F	Initial temperature of material, T, °F	Total Btu's in material
	(1)	(2)	(3) Col. 1 × Col. 2	(4)	(5) Col. 3 × Col. 4
Cement	564 (W_c)	0.22	124	150 (T_c)	18,600
Water	282 (W_w)	1.00	282	80 (T_w)	22,560
Total aggregate	3100 (W_a)	0.22	682	80 (T_a)	54,560
			1088		95,720

Initial concrete temperature $= \frac{95,720}{1088} = 88.0°F$

To achieve 1°F reduction in initial concrete temperature:

Cement temperature must be lowered $= \frac{1088}{124} = 9°F$

Or water temperature dropped $= \frac{1088}{282} = 3.8°F$

Or aggregate temperature cooled $= \frac{1088}{682} = 1.6°F$

B. Effect of Ice (75 lb) on Temperature of Concrete

Material	Weight, W, lb/cu yd	Specific heat	Btu to vary temperature 1°F	Initial temperature of material, T, °F	Total Btu's in material
	(1)	(2)	(3) Col. 1 × Col. 2	(4)	(5) Col. 3 × Col. 4
Cement	564 (W_c)	0.22	124	150 (T_c)	18,600
Water	207 (W_w)	1.00	207	80 (T_w)	16,560
Total aggregate	3100 (W_a)	0.22	682	80 (T_a)	54,560
Ice*	75 (W_i)	1.00	75	32 (T_i)	2,400
			1088		
minus	75 (W_i) × heat of fusion, 144 Btu/lb =				−10,800
					81,320

Concrete temperature $= \frac{81,320}{1088} = 74.7°F$

*$32 W_i - 144 W_i = -112 W_i$.

amount of water and ice must not exceed the total mixing-water requirements. Fig. 11-6 shows crushed ice being charged into a truck mixer prior to the addition of other materials.

Aggregates have a pronounced effect on the fresh concrete temperature because they represent 70% to 85% of the total weight of concrete. To lower the temperature of concrete 10°F requires only a 15°F reduction in the temperature of the aggregates.

There are several simple methods of keeping aggregates cool. Stockpiles should be shaded from the sun and kept moist by sprinkling. Since evaporation is a cooling process, sprinkling provides effective cooling, especially when the relative humidity is low.

Sprinkling of coarse aggregates should be adjusted to avoid producing excessive variations in the free moisture content and thereby causing a loss of slump uniformity. Refrigeration is another method of cooling materials. Aggregates can be immersed in cold-water tanks, or cooled air can be circulated through storage bins. Vacuum cooling can reduce aggregate temperatures to as low as 34°F.

Fig. 11-6. Substituting ice for part of the mixing water will substantially lower concrete temperature. A crusher delivers finely crushed ice to a truck mixer reliably and quickly.

Cement temperature has only a minor effect on the temperature of the freshly mixed concrete because of cement's low specific heat and the relatively small amount of cement in the mixture. A cement temperature change of 10°F generally will change the concrete temperature by only 1°F. Because cement loses heat slowly during storage, it may be warm when delivered. (This heat is produced in grinding the cement clinker during manufacture.) Since the temperature of cement does affect the temperature of the fresh concrete to some extent, some specifications place a limit on its temperature at the time of use. The ACI 305 recommended maximum temperature at which cement should be used in concrete is 170°F. However, it is preferable to specify a maximum temperature for freshly mixed concrete rather than place a temperature limit on individual ingredients.

PREPARATION BEFORE CONCRETING

Before concrete is placed, certain precautions should be taken during hot weather to maintain or reduce concrete temperature. Mixers, chutes, belts, hoppers, pump lines, and other equipment for handling concrete should be shaded, painted white, or covered with wet burlap to reduce solar heat.

Forms, reinforcing steel, and subgrade should be fogged or sprinkled with cool water just before the concrete is placed. Fogging the area during placing and finishing operations cools the contact surfaces and surrounding air and increases its relative humidity. This reduces the temperature rise of the concrete and minimizes the rate of evaporation of water from the concrete after placement. For slabs on ground, it is a good practice to moisten the subgrade the evening before concreting. There should be no standing water or puddles on forms or subgrade at the time concrete is placed.

During extremely hot periods, improved results can be obtained by restricting concrete placement to early morning, evening, or nighttime hours, especially in arid climates. This practice has resulted in less thermal shrinkage and cracking of thick slabs and pavements.

TRANSPORTING, PLACING, FINISHING

Transporting and placing concrete should be done as quickly as is practical during hot weather. Delays contribute to loss of slump and an increase in concrete temperature. Sufficient labor and equipment must be available at the jobsite to handle and place concrete immediately upon delivery.

Prolonged mixing, even at agitating speed, should be avoided. If delays occur, the heat generated by mixing can be minimized by stopping the mixer and then agitating intermittently. The Standard Specification for Ready Mixed Concrete (ASTM C94) requires that discharge of concrete be completed within 1½ hours or before the drum has revolved 300 times, whichever occurs first. During hot weather the time limit can be reasonably reduced to 1 hour or even 45 minutes.

Since concrete hardens more rapidly in hot weather, extra care must be taken with placement techniques to avoid cold joints. For placement of walls, shallower layers can be specified to assure enough time for consolidation with the previous lift. Temporary sunshades and windbreaks help to minimize cold joints.

Floating should be done promptly after the water sheen disappears from the surface or when the concrete can support the weight of a finisher. Finishing on dry and windy days requires extra care. Rapid drying of the concrete at the surface may cause plastic shrinkage cracking.

PLASTIC SHRINKAGE CRACKING

Plastic shrinkage cracks are cracks that sometimes occur in the surface of freshly mixed concrete soon after it has been placed and while it is being finished (Fig. 11-7). These cracks appear mostly on horizontal surfaces and can be substantially eliminated if preventive measures are taken.

Plastic shrinkage cracking is usually associated with hot-weather concreting; however, it can occur any time ambient conditions produce rapid evaporation of moisture from the concrete surface. Such cracks occur when water evaporates from the surface faster than it can appear at the surface during the bleeding process. This creates rapid drying shrinkage and tensile stresses in the surface that often result in short, irregular cracks. The following conditions, singly or collectively, increase evaporation of surface moisture and increase the possibility of plastic shrinkage cracking:

1. High air and concrete temperature
2. Low humidity
3. High winds

Fig. 11-7. Typical plastic shrinkage cracks.

The crack length is generally from a few inches to a few feet in length and they are usually spaced in an irregular pattern from a few inches to 2 ft apart. Fig. 11-8 is useful for determining when precautionary measures should be taken. There is no way to predict with certainty when plastic shrinkage cracking will occur. However, when the rate of evaporation exceeds 0.2 lb per square foot per hour, precautionary measures are almost mandatory. Cracking is possible if the rate of evaporation exceeds 0.1 lb per square foot per hour.

The simple precautions listed below can minimize the possibility of plastic shrinkage cracking. They should be considered while planning for hot-weather-concrete construction or while dealing with the problem after construction has started. They are listed in the order in which they should be done during construction.

1. Moisten the subgrade and forms.
2. Moisten concrete aggregates that are dry and absorptive.
3. Erect temporary windbreaks to reduce wind velocity over the concrete surface.
4. Erect temporary sunshades to reduce concrete surface temperatures.
5. Keep the freshly mixed concrete temperature low by cooling the aggregates and mixing water.

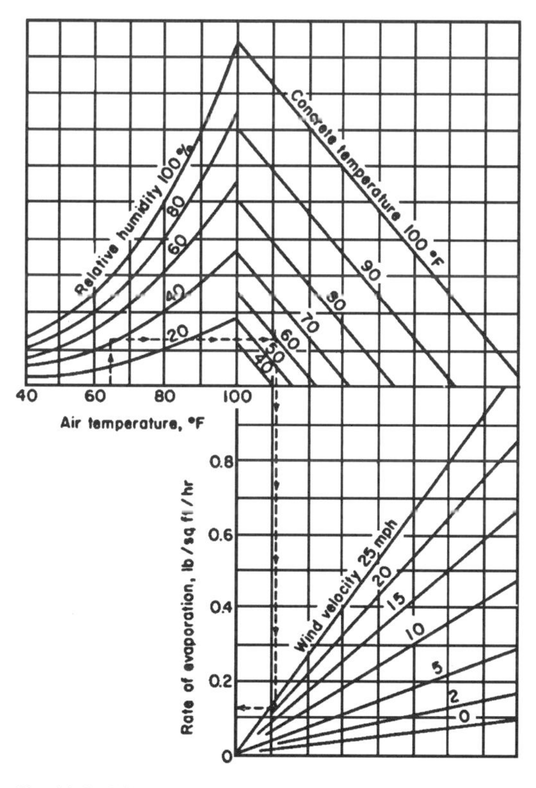

Fig. 11-8. Effect of concrete and air temperatures, relative humidity, and wind velocity on rate of evaporation of surface moisture from concrete.

6. Protect the concrete with temporary coverings, such as polyethylene sheeting, during any appreciable delay between placing and finishing. Evaporation retarders (usually polymers) can be spray-applied immediately after screeding to retard water evaporation before final finishing operations and curing commence. These materials are floated and troweled into the surface during finishing and should have no adverse effect on the concrete or inhibit the adhesion of membrane-curing compounds.
7. Reduce time between placing and start of curing by eliminating delays during construction.
8. Protect the concrete immediately after final finishing to minimize evaporation. This is most important to avoid cracking. Use of a fog spray to raise the relative humidity of the ambient air is an effective means of preventing evaporation from the concrete. Fogging should be continued until a suitable curing material such as a curing compound, wet burlap, or curing paper can be applied.

If plastic shrinkage cracks should appear during finishing, the cracks can be closed by striking each side of the crack with a float and refinishing. However, the cracking may recur unless the causes are corrected.

CURING AND PROTECTION

Curing and protection are more critical in hot and cold weather than in temperate periods. Retaining forms in place cannot be considered a satisfactory substitute for curing in hot weather; they should be loosened as soon as practical without damage to the concrete. Water should then be applied at the top exposed concrete surfaces, for example, with a soil-soaker hose, and allowed to run down inside the forms. On hardened concrete and on flat concrete surfaces in particular, curing water should not be excessively cooler than the concrete. This will minimize cracking caused by thermal stresses due to temperature differentials between the concrete and curing water.

The need for moist curing is greatest during the first few hours after finishing. To prevent the drying of exposed concrete surfaces, moist curing should commence as soon as the surfaces are finished and continue for at least 24 hours. In hot weather, continuous moist curing for the entire curing period is preferred. However, if moist curing can not be continued beyond 24 hours, the concrete surfaces should be protected from drying with curing paper, heat-reflecting plastic sheets, or membrane-forming curing compounds while the surfaces are still damp. Moist-cured surfaces should dry out slowly after the curing period to reduce the possibility of surface crazing* and cracking.

*Crazing, a network pattern of fine cracks that do not penetrate much below the surface, is caused by minor surface shrinkage. Crazing cracks are very fine and barely visible except when the concrete is drying after the surface has been wet. The cracks encompass small concrete areas less than 2 in. in dimension, forming a chicken-wire pattern.

White-pigmented curing compounds can be used on horizontal surfaces. Application of a curing compound during hot weather should be preceded by 24 hours of moist curing. If this is not practical, the compound should be applied immediately after final finishing. The concrete surfaces should be moist.

ADMIXTURES

For unusual cases in hot weather and where careful inspection is maintained, a retarding admixture may be beneficial in delaying the setting time (Fig. 11-3), despite the somewhat increased rate of slump loss generally resulting from their use.

Retarding admixtures should conform to the requirements of ASTM C 494 Type B. Admixtures should be tested with job materials under job conditions before construction begins in order to determine their compatibility with the basic concrete ingredients and their ability under the particular conditions to produce the desired properties.

HEAT OF HYDRATION

Heat generated during cement hydration raises the temperature of the concrete to a greater or lesser extent depending on the size of the concrete placement, its surrounding environment, and the amount of cement in the concrete. ACI 211.1 states that as a general rule a 10°F to 15°F temperature rise per 100 lb of cement can be expected from the heat of hydration.* There may be instances in hot-weather-concrete work and massive concrete placements when measures must be taken to cope with the generation of heat and attendant thermal volume changes to control cracking (see Chapters 13 and 15).

CHAPTER 12

Cold-Weather Concreting

Concrete can be placed safely throughout the winter months in cold climates if certain precautions are taken. Cold weather is defined by ACI 306 as "a period when for more than 3 successive days the mean daily temperature drops below 40°F." Normal concreting practices can be resumed once the ambient temperature is above 50°F for more than half a day.

During cold weather, the concrete mixture and its temperature should be adapted to the construction procedure and ambient weather conditions. Preparations should be made to protect the concrete; enclosures, windbreaks, portable heaters, insulated forms, and blankets should be ready to maintain the concrete temperature. Forms, reinforcing steel, and embedded fixtures must be clear of snow and ice at the time concrete is placed. Thermometers and proper storage facilities for test cylinders should be available to verify that precautions are adequate.

Fig. 12-1. When suitable preparations have been made, cold weather is no obstacle to concrete construction.

EFFECT OF FREEZING FRESH CONCRETE

Concrete gains very little strength at low temperatures. Freshly mixed concrete must be protected against the disruptive effects of freezing until the degree of saturation of the concrete has been sufficiently reduced by the process of hydration. The time at which this reduction is accomplished corresponds roughly to the time required for the concrete to attain a compressive strength of 500 psi. At normal temperatures, this occurs within the first 24 hours after placement. Significant ultimate strength reductions, up to about 50%, can occur if concrete is frozen within a few hours after placement or before it attains a compressive strength of 500 psi.

Concrete that has been frozen just once at an early age can be restored to nearly normal strength by providing favorable subsequent curing conditions. Such concrete, however, will not be as resistant to weathering nor as watertight as concrete that had not been frozen. The critical period after which concrete is not seriously damaged by one or two freezing cycles is dependent upon the concrete ingredients and conditions of mixing, placing, curing, and subsequent drying. For example, air-entrained concrete is less susceptible to damage by early freezing than non-air-entrained concrete. See Chapter 5 on "Freeze-Thaw Resistance" and "Resistance to Deicers and Salts" for more information.

STRENGTH GAIN OF CONCRETE AT LOW TEMPERATURES

Temperature affects the rate at which hydration of cement occurs—low temperatures retard hydration and consequently retard the hardening and strength gain of concrete.

If concrete is frozen and kept frozen above about 14°F, it will gain strength slowly. Below that temperature, cement hydration and concrete strength gain cease. Fig. 12-3 illustrates the effect of cool temperatures on setting time. Figs. 12-4 and 12-5 show the age-compressive strength relationship for concrete that has been

Fig. 12-2. Insulated column forms permit concreting even when the air temperature is well below the freezing mark. For more details see Figs. 12-20 and 12-21.

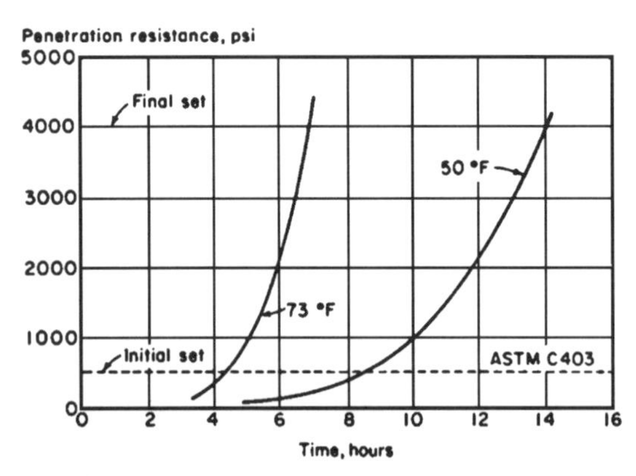

Fig. 12-3. Effect of cold temperature on rate of hardening.

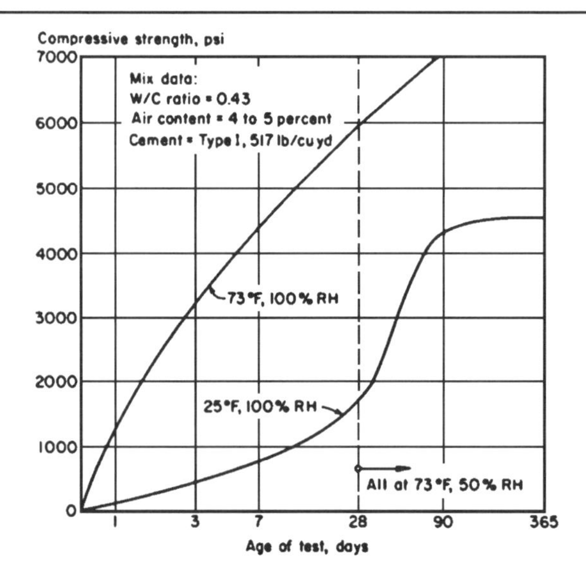

Fig. 12-4. Effect of temperature conditions on the strength development of concrete. Specimens for the lower curve were made at 40°F and placed immediately in a curing room at 25°F. Both curves represent 100% relative-humidity curing for first 28 days followed by 50% relative-humidity curing.

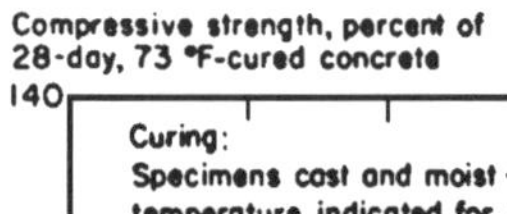

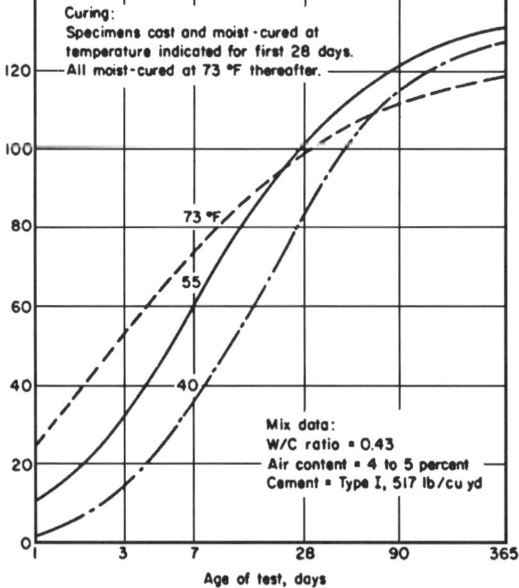

Fig. 12-5. Effect of low temperatures on concrete compressive strength at various ages.

cast and cured at various temperatures. Note in Fig. 12-5 that concrete cast and cured at 40°F and 55°F had relatively low strengths for the first week; but after 28 days—when all specimens were moist-cured at 73°F—strengths for the 40°F and 55°F concretes grew faster than the 73°F concrete and at one year they were slightly higher.

Higher early strengths can be achieved through use of Type III, high-early-strength, cement, as illustrated in Fig. 12-6. Principal advantages occur during the first 7 days. At a 40°F curing temperature, the advantages of Type III cement are more pronounced and persist longer than at higher temperatures.

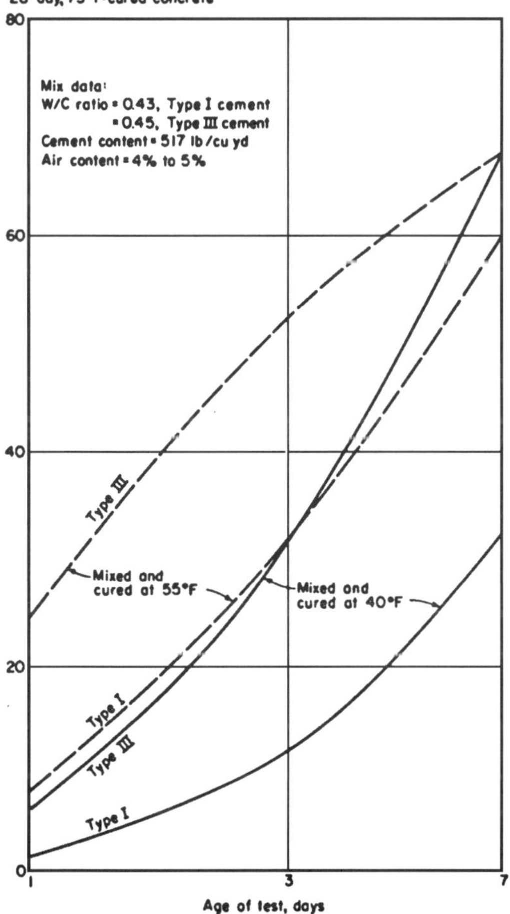

Fig. 12-6. Early-age compressive-strength relationships for Type I and Type III portland cements at low curing temperatures.

HEAT OF HYDRATION

Concrete generates heat during hardening as a result of the chemical process by which cement reacts with water to form a hard, stable paste. The heat generated is called heat of hydration; it varies in amount and rate for different portland cements. Heat generation and buildup are affected by dimensions of the concrete, ambient air temperature, initial concrete temperature, water-cement ratio, cement composition and fineness, amount of cement, and admixtures.

Heat of hydration is particularly useful in winter concreting as it often generates enough heat to provide a satisfactory curing temperature without other temporary heat sources, particularly in more massive elements.

Concrete must be delivered at the proper temperature and account must be taken of the temperature of forms, reinforcing steel, the ground, or other concrete on which the concrete is cast. Concrete should not be cast on frozen concrete or on frozen ground.

Fig. 12-7 shows a concrete pedestal on a spread footing being covered with a tarpaulin just after the concrete was placed. Thermometer readings of the concrete's temperature will tell whether the covering is adequate. The heat liberated during hydration will offset to a considerable degree the loss of heat during placing, finishing, and early curing operations.

Fig. 12-7. Concrete footing pedestal being covered with a tarpaulin to retain the heat of hydration.

SPECIAL CONCRETE MIXTURES

High strength at an early age is desirable in winter construction to reduce the length of time temporary protection is required. The additional cost of high-early-strength concrete is often offset by earlier reuse of forms and removal of shores, savings in the shorter duration of temporary heating, earlier finishing of flatwork, and earlier use of the structure. High-early-strength concrete can be obtained by using one or a combination of the following:

1. Type III or IIIA high-early-strength cement
2. Additional portland cement (100 to 200 lb per cu yd)
3. Chemical accelerators

Small amounts of an accelerator such as calcium chloride (at a maximum dosage of 2% by weight of portland cement) can be used to accelerate the setting and early-age strength development of concrete in cold weather. Accelerators containing chlorides should not be used where there is an in-service potential for corrosion, such as in prestressed concrete or where aluminum or galvanized inserts will be used. Chlorides are not recommended for concretes exposed to soil or water containing sulfates or for concrete susceptible to alkali-aggregate reactions. Other restrictions on accelerators containing chlorides are discussed in detail in Chapter 6.

Accelerators must not be used as a substitute for proper curing and frost protection. Also, the use of so-called antifreeze compounds to lower the freezing point of concrete must not be permitted. The quantity of these materials needed to appreciably lower the freezing point of concrete is so great that strength and other properties are seriously affected by their use.

Low water-cement ratio and low-slump concrete is particularly desirable for cold-weather flatwork. Evaporation is minimized and finishing can be accomplished quicker (Fig. 12-8).

Fig. 12-8. Finishing the concrete surface can proceed because a windbreak has been provided, there is adequate heat under the slab, and the concrete has low slump.

AIR-ENTRAINED CONCRETE

Entrained air is particularly desirable in any concrete placed during freezing weather. Concrete that is not air entrained will suffer a strength loss as a result of freezing and thawing (Fig. 12-9). Air entrainment provides the capacity to absorb stresses due to ice formation within the concrete. See Chapter 5, "Air-Entrained Concrete."

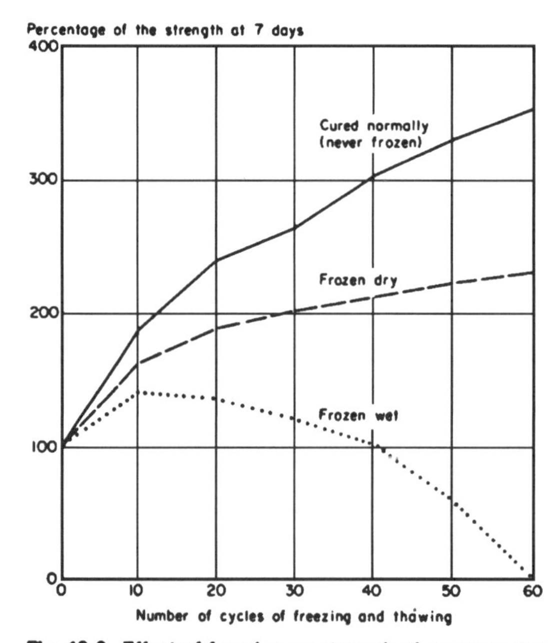

Fig. 12-9. Effect of freezing on strength of concrete that does not contain entrained air (cured 7 days before first freeze).

Cold-Weather Concreting implies that air entrainment should always be used for construction during the freezing months. The exception is concrete work done under roof where there is no chance that rain, snow, or water from other sources can saturate the concrete and where there is no chance of freezing.

The likelihood of water saturating a concrete floor during construction is very real. Fig. 12-10 shows conditions in the upper story of an apartment building during winter construction. Snow fell on the top deck. When heaters were used below to warm the deck, the snow melted. Water ran through floor openings down to a level that was not being heated. The water-saturated concrete froze, which caused a strength loss, particularly at the floor surface. This could also result in greater deflection of the floor and a surface that is less abrasion-resistant than it might have been.

Fig. 12-10. Example of a concrete floor that was saturated and then frozen, showing the need for air entrainment.

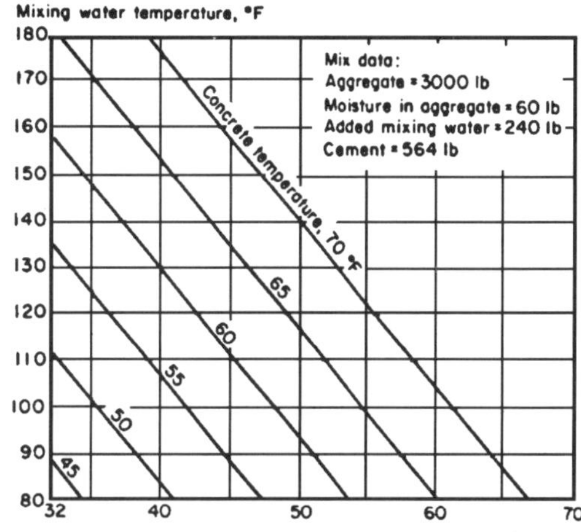

Fig. 12-11. Temperature of mixing water needed to produce heated concrete of required temperature. Temperatures are based on the mixture shown but are reasonably accurate for other typical mixtures.

TEMPERATURE OF CONCRETE

Temperature of Concrete As Mixed

The temperature of fresh concrete *as mixed* should not be less than shown in Lines 1, 2, or 3 of Table 12-1. Note that lower concrete temperatures are recommended for more massive concrete sections because heat generated during hydration is dissipated less rapidly in heavier sections. Also note that at lower air temperatures more heat is lost from concrete during transporting and placing; hence the recommended concrete temperatures are higher for colder weather.

There is little advantage in using fresh concrete at a temperature much above 70°F. Higher concrete temperatures do not afford proportionately longer protection from freezing because the rate of heat loss is greater. Also, high concrete temperatures are undesirable since they increase thermal shrinkage after hardening, require more mixing water for the same slump, and contribute to the possiblity of plastic-shrinkage cracking (caused by rapid moisture loss through evaporation). Therefore, the temperature of the concrete *as mixed* should not be more than 10°F above the minimums recommended in Table 12-1.

Aggregate temperature. The temperature of aggregates varies with weather and type of storage. Aggregates usually contain frozen lumps and ice when the temperature is below freezing. Frozen aggregates must be thawed to avoid pockets of aggregate in the concrete after batching, mixing, and placing. If thawing takes place in the mixer, excessively high water contents due to the ice melting must be avoided.

At temperatures above freezing it is seldom necessary to heat aggregates. At temperatures below freezing, often only the fine aggregate needs to be heated to produce concrete of the required temperature, provided the coarse aggregate is free of frozen lumps. If aggregate temperatures are above freezing, the desired concrete temperature can usually be obtained by heating only the mixing water.

Circulating steam through pipes over which aggregates are stockpiled is a recommended method for heating aggregates. Stockpiles can be covered with tarpaulins to retain and distribute heat and to prevent formation of ice. Live steam, preferably at pressures of 75 to 125 psi, can be injected directly into the aggregate pile to heat it, but the resultant variable moisture content in aggregates might result in erratic mixing-water control.

On small jobs aggregates can be heated by stockpiling over metal culvert pipes in which fires are maintained. Care should be taken to prevent scorching the aggregates.

Mixing-water temperature. Of the ingredients used to make concrete, mixing water is the easiest and most practical to heat. The weight of aggregates and cement in concrete is much greater than the weight of water; however, water can store about five times as much heat as can cement and aggregate of the same weight. For cement and aggregate the average specific heat (that is, heat units required to raise the temperature of 1 lb of material 1°F) is 0.22 Btu per pound per degree Fahrenheit compared to 1.0 for water.

Equation for concrete temperature. Fig. 12-11 shows graphically the effect of temperature of materials on temperature of freshly mixed concrete. The chart is based on the equation

$$T=\frac{0.22(T_aW_a+T_cW_c)+T_wW_w+T_{wa}W_{wa}}{0.22(W_a+W_c)+W_w+W_{wa}}$$

Table 12-1. Recommended Concrete Temperature for Cold-Weather Construction—Air-Entrained Concrete*

Line	Condition		Thickness of sections, in.: Less than 12	12 to 36	36 to 72	Over 72
1	Minimum temperature of fresh concrete as *mixed* for weather indicated, °F	Above 30°F	60	55	50	45
2		0°F to 30°F	65	60	55	50
3		Below 0°F	70	65	60	55
4	Minimum temperature of fresh concrete *as placed and maintained*, °F**		55	50	45	40
5	Maximum allowable *gradual* drop in temperature in first 24 hours after end of protection, °F		50	40	30	20

*Adapted from Table 1.4.1 of Reference 12-13.
**Placement temperatures listed are for normal-weight concrete. Lower temperatures can be used for lightweight concrete if justified by tests. For recommended duration of temperatures in Line 4, see Table 12-2.

where

T = temperature of the freshly mixed concrete
T_a, T_c, T_w, and T_{wa} = temperature of aggregates, cement, added mixing water, and free moisture on aggregates, respectively
W_a, W_c, W_w, and W_{wa} = weight of aggregates, cement, added mixing water, and free moisture on aggregates, respectively

If the weighted average temperature of aggregates and cement is above 32°F, the proper mixing-water temperature for the required concrete temperature can be selected from Fig. 12-11. The range of concrete temperatures in the chart corresponds with the recommended values given in Lines 1, 2, and 3 of Table 12-1.

To avoid the possibility of a quick or flash set of the concrete when either water or aggregates are heated to above 100°F, they should be combined in the mixer first before the cement is added. If this mixer-loading sequence is followed, water temperatures up to the boiling point can be used, provided the aggregates are cold enough to reduce the final temperature of the aggregates and water mixture to appreciably less than 100°F.

Fluctuations in mixing-water temperature from batch to batch should be avoided. The temperature of the mixing water can be adjusted by blending hot and cold water.

Temperature of Concrete As Placed and Maintained

There will be some temperature loss after mixing while the truck mixer is traveling to the construction site and waiting to discharge its load. The concrete should be placed in the forms before its temperature drops below that given on Line 4 of Table 12-1, and that concrete temperature should be maintained for the duration of the protection period given in Table 12-2.

CONTROL TESTS

Thermometers are needed to check the concrete temperatures as delivered, as placed, and as maintained. An inexpensive pocket thermometer is shown in Fig. 12-12.

After the concrete has hardened, temperatures can be checked with special surface thermometers or with an ordinary thermometer that is kept covered with insulating blankets. A simple way to check temperature below the concrete surface is shown in Fig. 12-13. Instead of filling the hole shown in Fig. 12-13 with a fluid, it can be fitted with insulation except at the bulb.

Concrete test cylinders must be maintained at a temperature between 60°F and 80°F at the jobsite for 24 hours until they are taken to a laboratory for curing. During this 24-hour period cylinders should be kept in a curing box with the temperature accurately controlled by a thermostat (Fig. 12-14). When stored in an insulated curing box outdoors, cylinders are less likely to be jostled by vibrations than if they were left on the floor of a trailer. If kept in a trailer or shanty where the heat may be turned off at night or over a weekend or holiday, the cylinders would not be at the prescribed curing temperatures during this critical period.

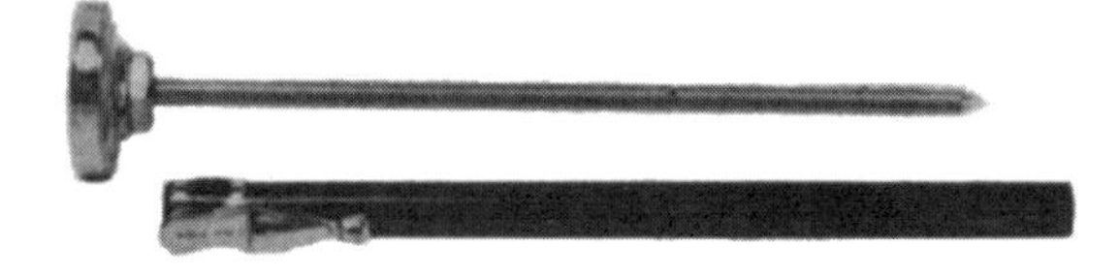

Fig. 12-12. A bimetallic pocket thermometer with a metal sensor suitable for checking concrete temperatures.

Table 12-2.

A. Recommended Duration of Concrete Temperature in Cold Weather—Air-Entrained Concrete*

Service category††	For durability: Conventional concrete,** days	For durability: High-early-strength concrete,† days	For safe stripping strength: Conventional concrete,** days	For safe stripping strength: High-early-strength concrete,† days
No load, not exposed, favorable moist-curing	2	1	2	1
No load, exposed, but later has favorable moist-curing	3	2	3	2
Partial load, exposed	3	2	6	4
Fully stressed, exposed	3	2	See Table B below	

B. Recommended Duration of Concrete Temperature for Fully Stressed, Exposed, Air-Entrained Concrete

Required percentage of design strength, f'_c	Days at 50°F			Days at 70°F		
	Type of portland cement			Type of portland cement		
	I	II	III	I	II	III
50	6	9	3	4	6	3
65	11	14	5	8	10	4
85	21	28	16	16	18	12
95	29	35	26	23	24	20

*Adapted from Table 1.4.2 of Reference 12-13. Cold weather is defined as that in which average daily temperature is less than 40°F for 3 successive days except that if temperatures above 50°F occur during at least 12 hours in any day, the concrete should no longer be regarded as winter concrete and normal curing practice should apply. For recommended concrete temperatures, see Table 12-1. For concrete that is *not* air entrained, ACI 306 states that protection for durability should be at least twice the number of days listed in Table A.

Part B was adapted from Table 7.7 of Reference 12-13. The values shown are approximations and will vary according to the thickness of concrete, mix proportions, etc. They are intended to represent the ages at which supporting forms can be removed. For recommended concrete temperatures, see Table 12-1.

**Made with ASTM Type I or II portland cement.

†Made with ASTM Type III portland cement, or an accelerator, or an extra 100 lb of cement per cu yd.

††"Exposed" means subject to freezing and thawing.

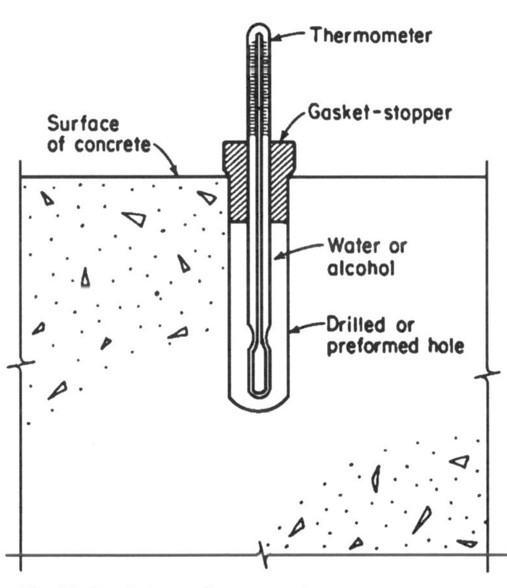

Fig. 12-13. Scheme for measuring concrete temperatures below the surface with a glass thermometer.

Fig. 12-14. Insulated curing box with thermostat for curing test cylinders. Heat is supplied by electric rubber heating mats on the bottom. A wide variety of designs are possible for curing boxes.

In addition to laboratory-cured cylinders, it is useful to field-cure some test cylinders in order to monitor actual curing conditions on the job in cold weather. It is sometimes difficult to find the right locations for field curing. A preferred location is in a boxout in a floor slab or wall with thermal insulation for cover. When placed on a formwork ledge just below a heated, suspended floor, possible high temperatures there will not duplicate the average temperature in the slab, nor the lowest temperature on top of the slab. Still, field-cured cylinders are more indicative of actual concrete strength than are laboratory-cured cylinders.

Molds should be stripped from the cylinders after the first 24 ±4 hours and the cylinders wrapped tightly in plastic bags. When cylinders are picked up for delivery to the laboratory, they must be maintained at 60°F to 80°F temperatures until they are placed in the laboratory curing room.

Cast-in-place cylinders (ASTM C 873) and nondestructive testing methods discussed in Chapter 14, as well as maturity techniques discussed later in this chapter, are helpful in monitoring in-place concrete strength.

Fig. 12-15. Even in the dead of winter, an outdoor swimming pool can be constructed if a heated enclosure is used.

CONCRETING ON GROUND

Concreting on ground during cold weather involves some extra effort and expense, but many contractors find that it more than pays for itself. In winter, the site may be frozen rather than a morass of mud. The concrete will furnish some if not all of the heat needed for proper curing. Insulated blankets or simple enclosures are easily provided. Embankments are frozen and require less bracing. With a good start during the winter months, construction gets above the ground before warmer weather arrives.

Placing concrete on the ground involves different procedures from those used at an upper level: (1) the ground must be thawed before placing concrete; (2) cement hydration will furnish some of the curing heat; (3) construction of enclosures is much simpler and use of insulating blankets may be sufficient; and (4) in the case of a floor slab, a *vented* heater is required if the area is enclosed.

Footings can be started if the ground is thawed with temporary heat and the footings backfilled as soon as possible. Concrete should never be placed on a frozen subgrade because when the subgrade thaws, uneven settlement may occur, causing cracking. Also, on a frozen subgrade, heat will migrate rapidly away from the bottom of the concrete and its rate of hardening will be retarded. Ideally, the temperature of the subgrade should be as close as practicable to the temperature of the concrete to be placed on it.

When the subgrade is frozen to a depth of only a few inches, this surface region can be thawed by (1) steaming; (2) spreading a layer of hot sand, gravel, or other granular material where the grade elevations allow it; (3) burning straw or hay if local air pollution ordinances permit it; or (4) covering the subgrade with insulation for a few days. Placing concrete for floor slabs and exposed footings should be delayed until the ground thaws and warms sufficiently to ensure that it will not freeze again during the curing period.

For slabs cast on ground at 35°F, ACI 306 gives two sets of tables and graphs showing required resistance to heat transfer (R) of insulating blankets. The data extends to a protection period of only one week and states that insulating blankets must be supplemented with heat for slabs 12 in. or less in thickness.

CONCRETING ABOVEGROUND

Working aboveground in cold weather usually involves several different approaches compared to working at ground level:

1. The concrete mixture need not necessarily be changed to generate more heat because portable heaters can be used to heat the undersides of floor and roof slabs. Nevertheless, there are advantages to having a mix that will produce a high strength at an early age; for example, artificial heat can be cut off sooner (see Table 12-2), and forms can be recycled faster.
2. Enclosures must be constructed to retain the heat under floor and roof slabs.
3. Portable heaters used to warm the underside of concrete can be direct-fired heating units (without venting).

Before concreting, the heaters under a formed deck should be turned on to preheat the forms and melt any snow or ice remaining on top. When slab finishing is completed, insulating blankets or other insulation must be placed on top of the slab to ensure that proper curing temperatures are maintained. The insulation *R* values necessary to maintain an adequate curing temperature above 50°F for 3 or 7 days may be estimated from Figs. 12-16 and 12-17. To maintain a temperature for longer periods, more insulation is required. Figs. 12-16 and 12-17 also apply to concrete walls. Insulation can be selected based on *R* values provided by insulation manufacturers or by using the information in Table 12-3. ACI 306 has additional graphs and tables for slabs on or above ground.

The period for the maintained temperature may be based on Table 12-2. However, the actual amount of insulation and length of the protection period should be determined from the monitored in-place concrete temperature and the desired strength. A correlation between curing temperature, curing time, and compressive strength can be determined from laboratory testing of the particular concrete mix used in the field (see maturity concept discussed in "Duration of Heating" in this chapter).

Since square and rectangular columns have twice as many heat-flow paths as a long, high wall, twice as much insulation is needed to maintain the same heat-loss characteristics. If the ambient temperature rises much above the temperature assumed in selecting insulation values, the temperature of the concrete may become excessive. This increases the probability of thermal shock and cracking when forms are removed. Temperature readings of insulated concrete should therefore be taken at regular intervals and should not be allowed to rise much above 80°F. In case of a sudden increase in concrete temperature, up to say 95°F, it may be necessary to remove some of the insulation or loosen the formwork. The maximum temperature differential between the concrete interior and the concrete surface should be about 35°F to minimize cracking. The weather forecast should be checked and appropriate action taken for expected temperature changes.

Columns and walls should not be cast on frozen foundations, because chilling of concrete in the bottom of the column or wall will retard strength development.

ENCLOSURES

Heated enclosures can be used for protecting concrete in cold weather. Enclosures can be of wood, canvas tarpaulins, or polyethylene film (Fig. 12-18). Prefabricated, rigid-plastic enclosures are also available. Plastic enclosures that admit daylight are the most popular (Fig. 12-18b).

When enclosures are being constructed below a deck, the framework can be extended above the deck to serve as a windbreak. A height of 6 ft will give good protection against biting winds.

Enclosures can be made to be moved with flying forms; more often, though, they must be removed so that the wind will not interfere with maneuvering the forms into position. Similarly, enclosures can be built in large panels like gang forms with the windbreak included (Fig. 12-1).

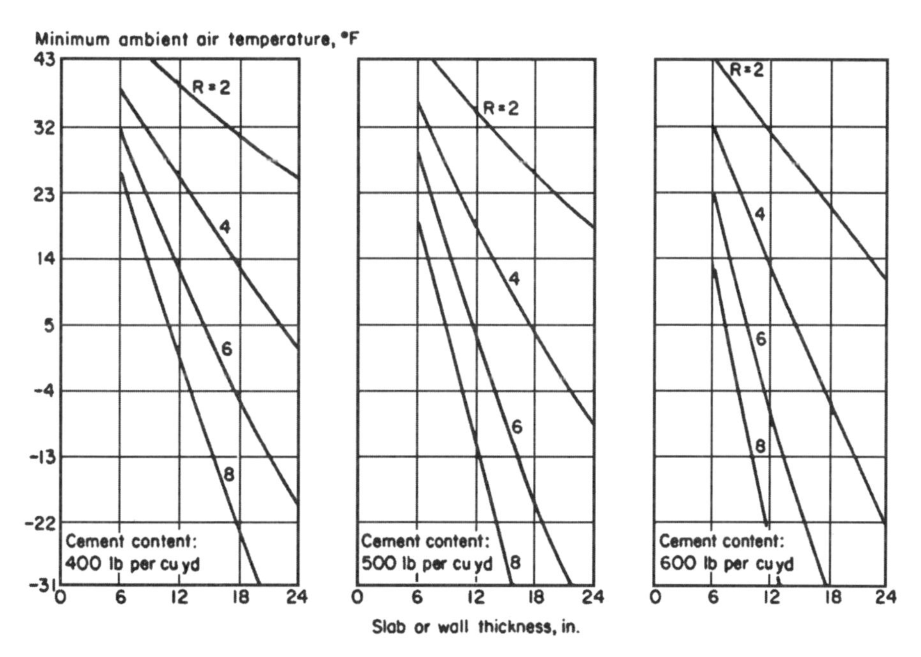

Fig. 12-16. Thermal resistance (*R*) of insulation required to maintain the concrete surface temperature of walls and slabs aboveground at 50°F or above for 3 days. Concrete temperature as placed: 50°F. Maximum wind velocity: 15 mph.

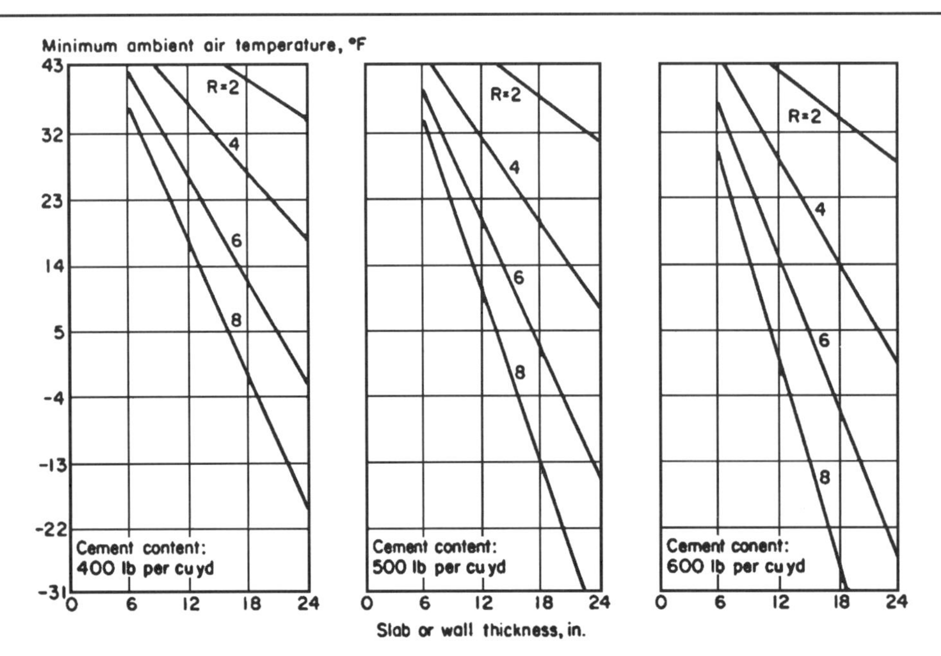

Fig. 12-17. Thermal resistance (*R*) of insulation required to maintain the concrete surface temperature of walls and slabs aboveground at 50°F or above for 7 days. Concrete temperatures as placed: 50°F. Maximum wind velocity: 15 mph. Note that in order to maintain a certain minimum temperature for a longer period of time, more insulation or a higher *R* value is required.

Table 12-3. Insulation Values of Various Materials

A.

Material	Density, pcf	Thermal resistance, *R*, for 1-in. thickness of material,* (°F · hr · ft²)/Btu
Board and Slabs		
Expanded polyurethane	1.5	6.25
Expanded polystyrene extruded smooth-skin surface	1.8 to 3.5	5.00
Expanded polystyrene extruded cut-cell surface	1.8	4.00
Glass fiber, organic bonded	4 to 9	4.00
Expanded polystyrene, molded beads	1	3.85
Mineral fiber with resin binder	15	3.45
Mineral fiberboard, wet felted	16 to 17	2.94
Vegetable fiberboard sheathing	18	2.64
Cellular glass	8.5	2.86
Laminated paperboard	30	2.00
Particle board (low density)	37	1.85
Plywood	34	1.24
Loose fill		
Wood fiber, soft woods	2.0 to 3.5	3.33
Perlite (expanded)	5.0 to 8.0	2.70
Vermiculite (exfoliated)	4.0 to 6.0	2.27
Vermiculite (exfoliated)	7.0 to 8.2	2.13
Sawdust or shavings	8.0 to 15.0	2.22

B.

Material	Material thickness, in.	Thermal resistance, *R*, for 1-in. thickness of material,* (°F · hr · ft²)/Btu
Mineral fiber blanket, fibrous form (rock, slag, or glass) 0.3 to 2 pcf	2 to 2.75	7
	3 to 3.5	11
	5.5 to 6.5	19
	6 to 7	22
	8.5 to 9	30
	12	38
Mineral fiber loose fill (rock, slag, or glass) 0.6 to 2 pcf	3.75 to 5	11
	6.5 to 8.75	19
	7.5 to 10	22
	10.25 to 13.75	30

*Values are from *ASHRAE Handbook of Fundamentals*, 1977 and 1981, American Society of Heating, Refrigerating, and Air-Conditioning Engineers, Inc., New York.

R values are the reciprocal of U values (conductivity).

a.

b.

Fig. 12-18. Heated enclosures maintain an adequate temperature for proper curing. a. Tarpaulin enclosure provides protection for a bridge during severe and prolonged winter weather. b. Polyethylene plastic sheets admitting daylight are used to fully enclose a building frame. The temperature inside is maintained at 50°F with space heaters.

Fig. 12-19. Insulating blankets trap heat and moisture in the concrete, providing beneficial curing.

Fig. 12-20. Blanket or batt insulation may be applied to the outside of job-built or prefabricated forms.

INSULATING MATERIALS

Heat and moisture can be retained in the concrete by covering it with commercial insulating blankets or batt insulation (Fig. 12-19). The effectiveness of insulation can be determined by placing a thermometer under it and in contact with the concrete. If the temperature falls below the minimum required, additional insulating material, or material with a higher *R* value, should be applied. Corners and edges of concrete are most vulnerable to freezing and the temperature should be checked there particularly.

The thermal resistance values for common insulating materials are given in Table 12-3. For maximum efficiency, insulating materials should be kept dry and in close contact with concrete or formwork.

Concrete pavements can be protected from cold weather by spreading 12 in. or more of dry straw or hay on the surface for insulation. Tarpaulins, polyethylene film, or waterproof paper should be used as a protective cover over the straw or hay to make the insulation more effective and prevent it from blowing away. The straw or hay should be kept dry or its insulation value will drop considerably.

Forms built for repeated use often can be economically insulated with commercial blanket or batt insulation (Fig. 12-20). The insulation should have a tough moistureproof covering to withstand handling abuse and exposure to the weather. Rigid insulation can also be used (Fig. 12-21).

Insulating blankets for construction are made of fiberglass, sponge rubber, open-cell polyurethane foam, vinyl foam, mineral wool, or cellulose fibers. The outer covers are made of canvas, woven polyethylene, or other tough fabrics that will take rough handling.

Fig. 12-21. With air temperatures down to −10°F, concrete was cast in this insulated column form made of ¾-in. high-density plywood inside, 1-in. rigid polystyrene in the middle, and ½-in. rough plywood outside. *R* value: 5.6 (°F · hr · ft²)/Btu.

The *R* value for a typical insulating blanket is about 7(°F · hr · ft²)/Btu, but since *R* values are not marked on the blankets, their effectiveness should be checked with a thermometer. If necessary, they can be used in two or three layers to attain the desired insulation.

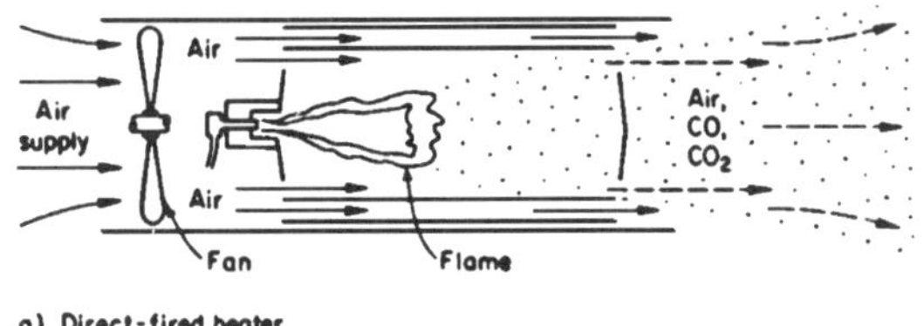

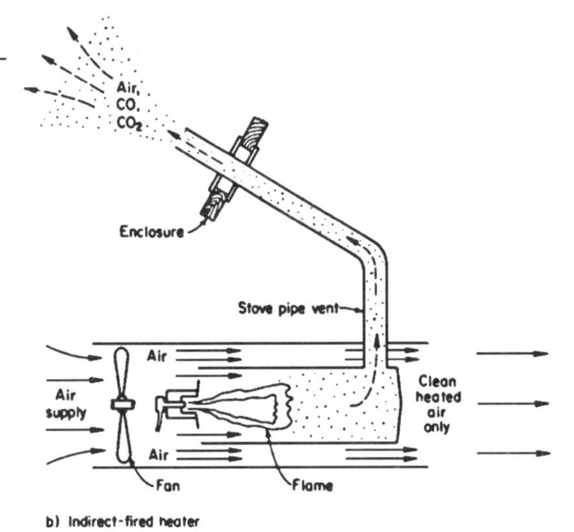

Fig. 12-22. Two basic types of heaters.

HEATERS

Two types of heaters are used in cold-weather concrete construction: direct fired and indirect fired (Fig. 12-22). Indirect-fired heaters are vented to remove the products of combustion. Where heat is to be supplied to the top of fresh concrete—for example, a floor slab—vented heaters are required. Carbon dioxide (CO_2) in the exhaust will be vented to the outside and prevented from reacting with the fresh concrete (Fig. 12-23). Direct-fired units can be used to heat enclosures beneath a floor or a roof deck (Fig. 12-24).

Any heater burning a fossil fuel produces CO_2 that will combine with calcium hydroxide on the surface of fresh concrete to form a weak layer of calcium carbonate that interferes with cement hydration. The result is a soft, chalky surface that will dust under traffic. Depth and degree of carbonation depend on concentration of CO_2, curing temperature, humidity, porosity of the concrete, length of exposure, and method of curing. Direct-fired heaters, therefore, should not be permitted to heat the air over concreting operations—at least until 24 hours have elapsed.

Carbon monoxide (CO), another product of combustion, is not usually a problem unless the heater is using recirculated air. Four hours of exposure to 200 parts per million of CO will produce headaches and nausea. Three hours of exposure to 600 ppm can be fatal. The American National Standard Safety Requirements for Temporary and Portable Space Heating Devices and Equipment Used in the Construction Industry (ANSI

Fig. 12-23. An indirect-fired heater. Notice vent pipe that carries combustion gases outside the enclosure.

Fig. 12-24. A direct-fired heater mounted outside the enclosure, thus using a fresh air supply.

A10.10) limits concentrations of CO to 50 ppm at worker breathing levels. The standard also establishes safety rules for ventilation and the stability, operation, fueling, and maintenance of heaters.

A salamander is an inexpensive combustion heater without a fan that discharges its combustion products into the surrounding air; heating is accomplished by radiation from its metal casing. Salamanders are fueled by coke, oil, wood, or liquid propane. They are but one form of a direct fired heater. A primary disadvantage of salamanders is the high temperature of their metal casing, a definite fire hazard. Salamanders should be placed so as not to overheat formwork or enclosure materials. When placed on floor slabs they should be elevated to avoid scorching the concrete.

Some heaters burn more than one type of fuel. The approximate heat values of fuels are as follows:

No. 1 fuel oil	135,000 Btu/gal
Kerosene	134,000 Btu/gal
Gasoline	128,000 Btu/gal
Liquid-propane gas	91,500 Btu/gal
Natural gas	1,000 Btu/ft³

The Btu rating of a portable heater is usually the heat content of the fuel consumed per hour. A rule of thumb is that about 36,000 Btu are required for each 10,000 cu ft of air to develop a 20°F heat rise.

Electricity can also be used to cure concrete in winter. The use of large electric blankets equipped with thermostats is one method. The blankets can also be used to thaw subgrades or concrete foundations.

Use of electric resistance wires that are cast into the concrete is another method. The power supplied is under 50 volts, and from 1.5 to 5 kilowatt-hours of electricity per cubic yard of concrete is required, depending on the circumstances. The method has been used in the Montreal, Quebec, area for many years. Where electrical resistance wires are used, insulation should be included during the initial setting period. If insulation is removed before the recommended time, the concrete should be covered with an impervious sheet and the power continued for the required time.

Steam is another source of heat for winter concreting. Live steam can be piped into an enclosure or supplied through radiant heating units. For accelerated steam curing at elevated temperatures, see Chapter 10, "Curing Concrete."

In choosing a heat source, it must be remembered that the concrete itself supplies heat through hydration of cement, and this is often enough for curing needs if the heat can be retained within the concrete with insulation.

DURATION OF HEATING

After concrete is in place, it should be protected and kept at a favorable curing temperature until it gains sufficient strength to withstand exposure to low temperatures, anticipated environment, and construction and service loads. Recommended periods of protection are given in Table 12-2. The columns headed "for durability" list the length of time in days required to provide adequate durability against exposure to freezing and thawing. The columns headed "for safe stripping strength" list the length of time in days after which bottom forms or reshores can be removed. This is based on the assumption that construction loads will be less than those provided for in the design and that normal curing conditions will enable the concrete to reach its full design strength before being put into service.

Maturity Concept

The maturity concept is based on the principle that strength gain is a function of curing time and temperature. The maturity concept, as described in ACI 306 and ASTM C1074 and C918, can be used to evaluate strength development when the prescribed curing temperatures have not been maintained for the required time or when curing temperatures have fluctuated. The concept is expressed by the equation

$$M = \Sigma(F - 14)\Delta t$$

where

M = maturity factor
Σ = summation
F = temperature, degrees Fahrenheit
Δt = duration of curing at temperature F, usually in hours

The equation is based on the premise that concrete gains strength (or that cement continues to hydrate) at temperatures as low as 14°F.

Before construction begins, a calibration curve is drawn plotting the relationship between compressive strength and the maturity factor for a series of test cylinders (of the particular concrete mixture proportions) cured in a laboratory and tested for strength at successive ages.

The maturity concept is not precise and has some limitations. The concept is useful only in checking the curing of concrete and estimating strength in relation to time and temperature. It presumes that all other factors affecting concrete strength have been properly controlled. Thus the maturity concept is another method for monitoring temperatures, but it is no substitute for quality control and proper concreting practices.

Moist Curing

Strength gain stops when moisture required for curing is no longer available. Concrete retained in forms or covered with insulation seldom loses enough moisture at 40°F to 55°F to impair curing. However, a positive means of providing moist curing is needed to offset drying when heated enclosures are used during cold weather.

Live steam exhausted into an enclosure around the concrete is an excellent method of curing because it provides both heat and moisture. Steam is especially practical in extremely cold weather because the moisture provided offsets the rapid drying that occurs when very cold air is heated.

Liquid membrane-forming compounds can be used for early curing of concrete surfaces within heated enclosures. These compounds are also helpful in reducing carbonation of the surface by unvented heaters.

Terminating the Heating Period

Rapid cooling of concrete at the end of the heating period should be avoided. Sudden cooling of the concrete surface while the interior is still warm may cause thermal cracking, especially in massive sections such as bridge piers, abutments, dams, and large structural members; thus cooling should be gradual. The maximum uniform drop in temperature throughout the first 24 hours after the end of protection should not be more than the amounts given in Line 5 of Table 12-1. Gradual cooling can be accomplished by lowering the heat or by simply shutting off the heat and allowing the enclosure to cool to outside air temperature.

FORM REMOVAL AND RESHORING

It is good practice in cold weather to leave forms in place as long as possible. Even within heated enclosures, forms serve to distribute heat more evenly and help prevent drying and local overheating.

Table 12-2A can be used to determine the length of time in days that vertical support for forms should be left in place. The time span is also the length of time that the required concrete temperature should be maintained. Table 12-2B lists length of time in days for heating and shoring or reshoring of members that will be fully stressed. The engineer in charge must determine what percentage of the design strength is required.

Side forms can be removed sooner than shoring and temporary falsework.

Nondestructive Test Methods

Various nondestructive tests can be used to evaluate the relative strength of hardened concrete. The most widely used are the rebound, penetration, pullout, and dynamic or vibration tests. Relatively new techniques being developed for testing the strength and other properties of hardened concrete include X-rays, gamma radiography, neutron moisture gages, magnetic cover meters, electricity, microwave absorption, and acoustic emissions. Each method has limitations, and caution should be exercised against acceptance of nondestructive test results as having a constant correlation to the traditional compression test, i.e., empirical correlations must be developed prior to use.

Rebound method. The Schmidt rebound hammer (Fig. 14-10) is essentially a surface-hardness tester that provides a quick, simple means of checking concrete uniformity. It measures the rebound of a spring-loaded plunger after it has struck a smooth concrete surface. The rebound number reading gives an indication of the compressive strength of the concrete.

The results of a Schmidt rebound hammer test (ASTM C 805) are affected by surface smoothness, size, shape, and rigidity of the specimen; age and moisture condition of the concrete; type of coarse aggregate; and carbonation of the concrete surface. When these limitations are recognized and the hammer is calibrated for the particular materials used in the concrete (Fig. 14-11), then this instrument can be useful for determining the relative compressive strength and uniformity of concrete in the structure.

Penetration method. The Windsor probe (ASTM

Fig. 14-10. The rebound hammer gives an indication of the compressive strength of concrete.

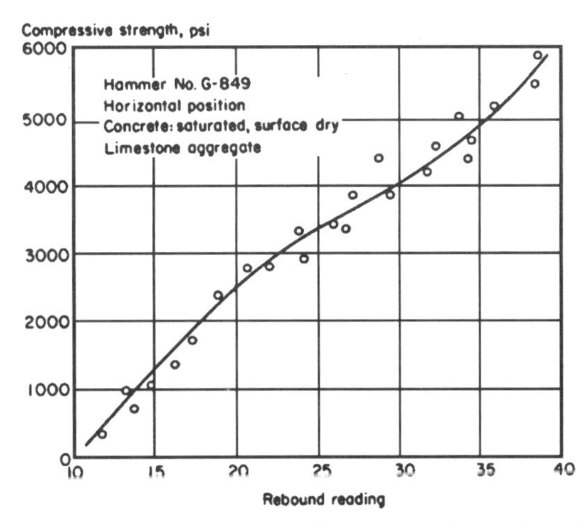

Fig. 14-11. Example of a calibration chart for an impact (rebound) test hammer.

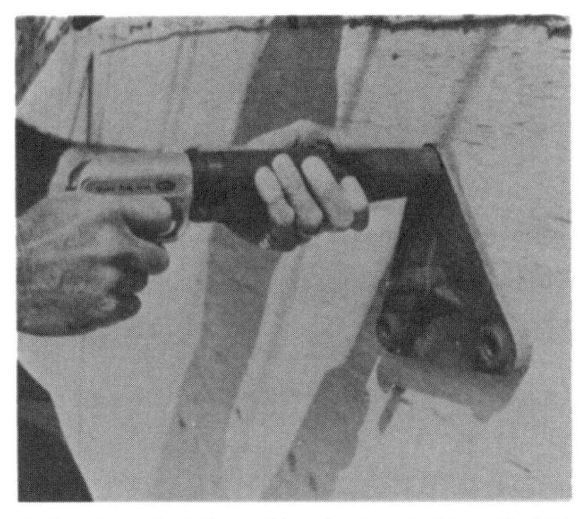

A. Powder-actuated gun drives hardened alloy probe into concrete.

B. Exposed length of probe is measured and relative compressive strength of the concrete then determined from calibration table.

Fig. 14-12. The Windsor-probe technique for determining the relative compressive strength of concrete.

C 803), like the rebound hammer, is basically a hardness tester that provides a quick means of determining the relative strength of the concrete. The equipment consists of a powder-actuated gun that drives a hardened alloy probe (needle) into the concrete (Fig. 14-12). The exposed length of the probe is measured and related by a calibration table to the compressive strength of the concrete.

The results of the Windsor-probe test will be influenced by surface smoothness of the concrete and the type and hardness of aggregate used. Therefore, a calibration table or curve for the particular concrete to be tested should be made, usually from cores or cast specimens, to improve accuracy.

Both the rebound hammer and the probe damage the concrete surface to some extent. The rebound hammer leaves a small indentation on the surface; the probe leaves a small hole and may cause minor cracking and small craters similar to popouts.

Pullout tests. A pullout test (ASTM C 900) involves casting the enlarged end of a steel rod in the concrete to be tested and then measuring the force required to pull it out (Fig. 14-13). The test measures the strength of the concrete—the measured strength being the direct shear strength of the concrete. This in turn is correlated with the compressive strength and thus a measurement of the in-place strength is made.

Fig. 14-13. Pullout test equipment being used to measure the strength of the concrete.

Dynamic or vibration tests. A dynamic or vibration (pulse velocity) test (ASTM C 597) is based on the principle that velocity of sound in a solid can be measured by (1) determining the resonant frequency of a specimen or (2) recording the travel time of short pulses of vibrations through a sample. High velocities are indicative of good concrete and low velocities are indicative of poor concrete.

Microseismic techniques employing low-frequency, mechanical energy can be used to detect, locate, and record discontinuities within solids. Modulus of elasticity as well as the presence and orientation of surface and internal cracking, can be determined. Fundamental transverse, longitudinal, and torsional frequencies of concrete specimens can be determined by ASTM C 215, a method frequently used in laboratory durability tests such as freezing and thawing (ASTM C 666).

Other tests. The use of X-rays for testing concrete properties is limited due to the costly and dangerous high-voltage equipment required as well as radiation hazards.

Gamma-radiography equipment can be used in the field to determine the location of reinforcement, density, and perhaps honeycombing in structural concrete units. ASTM C 1040 procedures use gamma radiation to determine the density of unhardened and hardened concrete in place.

Battery-operated magnetic detection devices like the pachometer or covermeter are available to measure the depth of reinforcement in concrete and to detect the position of rebars. Electrical-resistivity equipment is being developed to estimate the thickness of concrete pavement slabs.

A microwave-absorption method has been developed to determine the moisture content of porous building materials such as concrete. Acoustic-emission techniques show promise for studying load levels in structures and locating the origin of cracking.

Table 14-2 lists several nondestructive test methods along with main applications.

Evaluation of Compression Test Results

The Building Code Requirements for Reinforced Concrete (ACI 318) states that the compressive strength of concrete can be considered satisfactory if the averages of all sets of three consecutive strength tests equal or exceed the specified 28-day strength and if no individual strength test (average of two cylinders) is more than 500 psi below the specified strength.

If the strength of any laboratory-cured cylinder falls more than 500 psi below the specified strength, f'_c, the strength of the in-place concrete should be evaluated. The strength should also be evaluated if field-cured cylinders have a strength of less than 85% that of companion laboratory-cured cylinders. The 85% requirement may be waived if the field-cured strength exceeds f'_c by more than 500 psi.

When necessary, the in-place strength should be determined by testing three cores for each strength test where the laboratory-cured cylinders were more than 500 psi below f'_c. If the structure will be dry in service, prior to testing, the cores should be dried for 7 days at a temperature of 60 to 80°F and a relative humidity of less than 60%. The cores should be submerged in water for at least 40 hours before testing if the structure is to be used in a more than superficially wet state.

Nondestructive test methods are not a substitute for core tests (ASTM C 42). If the average strength of three cores is at least 85% of f'_c and if no single core is less than 75% of f'_c, the concrete in the area represented by the core is considered structurally adequate. If the results of properly made core tests are so low as to leave structural integrity in doubt, load tests as outlined in Chapter 20 of ACI 318 may be performed.

Table 14-2. Nondestructive Evaluation (NDE) Methods for Concrete Materials

Concrete properties	Recommended NDE methods	Possible NDE methods
Strength	Penetration probe Rebound hammer Pullout methods	
General quality and uniformity	Penetration probe Rebound hammer Ultrasonic pulse velocity Gamma radiography	Ultrasonic pulse echo
Thickness		Radar Gamma radiography Ultrasonic pulse echo
Stiffness	Ultrasonic pulse velocity	Proof loading (load-deflection)
Density	Gamma radiography Ultrasonic pulse velocity	Neutron density gage
Rebar size and location	Covermeter (pachometer) Gamma radiography	X-ray radiography Ultrasonic pulse echo Radar
Corrosion state of reinforcing steel	Electrical potential measurement	
Presence of subsurface voids	Acoustic impact Gamma radiography Ultrasonic pulse velocity	Thermal inspection X-ray radiography Ultrasonic pulse echo Radar
Structural integrity of concrete structures	Proof loading (load-deflection)	Proof testing using acoustic emission

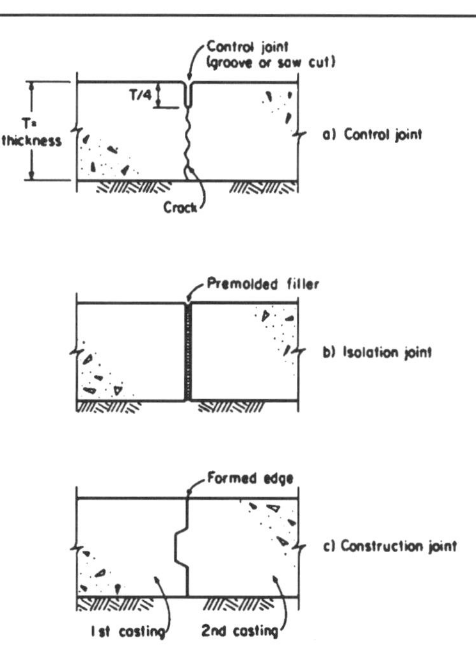

Fig. 1-16. The three basic types of joints used in concrete slab-on-ground construction.

Control of Cracking

Two basic causes of cracks in concrete are (1) stress due to applied loads and (2) stress due to drying shrinkage or temperature changes in restrained conditions.

Drying shrinkage is an inherent, unavoidable property of concrete; therefore, properly positioned reinforcing steel is used to reduce crack widths, or joints (Fig. 1-16) are used to predetermine and control the location of cracks. Thermal stress due to fluctuations in temperature can cause cracking, particularly at an early age.

Concrete shrinkage cracks occur because of restraint. When shrinkage occurs and there is no restraint, the concrete does not crack. Restraint comes from several sources. Drying shrinkage is always greater near the surface of concrete; the moist inner portions restrain the concrete near the surface, which can cause cracking. Other sources of restraint are reinforcing steel embedded in concrete, the interconnected parts of a concrete structure, and the friction of the subgrade on which concrete is placed.

Joints are the most effective method of controlling unsightly cracking. If a sizable expanse of concrete (a wall, slab, or pavement) is not provided with properly spaced joints to accommodate drying shrinkage and temperature contraction, the concrete will crack in a random manner.

Control joints are grooved, formed, or sawed into sidewalks, driveways, pavements, floors, and walls so that cracking will occur in these joints rather than in a random manner. Control joints permit movement in the plane of a slab or wall. They extend to a depth of approximately one-quarter the concrete thickness.

Isolation joints separate a slab from other parts of a structure and permit horizontal and vertical movements of the slab. They are placed at the junction of floors with walls, columns, footings, and other points where restraint can occur. They extend the full depth of the slab and include a premolded joint filler.

Construction joints occur where concrete work is concluded for the day; they separate areas of concrete placed at different times. In slabs-on-ground, construction joints usually align with and function as control or isolation joints.

Appendix

D

Western Wood Products Quality, Measurement Standards, and Lumber Definitions*

Method of measuring knots and evaluating acceptability in boards and timbers: pp. 205 to 210

Forces that act on wood structural members: pp. 210 to 214

Allowable, acceptable variations in cup, crook, twist: pp. 233 to 239

Definitions of various lumber terms: pp. 216 to 233

*This appendix is made up of pp. 205 to 214 and 216 to 239 from *Western Lumber Grading Rules '98*, Western Wood Products Association, 1998; reprinted by permission of WWPA.

■ 200.00 MEASUREMENTS OF CHARACTERISTICS

■ 210.00 KNOTS

In all Framing lumber 4″ and less in thickness, the size of a knot on a wide face is determined by its average dimension as in a line across the width of the piece. The size of knots on wide faces may be increased proportionately from the size permitted at the edge to the size permitted at the centerline. Knots appearing on narrow faces are limited to the same displacement as knots specified at edges of wide faces. Knots in Beams and Stringers and Posts and Timbers are measured differently than knots in 4″ and thinner material. Examples of these measurement methods are shown in Sections 212.00 and 213.00.

■ 211.00 WIDE FACE KNOTS 2″ to 4″ Thick Lumber

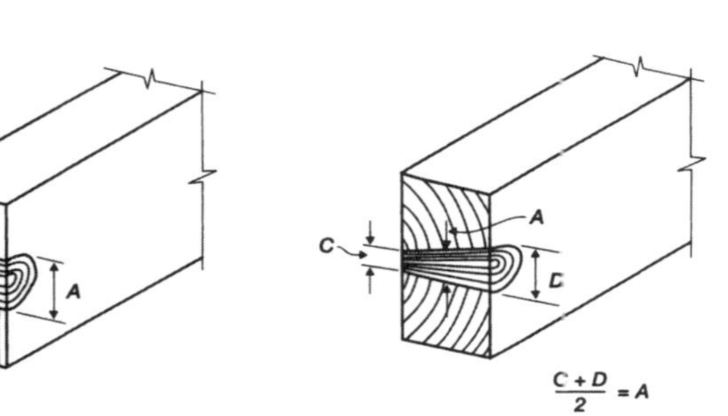

A – Equals allowable knot size.

■ 212.00 BEAM and STRINGER KNOTS 5″ and Thicker Lumber Width More Than 2″ Greater Than Thickness

B – Measure least dimension.
C – Measure between lines parallel with the edges.
D – Measure from edge of narrow face to line parallel with the edge.

■ 213.00 POST and TIMBER KNOTS 5″ x 5″ and Larger Lumber Width Not More Than 2″ Greater Than Thickness

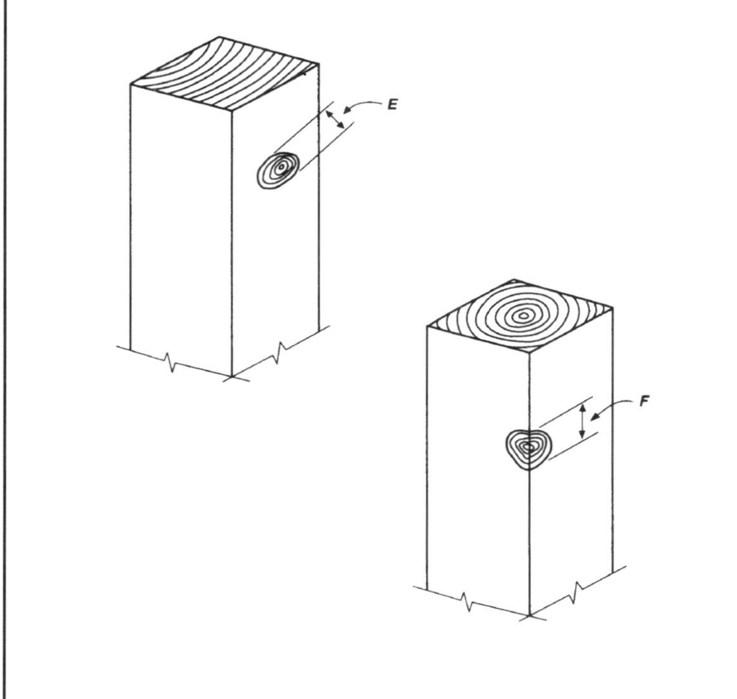

E – Measure least dimension.
F – Measure along corner or measure size most nearly representing diameter of branch causing the knot.

■ 220.00 SHAKES, CHECKS and SPLITS

In all grades of Beams and Stringers, these characteristics are measured only in the middle half of the width. Restrictions on checks apply for a distance from the ends equal to three times the width of the wide face. Shake is measured at the end between lines enclosing the shake and parallel to the wide face. Illustrations of how these characteristics are measured are as follows:

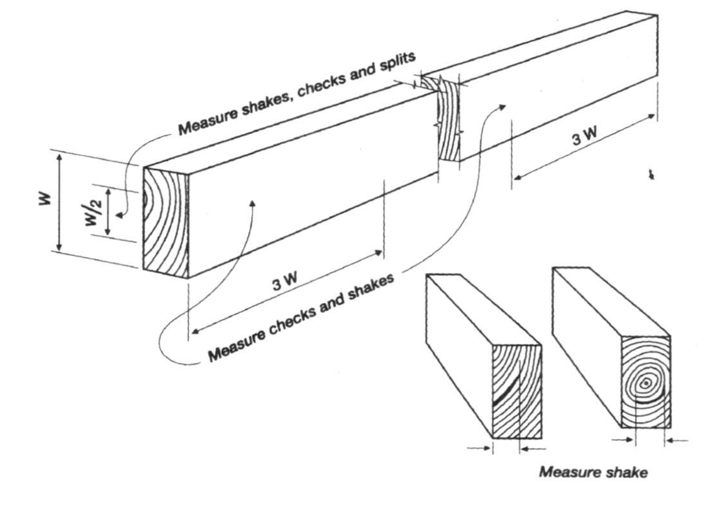

■ 221.00 SHAKES in Posts and Timbers are measured at the ends of pieces, between lines parallel with the two faces that give the least dimension.

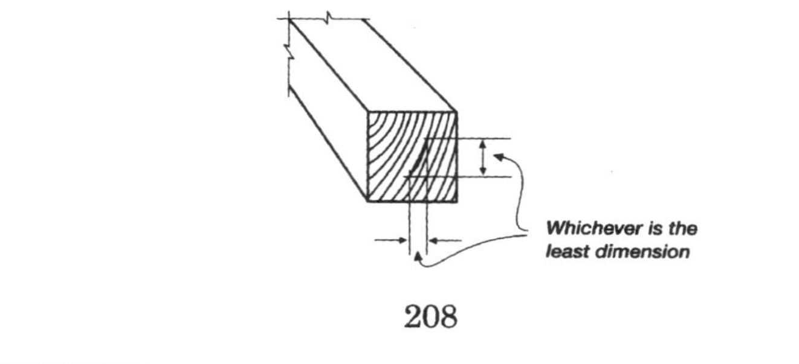

■ **222.00** CHECKS are measured as an average of the penetration perpendicular to the wide face. Where two or more checks appear on the same face, only the deepest one is measured. Where two checks are directly opposite each other, the sum of their depths is considered.

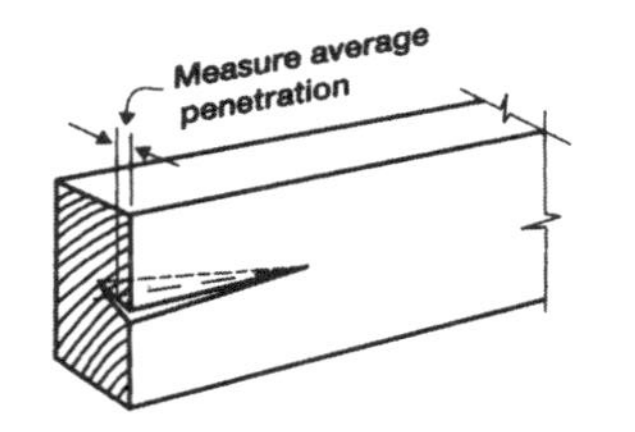

■ **223.00** SPLITS are measured as the penetration of a split from the end of the piece and parallel to the edges of the piece.

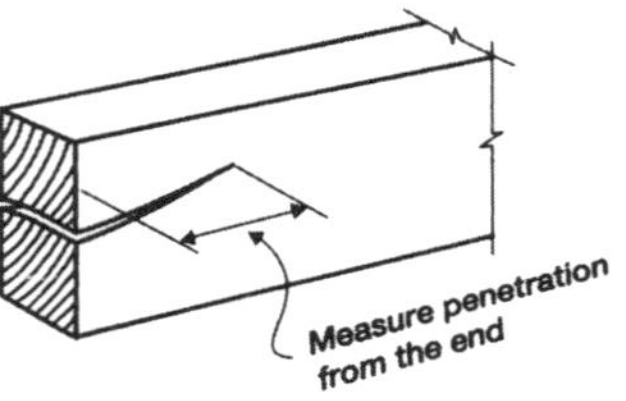

■ 230.00 SLOPE OF GRAIN

Slope of grain is the deviation of the wood fiber from a line parallel to the edges of a piece. The deviation is expressed as a ratio such as a slope of grain of 1 in 8, etc.

In lumber 2″ nominal and thicker and 4″ nominal and wider, slope of grain is measured over a sufficient length and area to be representative of the general slope of the fibers. Local deviations around knots and elsewhere are disregarded in the general slope measurement.

In thinner or narrower lumber, areas of local slope of grain exceeding the slope of grain provisions of the grade shall be graded like knots. Such areas are limited to the permitted knot displacement for the grade.

In lumber less than 1″ net in thickness, the average slope of grain anywhere in the length shall not pass completely through the thickness of the piece in a length less than the allowable slope.

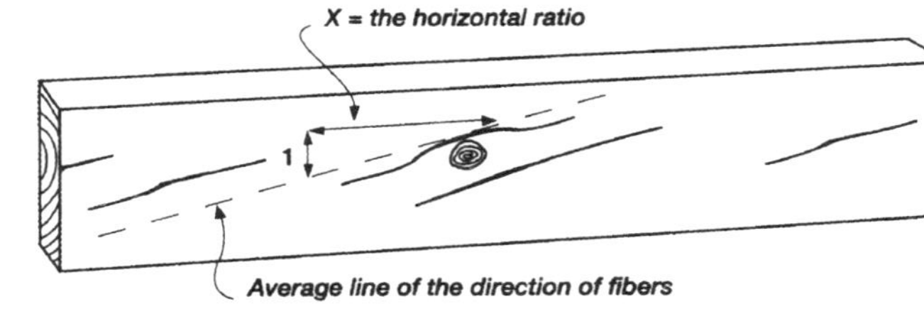

■ 240.00 STRESSES ILLUSTRATED

■ 241.00 EXTREME FIBER IN BENDING – (Fb) AND HORIZONTAL SHEAR – (Fv)

Structural members may carry loads on spans between supports and the lumber is stressed internally to the extent required to resist the external load. The loads cause pieces to bend, producing tension in the extreme fibers along the face farthest from the applied

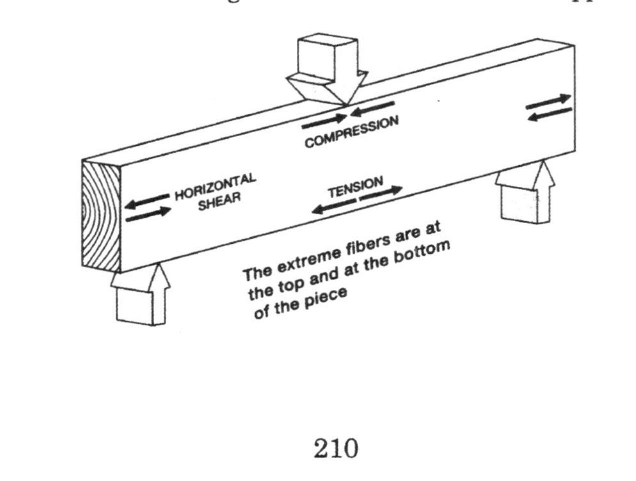

load and compression in the extreme fibers along the face nearest to the applied load. At the same time, over each support, there is a stress that tends to slide the fibers over each other horizontally. This action is similar to the way the ends of playing cards slide over each other when a deck is sharply bent. The internal force that resists this action is the horizontal shear value of the wood. The shearing stress is maximum at the center of the depth of the piece.

■ 242.00 MODULUS OF ELASTICITY – (E)

The relationship between the amount a piece deflects and the load causing the deflection determines its stiffness. This is called the modulus of elasticity of the species. A piece may deflect slightly or a lot depending on its size, the span, the load and the modulus of elasticity for the particular species. A large deflection is not necessarily a sign of insufficient strength. For example: the floors of a residence are usually limited to a deflection of 1/360 of the span, or less, while a Scaffold Plank may deflect substantially more.

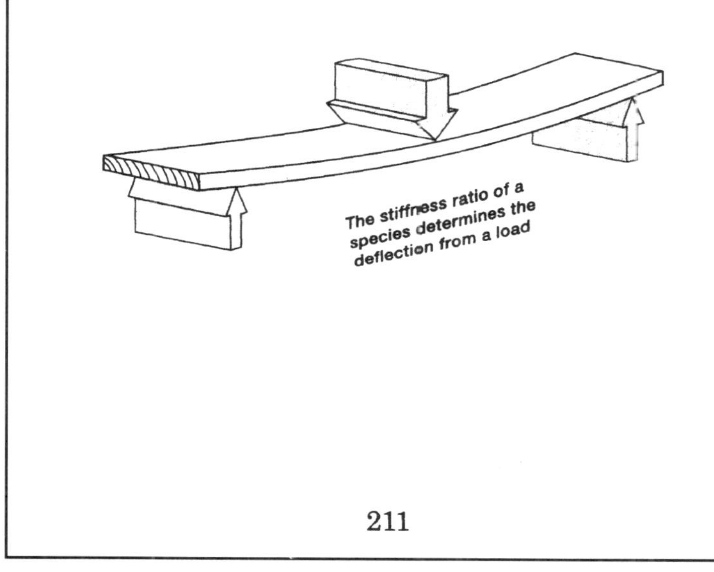

The stiffness ratio of a
species determines the
deflection from a load

■ 243.00 COMPRESSION PERPENDICULAR TO GRAIN – ($Fc_{\perp}$)

Where a joist, beam or similar piece of lumber bears on supports, the loads tend to compress the fibers. It is therefore necessary that the bearing area is sufficient to prevent side grain crushing.

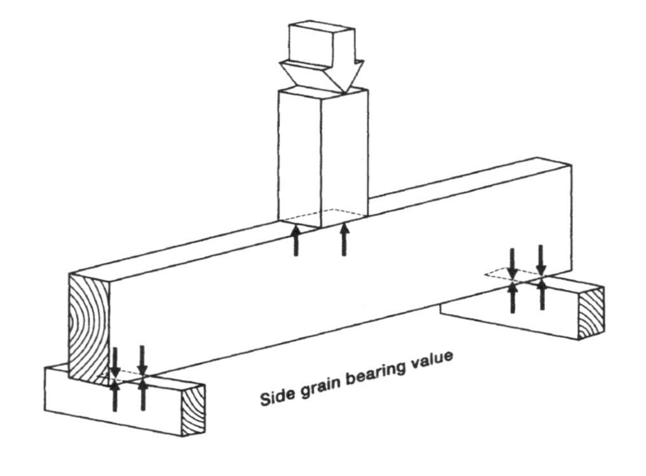

■ 244.00 COMPRESSION PARALLEL TO GRAIN – (Fc//)

In many parts of a structure, stress-grades are used with the loads supported on the ends of the pieces. Such uses are as studs, posts, columns and struts. The internal stress induced by this kind of loading is the same across the whole cross section and the fibers are uniformly stressed parallel to and along the full length of the piece.

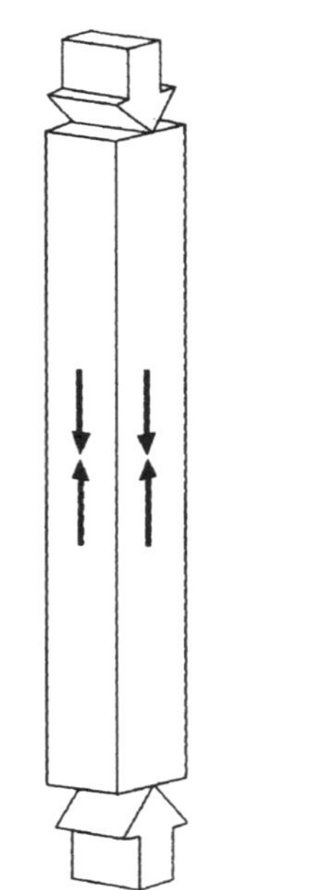

213

■ 245.00 FIBER STRESS IN TENSION – (Ft)

Tensile stresses are similar to compression parallel to grain in that they act across the full cross section and tend to stretch the piece. Length does not affect tensile stresses.

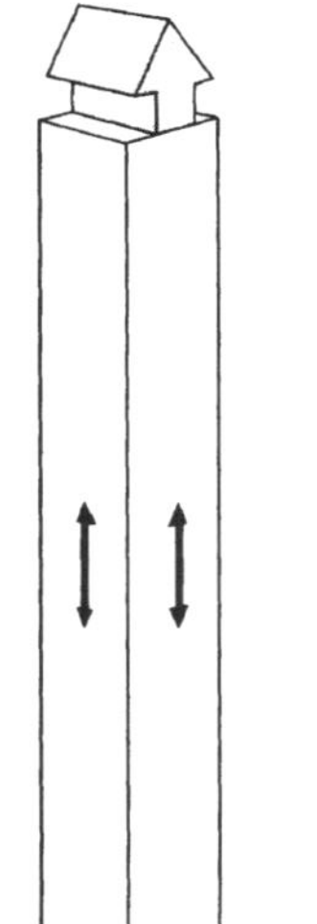

214

CUP TABLE

Cup	Face Width 2″&3″	4″	5″&6″	8″	10″	12″
Very Light	1/32	1/32	1/32	1/16	3/32	1/8
Light	1/32	1/32	1/16	1/8	3/16	1/4
Medium	1/32	1/16	1/8	3/16	1/4	3/8
Heavy	1/16	1/8	3/16	1/4	3/8	1/2

14″ and wider proportionately more.

CROOK TABLE for FRAMING and ALTERNATE BOARD RULES

Length in Feet	Crook	Face Width 2″	3″	4″	5″& 6″	8″	10″	12″
4 & 6	Very Light	1/8	1/8	1/8	1/8	1/16	1/16	1/16
	Light	1/4	1/4	1/4	3/16	1/8	1/16	1/16
	Medium	3/8	3/8	3/8	1/4	3/16	1/8	1/8
	Heavy	1/2	1/2	1/2	3/8	1/4	3/16	3/16
8	Very Light	1/4	1/4	3/16	1/8	1/8	1/16	1/16
	Light	3/8	3/8	3/8	5/16	1/4	3/16	1/8
	Medium	1/2	1/2	1/2	1/2	3/8	1/4	3/16
	Heavy	3/4	3/4	3/4	5/8	1/2	3/8	1/4
10	Very Light	3/8	5/16	1/4	3/16	3/16	1/8	1/8
	Light	3/4	5/8	1/2	7/16	3/8	1/4	3/16
	Medium	1 3/8	1	3/4	5/8	1/2	7/16	3/8
	Heavy	1 3/4	1 1/4	1 1/8	1	7/8	3/4	5/8
12	Very Light	1/2	3/8	3/8	5/16	1/4	1/4	3/16
	Light	1	3/4	11/16	5/8	1/2	7/16	3/8
	Medium	1 1/2	1 1/8	1	7/8	13/16	3/4	9/16
	Heavy	2	1 1/2	1 3/8	1 1/4	1 1/8	1	13/16
14	Very Light	5/8	1/2	7/16	3/8	5/16	1/4	3/16
	Light	1 1/4	1	7/8	3/4	5/8	1/2	3/8
	Medium	2	1 1/2	1 1/4	1 1/8	1	7/8	3/4
	Heavy	2 3/4	2	1 3/4	1 1/2	1 1/4	1 1/8	1
16	Very Light	3/4	5/8	1/2	7/16	3/8	5/16	1/4
	Light	1 5/8	1 1/4	1	7/8	3/4	5/8	1/2
	Medium	2 1/2	1 7/8	1 1/2	1 3/8	1 1/8	1	7/8
	Heavy	3 1/4	2 1/2	2	1 3/4	1 1/2	1 1/4	1 1/8

(continued next page)

CROOK TABLE for FRAMING and ALTERNATE BOARD RULES, CONT.

Length in Feet	Crook	2″	3″	4″	5″& 6″	8″	10″	12″
18	Very Light	1	¾	⅝	½	7/16	⅜	5/16
	Light	2	1⅜	1⅛	1	⅞	¾	⅝
	Medium	3	2 1/16	1⅝	1½	1¼	1⅛	1
	Heavy	4	2¾	2¼	2	1¾	1½	1¼
20	Very Light	1⅛	⅞	¾	⅝	½	7/16	⅜
	Light	2¼	1½	1⅜	1¼	1	⅞	¾
	Medium	3⅜	2¼	2 1/16	1⅞	1½	1 5/16	1⅛
	Heavy	4½	3	2¾	2½	2	1¾	1½
22	Very Light	1¼	1	⅞	¾	⅝	½	7/16
	Light	2½	1¾	1⅝	1½	1¼	1	⅞
	Medium	3¾	2⅝	2 7/16	2¼	1⅞	1½	1¼
	Heavy	5	3½	3¼	3	2½	2	1¾
24	Very Light	1½	1⅛	1	⅞	¾	⅝	½
	Light	3	2	1⅞	1¾	1½	1¼	1
	Medium	4½	3	2¾	2⅝	2¼	1⅞	1⅝
	Heavy	6	4	3¾	3½	3	2½	2¼

235

TWIST TABLE

Length in Feet	Twist	2″	3″& 4″	5″& 6″	8″	10″	12″
4	Very Light	1/16	⅛	3/16	¼	5/16	⅜
	Light	⅛	¼	⅜	½	⅝	¾
	Medium	3/16	⅜	½	¾	⅞	1⅛
	Heavy	¼	½	¾	1	1¼	1½
6	Very Light	3/32	3/16	5/16	⅜	7/16	9/16
	Light	3/16	⅜	½	¾	⅞	1⅛
	Medium	9/32	½	¾	1⅛	1⅜	1⅝
	Heavy	⅜	¾	1⅛	1½	1⅞	2¼
8	Very Light	⅛	¼	⅜	½	⅝	¾
	Light	¼	½	¾	1	1¼	1½
	Medium	⅜	¾	1⅛	1½	1⅞	2¼
	Heavy	½	1	1½	2	2½	3
10	Very Light	5/32	5/16	7/16	⅝	¾	15/16
	Light	5/16	⅝	⅞	1¼	1½	1⅞
	Medium	½	⅞	1⅜	1⅞	2⅜	2¾
	Heavy	⅝	1¼	1⅞	2½	3⅛	3¾
12	Very Light	3/16	⅜	9/16	¾	15/16	1⅛
	Light	⅜	¾	1⅛	1½	1⅞	2¼
	Medium	9/16	1⅛	1⅝	2¼	2¾	3⅜
	Heavy	¾	1½	2¼	3	3¾	4½

(continued next page)

236

TWIST TABLE, CONT.

Length in Feet	Twist	Face Width 2″	3″& 4″	5″& 6″	8″	10″	12″
14	Very Light	7/32	7/16	5/8	7/8	1 1/16	1 5/16
	Light	7/16	7/8	1 1/4	1 3/4	2 1/8	2 5/8
	Medium	5/8	1 1/4	1 7/8	2 5/8	3 1/4	3 7/8
	Heavy	7/8	1 3/4	2 5/8	3 1/2	4 3/8	5 1/4
16	Very Light	1/4	1/2	3/4	1	1 1/4	1 1/2
	Light	1/2	1	1 1/2	2	2 1/2	3
	Medium	3/4	1 1/2	2 1/4	3	3 3/4	4 1/2
	Heavy	1	2	3	4	5	6
18	Very Light	5/16	9/16	13/16	1 1/8	1 7/16	1 11/16
	Light	9/16	1 1/8	1 5/8	2 1/4	2 3/4	3 3/8
	Medium	7/8	1 5/8	2 1/2	3 3/8	4 1/4	5
	Heavy	1 1/8	2 1/4	3 3/8	4 1/2	5 5/8	6 3/4
20 and Longer	Very Light	5/16	5/8	15/16	1 1/4	1 9/16	1 7/8
	Light	5/8	1 1/4	1 7/8	2 1/2	3 1/8	3 3/4
	Medium	1	1 7/8	2 3/4	3 3/4	4 5/8	5 5/8
	Heavy	1 1/4	2 1/2	3 3/4	5	6 1/4	7 1/2

Maximum twist is limited to the amount shown in the above table for the appropriate lengths and widths. Pieces differing in length and width may have twist proportionate to the amount shown. Maximum twist is limited to occasional pieces of an item.

CROOK TABLE for SELECTS and FINISH

Length		Face Width 4″	6″	8″	10″	12″
8 Feet	C Select & Btr., Superior	1/4	1/4	3/16	3/16	1/8
	D Select, Prime	3/8	3/8	5/16	5/16	1/4
10 Feet	C Select & Btr., Superior	3/8	5/16	5/16	1/4	3/16
	D Select, Prime	9/16	9/16	1/2	7/16	3/8
12 Feet	C Select & Btr., Superior	9/16	1/2	7/16	3/8	5/16
	D Select, Prime	7/8	3/4	11/16	5/8	9/16
14 Feet	C Select & Btr., Superior	3/4	11/16	9/16	1/2	3/8
	D Select, Prime	1 1/8	1 1/16	15/16	7/8	3/4
16 Feet	C Select & Btr., Superior	1	7/8	3/4	5/8	1/2
	D Select, Prime	1 1/2	1 3/8	1 1/4	1 1/8	1

In the grades of Selects and Finish, maximum crook is limited to the amount shown in the above table for the appropriate length, width and grade. Pieces differing in length and width from these basic sizes may have crook in proportion to the amounts shown. Maximum crook is limited to occasional pieces of any item.

CROOK TABLE for COMMON BOARDS

Length		Face Width 4″	6″	8″	10″	12″
8 Feet	2 & Btr. Com.	½	7/16	⅜	5/16	¼
	3 Com.	13/16	¾	11/16	⅝	½
	4 Com.	1	15/16	⅞	13/16	¾
10 Feet	2 & Btr. Com.	13/16	11/16	9/16	½	⅜
	3 Com.	1¼	1 3/16	1 1/16	1	13/16
	4 Com.	1 9/16	1 7/16	1⅜	1¼	1 3/16
12 Feet	2 & Btr. Com.	1⅛	1	⅞	11/16	9/16
	3 Com.	1 13/16	1 11/16	1 9/16	1 7/16	1⅛
	4 Com.	2¼	2⅛	2	1 13/16	1 11/16
14 Feet	2 & Btr. Com.	1 9/16	1 5/16	1⅛	15/16	¾
	3 Com.	2½	2 5/16	2⅛	1 5/16	1 9/16
	4 Com.	3 1/16	2⅞	2 11/16	2½	2 5/16
16 Feet	2 & Btr. Com.	2	1¾	1½	1¼	1
	3 Com.	3¼	3	2¾	2½	2
	4 Com.	4	3¾	3½	3¼	3

Maximum crook is limited to the amount shown in the above table for the appropriate length, width and grade. Pieces differing in length and width from these basic sizes may have crook in proportion to the amounts shown. Maximum crook is limited to occasional pieces of any item.

DEFINITIONS

■ **702.00 BURL** – A distortion of grain, usually caused by abnormal growth due to injury of the tree. The effect of burls is assessed in relation to knots.

■ **704.00 CHECKS** – A separation of the wood normally occurring across or through the rings of annual growth and usually as a result of seasoning.

(a) A surface check occurs on a face of a piece.

(b) A through check extends from one surface of a piece to the opposite or adjoining surface.

(c) Small checks are not over 1/32″ wide and not over 4″ long.

(d) Medium checks are not over 1/32″ wide and not over 10″ long.

(e) Large checks are more than 1/32″ wide or longer than 10″ or both.

(f) A roller check is a crack in the wood structure caused by a piece of cupped lumber being flattened in passing between the machine rollers.

A light roller check is a perceptible opening not over 2′ long.

A medium roller check is a perceptible opening over 2′ long but not exceeding 4′ in length.

A heavy roller check is over 4′ in length.

■ **706.00 COMPRESSION WOOD** – Abnormal wood that forms on the underside of leaning and crooked coniferous trees. It is characterized, aside from its distinguishing color, by being hard and brittle and by its relatively lifeless appearance. Compression wood shall be limited in effect to other appearance or strength reducing characteristics permitted in the grade.

■ **708.00 DECAY (UNSOUND WOOD)** – A disintegration of the wood substance due to action of wood-destroying fungi, and is also known as dote or rot. Some examples are as follows:

(a) Heart center decay is a localized decay developing along the pith in some species and is detected by visual inspection. The limitation for heart center decay applies to Southern Pine. Heart center decay develops in the living tree and does not progress further after the tree is cut.

(b) White specks are small white or brown pits or spots in wood caused by the fungus "Fomes pini". It develops in the living tree and does not develop further in wood in service. Where permitted in these rules it is so limited that it has no more effect on the intended use of the pieces than other characteristics permitted in the same grade. Pieces containing white speck are no more subject to decay than pieces which do not contain it. Note: "Firm" in relation to white speck infers that it will not crumble readily under thumb pressure and cannot be easily picked out.

(c) Honeycomb is similar to to white speck but the pockets are larger. Where permitted in the rules it is so limited that it has no more effect on the intended use of the piece than other characteristics permitted in the same grade. Pieces containing honeycomb are no more subject to decay than pieces which do not contain it. Note: "Firm" in relation to honeycomb infers that it will not crumble readily under thumb pressure and cannot be easily picked out.

(d) Incipient decay is an early stage of decay in which disintegration of the wood fibers has not proceeded far enough to soften or otherwise change the hardness of the wood perceptibly. It is usually accompanied by a slight discoloration or bleaching of the wood.

(e) Peck is channeled or pitted areas or pockets found in cedar and cypress. Wood tissue between pecky areas remains unaffected in appearance and strength. All further growth of the fungus causing peckiness ceases after the trees are felled.

■ **710.00 EDGE** – There are three meanings for edge: (1) The narrow face of rectangular-shaped pieces. (2) The corner of a piece at the intersection of two longitudinal faces. (3) In stress grades that part of the wide face nearest the corner of the piece.

(a) Eased edges means slightly rounded surfacing on pieces of lumber to remove sharp corners. The standard radius for 1″, 2″, 3″ and 4″ nominal thickness lumber shall not exceed 1/16″, 1/8″, 3/16″ and 1/4″ respectively. Note: Lumber 4″ or less in thickness is frequently shipped with eased edges unless otherwise specified. *[In 2″ thicknesses, a 3/16″ eased edge is allowed on STUD grade lumber only and must be indicated on the grade stamp as 3/16 EE.]

(b) Square edged means free from wane and without eased edges.

(c) Free of wane means without wane but has either eased or square edges. (See WANE definition.)

(d) Square corners means without eased edges but has an allowance for wane in certain grades.

(e) To "destroy the nailing edge" shall mean (1) the decay occupies more of the narrow face than the allowable maximum wane in thickness when in streak form, or (2) the decay occupies more than twice the length of the allowable knot hole when a spot occurs completely through the narrow face.

■ **712.00 GRAIN** – The fibers in wood and their direction, size, arrangement, appearance or quality.

(a) For requirements and method of measuring medium grain, close grain and dense material, see Section 170.00.

(b) Slope of grain is the deviation of the line of fibers from a straight line parallel to the sides of the piece. For method of measurement, see Section 230.00.

---

*Provisions in brackets are only for lumber graded under WWPA rules.

(c) Summerwood is the portion of the annual growth ring formed during the latter part of the yearly growth ring. It is darker in color, more dense, and stronger mechanically than springwood.
(d) Springwood is the portion of the annual growth ring formed during the early part of the yearly growth period. It is lighter in color, less dense, and not as strong mechanically as summerwood.
(e) Vertical grain (VG) (Edge grain EG) (Rift grain) lumber is a piece or pieces sawn at approximately right angles to the annual growth rings so that the rings form an angle of 45 degrees or more with the surface of the piece.
(f) Flat grain (FG) (Slash grain SG) lumber is a piece or pieces sawn approximately parallel to the annual growth rings so that all or some of the rings form an angle of less than 45 degrees with the surface of the piece.
(g) Mixed grain (MG) lumber includes either or both vertical and flat grain pieces.
(h) Spiral grain is a deviation in the slope of grain caused when the fibers in a tree take a spiral course around the trunk of the tree, instead of the normal vertical course.
(i) Diagonal grain is a deviation in the slope of grain caused by sawing at an angle with the bark of the tree. See slope of grain.

■ **714.00 HEART** – (Heartwood) Inner core of the tree trunk comprising the annual rings containing nonliving elements. In some species, heartwood has a prominent color different from sapwood.

(a) Boxed heart means with the pith enclosed in the piece.
(b) Heart center is the pith or center core of the log.
(c) Free of heart centers (FOHC) means without pith (side cut). An occasional piece (see Section 726.00) when showing pith for not more than 1/4 the length on the surface shall be accepted.

(d) Firm red heart is a stage of incipient decay characterized by a reddish color in the heartwood, which does not render the wood unfit for the majority of yard purposes.
(e) Heartwood and sapwood of equivalent character are of equal strength. No requirement of heartwood is made when strength alone is the governing factor.
(f) Heartwood is more durable than sapwood. When wood is to be exposed to decay-producing conditions without preservative treatment, it shall be permitted to specify the minimum percentage of heartwood to be present in all pieces in a shipment.* [Decking grades (Sections 55.00 and 55.20) or National Grading Rule grades (Sections 40.00, 41.00, 42.00 and 62.00) of Cedar may be grade-stamped as heartwood. In so stamped, it shall be heartwood except that an occasional piece may contain sapwood showing on one side only providing 90 percent or more of the side on which it appears is heartwood.]
(g) Sapwood takes preservative treatment more readily than heartwood.

■ **716.00 HOLES** – Holes either extend partially or wholly through the piece. An alternate designation for holes which extend only partially through the piece is surface pits. Unless otherwise specified, holes are measured the same as knots. Holes are classified by size as follows:

(a) A pin hole is not over 1/16" in diameter.
(b) A medium (small) hole is not over 1/4" in diameter.
(c) A large hole is not over 1" in diameter.
(d) A very large hole is over 1" in diameter.

■ **718.00 KNOTS** – A portion of a branch or limb that has become incorporated in a piece of lumber. In lumber,

*Provisions in brackets are only for lumber graded under WWPA rules.

knots are classified as to form, size, quality and occurrence. A red knot is one that results from a live branch growth in the tree and is intergrown with the surrounding wood. A black knot is one that results from a dead branch which the wood growth of the tree has surrounded.

(a) A round knot is produced when the limb is cut at approximately a right angle to its long axis.
(b) An oval knot is produced when the limb is cut at slightly more than a right angle to the long axis.
(c) A spike knot is produced when the limb is cut either lengthwise or diagonally.
(d) A pin knot is not over ½″.
(e) A small knot is not over ¾″.
(f) A medium knot is not over 1½″.
(g) A large knot is over 1½″.
(h) A sound knot contains no decay.
(i) A pith knot is sound in all respects except it contains a pith hole not over ¼″ in diameter.
(j) A hollow knot is a sound knot containing a hole greater than ¼″ in diameter. Through opening of a hollow knot is limited to the size of other holes permitted.
(k) An unsound knot contains decay.
(l) A "firm" knot is solid across its face but contains incipient decay.
(m) A tight knot is so fixed by growth, shape or position that it retains its place in the piece.
(n) An intergrown knot is one whose growth rings are partially or completely intergrown on one or more faces with the growth rings of the surrounding wood.
(o) A watertight knot has its annual rings completely intergrown with those of the surrounding wood on one surface of the piece, and it is sound on that surface.
(p) An encased knot is one which is not intergrown with the growth rings of the surrounding wood.
(q) A "loose" or "not firmly fixed" knot is one not held tightly in place by growth, shape or position.

(r) A "fixed" knot will retain its place in dry lumber under ordinary conditions but is movable under pressure though not easily pushed out.
(s) A knot cluster is two or more knots grouped together as a unit with the fibers of the wood deflected around the entire unit. A group of single knots is not a knot cluster.
(t) A star-checked knot has radial checks.
(u) Well-scattered knots are not in clusters and each knot is separated from any other by a distance at least equal to the diameter of the smaller of the two.
(v) Well-spaced knots means that the sum of the sizes of all knots in any 6″ of length of a piece must not exceed twice the size of the largest knot permitted. More than one knot of maximum permissible size must not be in same 6″ of length and the combination of knots must not be serious.

■ **720.00 MANUFACTURING IMPERFECTIONS** – Means all imperfections or blemishes which are the result of surfacing, such as the following:

(a) Chipped grain is a barely perceptible irregularity in the surface of a piece caused when particles of wood are chipped or broken below the line of cut. It is too small to be classed as torn grain and is not considered unless in excess of 25% of the surface is involved.
(b) Torn grain is an irregularity in the surface of a piece where wood has been torn or broken out by surfacing. Torn grain is described as follows:
    Very light torn grain – not over 1/64″ deep.
    Light torn grain – not over 1/32″ deep.
    Medium torn grain – not over 1/16″ deep.
    Heavy torn grain – not over ⅛″ deep.
    Very heavy torn grain – over ⅛″ deep.
(c) Raised grain is a roughened condition of the surface of dressed lumber in which the hard summerwood is raised above the softer springwood, but not torn loose from it.

*[Slight raised grain is an unevenness somewhat less than 1/64".]
Very light raised grain is not over 1/64".
Light raised grain is not over 1/32".
Medium raised grain is not over 1/16".
Heavy raised grain is not over 1/8".

(d) Loosened grain is a grain separation or loosening between springwood and summerwood without displacement.
Very light loosened grain is not over 1/64" separation.
Light loosened grain is not over 1/32" separation.
Medium loosened grain is not over 1/16" separation.
Heavy loosened grain is not over 1/8" separation.
Very heavy loosened grain is over 1/8" separation.

(e) Skips are areas on a piece that failed to surface clean. Skips are described as follows:
Very light skip is not over 1/64" deep, *[and approximately 6" in length.]
Light skip is not over 1/32" deep, *[on face, may be 12" in length and on edge may be 2' long.]
Medium skip is not over 1/16" deep, *[on face, may be 12" in length and on edge may be 2' long.]
Heavy skip is not over 1/8" deep.

(f) Hit and miss is a series of skips not over 1/16" deep with surfaced areas between.

(g) Hit or miss means completely or partly surfaced or entirely rough. Scantness may be 1/16".

(h) Mismatch is an uneven fit in worked lumber when adjoining pieces do not meet tightly at all points of contact or when the surfaces of adjoining pieces are not in the same plane.
Slight mismatch is a barely evident trace of mismatch.
Very light mismatch is not over 1/64".
Light mismatch is not over 1/32".

*Provisions in brackets apply to all except dimension lumber.

Medium mismatch is not over 1/16".
Heavy mismatch is not over 1/8".

(i) Machine burn is a darkening of the wood due to overheating by machine knives or rolls when pieces are stopped in machine.

(j) Machine bite is a depressed cut of the machine knives at the end of the piece.
Very light machine bite is not over 1/64" deep.
Light machine bite is not over 1/32" deep.
Medium machine bite is not over 1/16" deep.
Heavy machine bite is not over 1/8" deep.
Very heavy machine bite is over 1/8" deep.

(k) Machine gouge is a groove cut by the machine below the desired line.
*[Slight machine gouge is somewhat less than 1/64" deep.]
Very light machine gouge is not over 1/64" deep.
Light machine gouge is not over 1/32" deep.
Medium machine gouge is not over 1/16" deep.
Heavy machine gouge is not over 1/8" deep.
Very heavy machine gouge is over 1/8" deep.

(l) A machine offset is an abrupt dressing variation in the edge surface which usually occurs near the end of the piece and without reducing the width or without changing the plane of the wide surface.
Very light machine offset is a variation not over 1/64".
Light machine offset is a variation not over 1/32".
Medium machine offset is a variation not over 1/16".
Heavy machine offset is a variation not over 1/8".
Very heavy machine offset is a variation over 1/8".

(m) Chip marks are shallow depressions or indentations on or in the surface of dressed lumber caused by shavings or chips getting embedded in the surface during dressing.

*Provisions in brackets apply to all except dimension lumber.

*[Slight chip marks are somewhat less than 1/64" deep.]
Very light chip marks are not over 1/64" deep.
Light chip marks are not over 1/32" deep.
Medium chip marks are not over 1/16" deep.
Heavy chip marks are not over 1/8" deep.

(n) Knife marks are the imprints or markings of the machine knives on the surface of dressed lumber.
Very slight knife marks are visible only from a favorable angle and are perfectly smooth to the touch.
Slight knife marks are readily visible but evidence no unevenness to the touch.

(o) Wavy dressing involves more uneven dressing than knife marks.
*[Very slight wavy dressing evidences unevenness that is barely perceptible to the touch.]
*[Slight wavy dressing evidences perceptible unevenness that is somewhat less than 1/64" deep.]
Very light wavy dressing is not over 1/64" deep.
Light wavy dressing is not over 1/32" deep.
Medium wavy dressing is not over 1/16" deep.
Heavy wavy dressing is not over 1/8" deep.
Very heavy wavy dressing is over 1/8" deep.

■ 722.00 MANUFACTURING IMPERFECTIONS CLASSIFICATION –

(a) Standard "A" Manufacture admits: Very light torn grain; *[occasional slight chip marks]; occasional very light chip marks; very slight knife marks.

(b) Standard "B" Manufacture admits: Very light torn grain; very light raised grain; very light loosened grain; *[Slight chip marks]; very light chip

*Provisions in brackets apply to all except dimension lumber.

marks; average of one *[Slight chip mark] very light chip mark per lineal foot but not more than two in any lineal foot; very slight knife marks; slight mismatch.

(c) Standard "C" Manufacture admits: Medium torn grain; light raised grain; light loosened grain; very light machine bite; very light machine gouge; very light machine offset; light chip marks if well-scattered; occasional medium chip marks; very slight knife marks; slight mismatch.

(d) Standard "D" Manufacture admits: Heavy torn grain; medium raised grain; very heavy loosened grain; light machine bite; light machine gouge; light machine offset; medium chip marks; slight knife marks; very light mismatch.

(e) Standard "E" Manufacture admits: Very heavy torn grain; raised grain; very heavy loosened grain; medium machine bite; machine gouge; medium machine offset; chip marks, knife marks; light wavy dressing; light mismatch.

(f) Standard "F" Manufacture admits: Very heavy torn grain; raised grain; very heavy loosened grain; heavy machine bite; machine gouge; heavy machine offset; chip marks; knife marks; medium wavy dressing; medium mismatch.

■ 723.00 METRIC CONVERSION – Metric dimensions listed in these rules are calculated at 25.4 millimeters (mm) times the actual dimension in inches, rounded to the nearest millimeter. In case of a dispute on size measurements, the conventional (inch) method of measurement shall take precedence.

(a) To round to the nearest millimeter: If the digit in the tenths of mm position (the digit after the decimal point) is less than 5, drop all fractional mm digits; if greater than 5 or it is 5 followed by at least one non-zero digit, round one mm higher; if

* Provisions in brackets apply to all except dimension lumber.

5 followed by only zeroes, retain the digit in the unit position (the digit before the decimal point) if it is even or increase it one mm if it is odd.

■ **724.00 MOISTURE CONTENT** – The weight of the water in wood expressed in percentage of the weight of the oven-dry wood.

■ **726.00 OCCASIONAL PIECES** – Means not more than 10% of the pieces in a parcel or shipment.

■ **728.00 PITCH** – Is an accumulation of resinous material.

(a) Light pitch is the light but evident presence of pitch.
(b) Medium pitch is a somewhat more evident presence of pitch than is the light.
(c) Heavy pitch is a very evident accumulation of pitch showing by its color and consistency.
(d) Massed pitch is a clearly defined accumulation of solid pitch in a body by itself.

■ **730.00 PITCH STREAK** – Is a well-defined accumulation of pitch in the wood cells in a streak. Pitch streaks are described as follows, with equivalent areas being permissible:

(a) Very small pitch streak ⅜″ in width and 15″ in length.
(b) Small pitch streak 1/12 the width and 1/6 the length of the piece.
(c) Medium pitch streak 1/6 the width and 1/3 the length of the piece.
(d) A large pitch streak is not over 1/4 the width by 1/2 the length of the surface.
(e) A very large pitch streak is over 1/4 the width by 1/2 the length of the surface.
(f) A pitch seam is a shake or check which contains pitch.

■ **732.00 PITH** – Pith is the small soft core in the structural center of the log.

(a) Very small pith is not over ⅛″ wide and occupies on face surface not over ¼ square inch (⅛″ x 2″, or 1/16″ x 4″).
(b) Small pith occupies not over ¾ square inch (¼″ x 3″, 3/16″ x 4″, ⅛″ x 6″, or 1/16″ x 12″).
(c) Free of pith means that pith on or within the body of the piece is prohibited.

■ **734.00 POCKET** – A well-defined opening between the rings of annual growth which develops during the growth of the tree. It usually contains pitch or bark. Pockets are described as follows with equivalent areas being permissible:

(a) Very small pocket – 1/16″ in width and 3″ in length, or ⅛″ in width and 2″ in length.
(b) Small pocket – 1/16″ in width and 6″ in length, or ⅛″ in width and 4″ in length, or ¼″ in width and 2″ in length.
(c) Medium pocket – 1/16″ in width and 12″ in length, or ⅛″ in width and 8″ in length, or ⅜″ in width and 4″ in length.
(d) A large pocket is not over 4 square inches in area.
(e) A very large pocket is over 4 inches in area.
(f) A closed pocket has an opening on one surface only.
(g) A through or open pocket has an opening on opposite surfaces, and the through opening is considered the same as a through hole of equal size.

■ **736.00 PLUGS AND FILLERS** – Wood plugs and fillers are inserted into pieces of lumber to improve their appearance and usefulness. Lumber containing plugs and fillers shall only be shipped when the order, acknowledgement and invoice carry reference to the inserts. Quality of the inserts and workmanship must be in keeping with the quality of the grade. In Dimension and other lumber graded for strength, inserts are limited to the same size and location as knots.

■ **738.00 SAPWOOD** – Outer layers of growth between the bark and the heartwood which contain the sap.

(a) Bright sapwood shows no stain and is not limited in any grade unless specifically stated in the grade description.

(b) Sapwood restrictions waived means that any restriction in a rule on the amount of sapwood permitted in pieces graded under that rule are not to apply.

(c) Bright sapwood no defect (BSND) means that bright sapwood is permitted in each piece in any amount.

■ **740.00 SHAKE** – A lengthwise separation of the wood which occurs between or through the rings of annual growth.

(a) A light shake is not over 1/32" wide.

(b) A medium shake is not over 1/8" wide.

(c) A surface shake occurs on only one surface of a piece.

(d) A through shake extends from one surface of a piece to the opposite or to an adjoining surface.

(e) A pith shake (or heart shake or heart check) extends through the growth rings from or through the pith towards the surface of a piece, and is distinguished from a seasoning check by the fact that its greatest width is nearest the pith, whereas the greatest width of a seasoning check in a pith-centered piece is farthest from the pith.

(f) A ring shake occurs between the growth rings to partially or wholly encircle the pith.

■ **742.00 SPLITS** – A separation of the wood through the piece to the opposite surface or to an adjoining surface due to the tearing apart of the wood cells.

(a) A very short split is equal in length to 1/2 the width of the piece.

(b) A short split is equal in length to the width of the piece and in no case exceeds 1/6 the length.

(c) A medium split is equal in length to twice the width of the piece and in no case exceeds 1/6 the length.

(d) A long split is longer than a medium split.

■ **744.00 STAINED WOOD** –

(a) Stained Heartwood and Firm Red Heart – Stained Heartwood or Firm Red Heart is a marked variation from the natural color. Note: It ranges from pink to brown. It is not to be confused with natural red heart. Natural color is usually uniformly distributed through certain annual rings, whereas stains are usually in irregular patches. In grades where it is permitted, it has no more effect on the intended use of the piece than other characteristics permitted in the grade.

(b) Stained Sapwood – Stained Sapwood similarly has no effect on the intended use of the pieces in which it is permitted but affects appearance in varying degrees.

(1) Light stained sapwood is so slightly discolored that it does not affect natural finishes.

(2) Medium stained sapwood has a pronounced difference in coloring. Note: Sometimes the usefulness for natural finishes but not for paint finishes is affected.

(3) Heavy stained sapwood has so pronounced a difference in color as to obscure the grain of the wood but the lumber containing it is acceptable for paint finishes.

(c) Discoloration through exposure to the elements is admitted in all grades of framing and sheathing lumber.

■ **746.00 STRESS GRADES** – Lumber grades having assigned working stress and modulus of elasticity values in accordance with accepted basic principles of strength grading, and the provisions of sections 6.3.2.1 and 6.3.2.2 of Voluntary Product Standard 20-94.

■ **748.00 TRIM –**

(a) Trimming of lumber is the act of cross-cutting a piece to a given length.

(b) Double end trimmed (DET). Note: It is intended that DET lumber be trimmed square on both ends. Tolerances are found in certified grading rules.

(c) Precision end trimmed (PET) lumber is trimmed square on both ends to uniform lengths with a manufacturing tolerance of 1/16″ over or under in length in 20% of the pieces.

(d) Square end trimmed lumber is trimmed square having a manufacturing tolerance of 1/64″ for each nominal 2″ of thickness or width.

(e)* [Pencil Trimmed lumber is lumber that is not physically trimmed by saws, but is marked in some manner to indicate the piece needs to be trimmed to make a particular grade. Pencil trim is permitted on either or both ends subject to agreement between buyer and seller. Shipments of pencil trimmed lumber are permitted only when ordered, acknowledged and invoiced as such. It is not permissible to gradestamp pencil trimmed lumber.]

■ **750.00 WANE** – Bark or lack of wood from any cause, except eased edges, on the edge or corner of a piece of lumber.

Wane away from ends extending partially or completely across any face is permitted for 1′ if no more serious than skips in dressing allowed or across a narrow face if no more damaging than the knot hole allowed (not to exceed in length twice the diameter of the maximum knot hole allowed in the grade) and is limited to one occurrence in each piece. These variations shall not be allowed in more than 5% of the pieces. (This provision applies only to the National Grading Rule for Dimension Lumber).

*Provisions in brackets are only for lumber graded under WWPA rules.

■ **752.00 WARP** – Any deviation from a true or plane surface, including bow, crook, cup and twist or any combination thereof. Warp restrictions are based on the average form of warp as it occurs normally, and any variation from this average form, such as short kinks, shall be appraised according to its equivalent effect. Pieces containing two or more forms shall be appraised according to the combined effect in determining the amount permissible. In these rules warp is classified as very light, light, medium, and heavy, and applied to each width and length as set forth in the various grades in accordance with the following provisions and tables:

(a) Bow is a deviation flatwise from a straight line drawn from end to end of a piece. It is measured at the point of greatest distance from the straight line. The maximum amount of bow allowed in a grade is as follows: If under 2″ thick, three times as much as crook for 2″ faces. If 2″ thick and under 3″, twice as much as crook for 2″ faces. If 3″ thick and over, the same as the amount of crook for that thickness.

(b) Crook is a deviation edgewise from a straight line drawn from end to end of a piece. It is measured at the point of greatest distance from the straight line. The maximum amount of crook allowed shall be that shown in the crook tables on pages 234, 235, 238 and 239.

(c) Cup is a deviation in the face of a piece from a straight line drawn from edge to edge of a piece. It is measured at the point of greatest distance from the straight line. The maximum amount of cup allowed shall be that shown in the cup table.

(d) Twist is a deviation flatwise, or a combination of flatwise and edgewise, in the form of a curl or spiral, and the amount is the distance an edge of a piece at one end is raised above a flat surface against which both edges at the opposite end are resting snugly. The maximum amount of twist allowed shall be that shown in the twist tables on pages 236 and 237.

■ **754.00 COMBINATION GRADES** – Product Standard PS 20 permits grouping the highest two grades in a grade category, and grade marking the combination as an "& Better" grade. The combined grade is assigned the allowable property values of the lower grade unless allowable property values have been assigned to the combination. [In the case of "No. 1 & Better," data collected for Douglas Fir, Larch, Douglas Fir-Larch, and Hem-Fir during the U. S. In-grade testing program permits development of allowable property values specific to the combination grade. When the "No. 1 & Better" grade combination is assigned specific allowable properties, as for Douglas Fir, Larch, Douglas Fir-Larch, and Hem-Fir, the material is required to be stamped with a "No. 1 & Better" grade stamp. If the lumber is gradestamped as "Select Structural" and "No. 1" rather than "No. 1 & Better," the values assigned to the individual grades apply.]

Appendix

E

Useful Tables and Formulas*

*This appendix is made up of pp. 306, 309, 358, 359, 586 to 588, 591, 596 to 600, and 603 from Sidney M. Levy, *Construction Site Work, Site Utilities, and Substructures Databook*, McGraw-Hill, 2001; reprinted by permission of the source organization.

9.0.0 Nails: Penny Designation ("d") and Lengths (U.S. and Metric)

Nail—penny size	Length in inches	Length in millimeters
2d	1	25.40
3d	1 1/4	31.75
4d	1 1/2	38.10
5d	1 3/4	44.45
6d	2	50.80
7d	2 1/4	57.15
8d	2 1/2	63.50
9d	2 3/4	69.85
10d	3	76.20
12d	3 1/4	82.55
16d	3 1/2	88.90
20d	3 3/4	95.25
30d	4 1/2	114.30
40d	5	127.00
50d	5 1/2	139.70
60d	6	152.40

9.1.0 Stainless Steel Sheets (Thickness and Weights)

Gauge	Thickness inches	mm.	Weight lb/ft^2	kg/m^2
8	0.17188	4.3658	7.2187	44.242
10	0.14063	3.5720	5.9062	28.834
11	0.1250	3.1750	5.1500	25.6312
12	0.10938	2.7783	4.5937	22.427
14	0.07813	1.9845	3.2812	16.019
16	0.06250	1.5875	2.6250	12.815
18	0.05000	1.2700	2.1000	10.252
20	0.03750	0.9525	1.5750	7.689
22	0.03125	0.7938	1.3125	6.409
24	0.02500	0.6350	1.0500	5.126
26	0.01875	0.4763	0.7875	3.845
28	0.01563	0.3970	0.6562	3.1816
Plates				
3/16"	0.1875	4.76	7.752	37.85
1/4"	0.25	6.35	10.336	50.46
5/16"	0.3125	7.94	12.920	63.08
3/8"	0.375	9.53	15.503	75.79
1/2"	0.50	12.70	20.671	100.92
5/8"	0.625	15.88	25.839	126.15
3/4"	0.75	19.05	31.007	151.38
1"	1.00	25.4	41.342	201.83

4.22.0 Estimating Concrete Masonry

NOMINAL LENGTH OF CONCRETE MASONRY WALLS BY STRETCHERS

(Based on units 15⅝" long and half units 7⅝" long with ⅜" thick head joints)

LENGTH OF WALL	NO. OF UNITS	LENGTH OF WALL	NO. OF UNITS	LENGTH OF WALL	NO. OF UNITS	LENGTH OF WALL	NO. OF UNITS	LENGTH OF WALL	NO. OF UNITS	LENGTH OF WALL	NO. OF UNITS
0'-8"	½	20'-8"	15½	40'-8"	30½	60'-8"	45½	80'-8"	60½	100'-8"	75½
1'-4"	1	21'-4"	16	41'-4"	31	61'-4"	46	81'-4"	61	101'-4"	76
2'-0"	1½	22'-0"	16½	42'-0"	31½	62'-0"	46½	82'-0"	61½	102'-0"	76½
2'-8"	2	22'-8"	17	42'-8"	32	62'-8"	47	82'-8"	62	102'-8"	77
3'-4"	2½	23'-4"	17½	43'-4"	32½	63'-4"	47½	83'-4"	62½	103'-4"	77½
4'-0"	3	24'-0"	18	44'-0"	33	64'-0"	48	84'-0"	63	104'-0"	78
4'-8"	3½	24'-8"	18½	44'-8"	33½	64'-8"	48½	84'-8"	63½	104'-8"	78½
5'-4"	4	25'-4"	19	45'-4"	34	65'-4"	49	85'-4"	64	105'-4"	79
6'-0"	4½	26'-0"	19½	46'-0"	34½	66'-0"	49½	86'-0"	64½	106'-0"	79½
6'-8"	5	26'-8"	20	46'-8"	35	66'-8"	50	86'-8"	65	106'-8"	80
7'-4"	5½	27'-4"	20½	47'-4"	35½	67'-4"	50½	87'-4"	65½	107'-4"	80½
8'-0"	6	28'-0"	21	48'-0"	36	68'-0"	51	88'-0"	66	108'-0"	81
8'-8"	6½	28'-8"	21½	48'-8"	36½	68'-8"	51½	88'-8"	66½	108'-8"	81½
9'-4"	7	29'-4"	22	49'-4"	37	69'-4"	52	89'-4"	67	109'-4"	82
10'-0"	7½	30'-0"	22½	50'-0"	37½	70'-0"	52½	90'-0"	67½	110'-0"	82½
10'-8"	8	30'-8"	23	50'-8"	38	70'-8"	53	90'-8"	68	110'-8"	83
11'-4"	8½	31'-4"	23½	51'-4"	38½	71'-4"	53½	91'-4"	68½	111'-4"	83½
12'-0"	9	32'-0"	24	52'-0"	39	72'-0"	54	92'-0"	69	112'-0"	84
12'-8"	9½	32'-8"	24½	52'-8"	39½	72'-8"	54½	92'-8"	69½	112'-8"	84½
13'-4"	10	33'-4"	25	53'-4"	40	73'-4"	55	93'-4"	70	113'-4"	85
14'-0"	10½	34'-0"	25½	54'-0"	40½	74'-0"	55½	94'-0"	70½	114'-0"	85½
14'-8"	11	34'-8"	26	54'-8"	41	74'-8"	56	94'-8"	71	114'-8"	86
15'-4"	11½	35'-4"	26½	55'-4"	41½	75'-4"	56½	95'-4"	71½	115'-4"	86½
16'-0"	12	36'-0"	27	56'-0"	42	76'-0"	57	96'-0"	72	116'-0"	87
16'-8"	12½	36'-8"	27½	56'-8"	42½	76'-8"	57½	96'-8"	72½	116'-8"	87½
17'-4"	13	37'-4"	28	57'-4"	43	77'-4"	58	97'-4"	73	117'-4"	88
18'-0"	13½	38'-0"	28½	58'-0"	43½	78'-0"	58½	98'-0"	73½	118'-0"	88½
18'-8"	14	38'-8"	29	58'-8"	44	78'-8"	59	98'-8"	74	118'-8"	89
19'-4"	14½	39'-4"	29½	59'-4"	44½	79'-4"	59½	99'-4"	74½	119'-4"	89½
20'-0"	15	40'-0"	30	60'-0"	45	80'-0"	60	100'-0"	75	120'-0"	90

NOMINAL HEIGHT OF CONCRETE MASONRY WALLS BY COURSES

(Based on units 7⅝" high ⅜" thick mortar joints)

HEIGHT OF WALL	NO. OF UNITS	HEIGHT OF WALL	NO. OF UNITS	HEIGHT OF WALL	NO. OF UNITS	HEIGHT OF WALL	NO. OF UNITS
0'-8"	1	8'-8"	13	16'-8"	25	24'-8"	37
1'-4"	2	9'-4"	14	17'-4"	26	25'-4"	38
2'-0"	3	10'-0"	15	18'-0"	27	26'-0"	39
2'-8"	4	10'-8"	16	18'-8"	28	26'-8"	40
3'-4"	5	11'-4"	17	19'-4"	29	27'-4"	41
4'-0"	6	12'-0"	18	20'-0"	30	28'-0"	42
4'-8"	7	12'-8"	19	20'-8"	31	28'-8"	43
5'-4"	8	13'-4"	20	21'-4"	32	29'-4"	44
6'-0"	9	14'-0"	21	22'-0"	33	30'-0"	45
6'-8"	10	14'-8"	22	22'-8"	34	30'-8"	46
7'-4"	11	15'-4"	23	23'-4"	35	31'-4"	47
8'-0"	12	16'-0"	24	24'-0"	36	30'-0"	48

HOW TO USE THESE TABLES

The tables on this page are an aid to estimating and designing with standard concrete masonry units. The following are examples of how they can be used to advantage.

Example:

Estimate the number of units required for a wall 76' long and 12' high.

From table: 76' = 57 units

12' = 18 courses

57 × 18 = 1026 = No. masonry units required

Example:

Estimate the number of units required for a foundation 24' × 30' = 11 courses high.

2 (24 + 30) = 108' = distance for a foundation

From table: 108' = 81 units

81 × 11 = 891 = No. masonry units required.

This table can also be useful in the layout of a building on a modular basis to eliminate cutting of units. Example: If design calls for a wall 41' long it can be found from the table that making wall 41'-4", will eliminate cutting units and consequent waste. Example: If the distance between two openings has been tentatively established at 2'-9", consulting the table will show that 2'-8" dimension would eliminate cutting of units.

4.23.0 Nominal Height of Brick and Block Walls by Coursing

COURSES	REGULAR 4 2¼" bricks + 4 equal joints =					MODULAR 3 bricks + 3 joints =	CONCRETE BLOCKS	
	10" ¼" joints	10½" ⅜" joints	11" ½" joints	11½" ⅝" joints	12" ¾" joints	8"	3⅝" blocks ⅜" joints	7⅝" blocks ⅜" joints
1	2½"	2⅝"	2¾"	2⅞"	3"	2 11/16"	4"	8"
2	5"	5¼"	5½"	5¾"	6"	5 5/16"	8"	1'4"
3	7½"	7⅞"	8¼"	8⅝"	9"	8"	1'0"	2'0"
4	10"	10½"	11"	11½"	1'0"	10 11/16"	1'4"	2'8"
5	1'0½"	1'1⅛"	1'1¾"	1'2⅜"	1'3"	1'1 5/16"	1'8"	3'4"
6	1'3"	1'3¾"	1'4½"	1'5¼"	1'6"	1'4"	2'0"	4'0"
7	1'5½"	1'6⅜"	1'7¼"	1'8⅛"	1'9"	1'6 11/16"	2'4"	4'8"
8	1'8"	1'9"	1'10"	1'11"	2'0"	1'9 5/16"	2'8"	5'4"
9	1'10½"	1'11⅝"	2'0¾"	2'1⅞"	2'3"	2'0"	3'0"	6'0"
10	2'1"	2'2¼"	2'3½"	2'4¾"	2'6"	2'2 11/16"	3'4"	6'8"
11	2'3½"	2'4⅞"	2'6¼"	2'7⅝"	2'9"	2'5 5/16"	3'8"	7'4"
12	2'6"	2'7½"	2'9"	2'10½"	3'0"	2'8"	4'0"	8'0"
13	2'8½"	2'10⅛"	2'11¾"	3'1⅜"	3'3"	2'10 11/16"	4'4"	8'8"
14	2'11"	3'0¾"	3'2½"	3'4¼"	3'6"	3'1 5/16"	4'8"	9'4"
15	3'1½"	3'3⅜"	3'5¼"	3'7⅛"	3'9"	3'4"	5'0"	10'0"
16	3'4"	3'6"	3'8"	3'10"	4'0"	3'6 11/16"	5'4"	10'8"
17	3'6½"	3'8⅝"	3'10¾"	4'0⅞"	4'3"	3'9 5/16"	5'8"	11'4"
18	3'9"	3'11¼"	4'1½"	4'3¾"	4'6"	4'0"	6'0"	12'0"
19	3'11½"	4'1⅞"	4'4¼"	4'6⅝"	4'9"	4'2 11/16"	6'4"	12'8"
20	4'2"	4'4½"	4'7"	4'9½"	5'0"	4'5 5/16"	6'8"	13'4"
21	4'4½"	4'7⅛"	4'9¾"	5'0⅜"	5'3"	4'8"	7'0"	14'0"
22	4'7"	4'9¾"	5'0½"	5'3¼"	5'6"	4'10 11/16"	7'4"	14'8"
23	4'9½"	5'0⅜"	5'3¼"	5'6⅛"	5'9"	5'1 5/16"	7'8"	15'4"
24	5'0"	5'3"	5'6"	5'9"	6'0"	5'4"	8'0"	16'0"
25	5'2½"	5'5⅝"	5'8¾"	5'11⅞"	6'3"	5'6 11/16"	8'4"	16'8"
26	5'5"	5'8¼"	5'11½"	6'2¾"	6'6"	5'9 5/16"	8'8"	17'4"
27	5'7½"	5'10⅞"	6'2¼"	6'5⅝"	6'9"	6'0"	9'0"	18'0"
28	5'10"	6'1½"	6'5"	6'8½"	7'0"	6'2 11/16"	9'4"	18'8"
29	6'0½"	6'4⅛"	6'7¾"	6'11⅜"	7'3"	6'5 5/16"	9'8"	19'4"
30	6'3"	6'6¾"	6'10½"	7'2¼"	7'6"	6'8"	10'0"	20'0"
31	6'5½"	6'9⅜"	7'1¼"	7'5⅛"	7'9"	6'10 11/16"	10'4"	20'8"
32	6'8"	7'0"	7'4"	7'8"	8'0"	7'1 5/16"	10'8"	21'4"
33	6'10½"	7'2⅝"	7'6¾"	7'10⅞"	8'3"	7'4"	11'0"	22'0"
34	7'1"	7'5¼"	7'9½"	8'1¾"	8'6"	7'6 11/16"	11'4"	22'8"
35	7'3½"	7'7⅞"	8'0¼"	8'4⅝"	8'9"	7'9 5/16"	11'8"	23'4"
36	7'6"	7'10½"	8'3"	8'7½"	9'0"	8'0"	12'0"	24'0"
37	7'8½"	8'1⅛"	8'5¾"	8'10⅜"	9'3"	8'2 11/16"	12'4"	24'8"
38	7'11"	8'3¾"	8'8½"	9'1¼"	9'6"	8'5 5/16"	12'8"	25'4"
39	8'1½"	8'6⅜"	8'11¼"	9'4⅛"	9'9"	8'8"	13'0"	26'0"
40	8'4"	8'9"	9'2"	9'7"	10'0"	8'10 11/16"	13'4"	26'8"
41	8'6½"	8'11⅝"	9'4¾"	9'9⅞"	10'3"	9'1 5/16"	13'8"	27'4"
42	8'9"	9'2¼"	9'7½"	10'0¾"	10'6"	9'4"	14'0"	28'0"
43	8'11½"	9'4⅞"	9'10¼"	10'3⅝"	10'9"	9'6 11/16"	14'4"	28'8"
44	9'2"	9'7½"	10'1"	10'6½"	11'0"	9'9 5/16"	14'8"	29'4"
45	9'4½"	9'10⅛"	10'3¾"	10'9⅜"	11'3"	10'0"	15'0"	30'0"
46	9'7"	10'0¾"	10'6½"	11'0¼"	11'6"	10'2 11/16"	15'4"	30'8"
47	9'9½"	10'3⅜"	10'9¼"	11'3⅛"	11'9"	10'5 5/16"	15'8"	31'4"
48	10'0"	10'6"	11'0"	11'6"	12'0"	10'8"	16'0"	32'0"
49	10'2½"	10'8⅝"	11'2¾"	11'8⅞"	12'3"	10'10 11/16"	16'4"	32'8"
50	10'5"	10'11¼"	11'5½"	11'11¾"	12'6"	11'1 5/16"	16'8"	33'4"

9.2.0 Comparable Thicknesses and Weights of Stainless Steel, Aluminum, and Copper

STAINLESS STEEL			ALUMINUM			COPPER		
Thickness (inch)	Gauge (U.S. Standard)	Lb/sq ft	Thickness (inch)	Gauge (B&S)	Lb/sq ft	Thickness (inch)	Oz sq ft	Lb/sq ft
.010	32	.420	.010	30	.141	.0108	8	.500
.0125	30	.525	.0126	28	.177	.0121	9	.563
						.0135	10	.625
.0156	28	.656	.0156		.220	.0148	11	.688
			.0179	25	.253	.0175	13	.813
.0187	26	.788						
.0219	25	.919	.020	24	.282	.021	16	1.000
.025	24	1.050	.0253	22	.352			
						.027	20	1.250
.031	22	1.313	.0313	—	.441	.032	24	1.500
.0375	20	1.575	.032	20	.451	.0337	28	1.750
			.0403	18	.563	.0431	32	2.000
			.0453	17	.100			
.050	18	2.100	.0506	16	.126			

Note that U.S. Standard Gauge (stainless sheet) is not directly comparable with the B&S Gauge (aluminum). A 20-gauge stainless averages .0375" thick; while a 20-gauge aluminum averages .032" thick; and 20-ounce copper is .027" thick. The higher strength of stainless steel permits use of thinner gauges than required for aluminum or copper, which makes stainless more competitive with aluminum on a weight-to-coverage basis and provides stainless with a substantial weight saving compared to copper. For example, 100 sq ft of .032" aluminum will weigh about 45 pounds, .021" (16-ounce) copper will weigh about 100 pounds, and .015" stainless will weigh about 66 pounds.

9.3.0 Wire and Sheetmetal Gauges and Weights

Name of Gage	*United States Standard Gage		The United States Steel Wire Gage	American or Brown & Sharpe Wire Gage	New Birmingham Standard Sheet & Hoop Gage	British Imperial or English Legal Standard Wire Gage	Birmingham or Stubs Iron Wire Gage	Name of Gage
Principal Use	Uncoated Steel Sheets and Light Plates		Steel Wire except Music Wire	Non-Ferrous Sheets and Wire	Iron and Steel Sheets and Hoops	Wire	Strips, Bands, Hoops and Wire	Principal Use
Gage No.	Weight Oz. per Sq. Ft.	Approx. Thickness Inches	Thickness, Inches					Gage No.
7/0's			.4900		.6666	.500		7/0's
6/0's			.4615	.5800	.625	.464		6/0's
5/0's			.4305	.5165	.5883	.432	.550	5/0's
4/0's			.3938	.4600	.5416	.400	.454	4/0's
3/0's			.3625	.3648	.500	.372	.425	3/0's
2/0's			.3310	.3249	.4452	.348	.380	2/0's
1/0			.3065	.2893	.3964	.324	.340	1/0
1			.2830	.2576	.3532	.300	.300	1
2			.2625	.2294	.3147	.276	.284	2
3	160	.2391	.2437	.2043	.2804	.252	.259	3
4	150	.2242	.2253	.1819	.250	.232	.238	4
5	140	.2092	.2070	.1620	.2225	.212	.220	5
6	130	.1943	.1920	.1443	.1981	.192	.203	6
7	120	.1793	.1770	.1285	.1764	.176	.180	7
8	110	.1644	.1620	.1144	.1570	.160	.165	8
9	100	.1495	.1483	.1019	.1398	.144	.148	9
10	90	.1345	.1350	.0907	.1250	.128	.134	10
11	80	.1196	.1205	.0808	.1113	.116	.120	11
12	70	.1046	.1055	.0720	.0991	.104	.109	12
13	60	.0897	.0915	.0641	.0882	.092	.095	13
14	50	.0747	.0800	.0571	.0785	.080	.083	14
15	45	.0673	.0720	.0508	.0699	.072	.072	15
16	40	.0598	.0625	.0453	.0625	.064	.065	16
17	36	.0538	.0540	.0403	.0556	.056	.058	17
18	32	.0478	.0475	.0359	.0495	.048	.049	18
19	28	.0418	.0410	.0320	.0440	.040	.042	19
20	24	.0359	.0348	.0285	.0392	.036	.035	20
21	22	.0329	.0317	.0253	.0349	.032	.032	21
22	20	.0299	.0286	.0226	.0313	.028	.028	22
23	18	.0269	.0258	.0201	.0278	.024	.025	23
24	16	.0239	.0230	.0179	.0248	.022	.022	24
25	14	.0209	.0204	.0159	.0220	.020	.020	25
26	12	.0179	.0181	.0142	.0196	.018	.018	26
27	11	.0164	.0173	.0126	.0175	.0164	.016	27
28	10	.0149	.0162	.0113	.0156	.0148	.014	28
29	9	.0135	.0150	.0100	.0139	.0136	.013	29
30	8	.0120	.0140	.0089	.0123	.0124	.012	30
31	7	.0105	.0132	.0080	.0110	.0116	.010	31
32	6.5	.0097	.0128	.0071	.0098	.0108	.009	32
33	6	.0090	.0118	.0063	.0087	.0100	.008	33
34	5.5	.0082	.0104	.0056	.0077	.0092	.007	34
35	5	.0075	.0095	.0050	.0069	.0084	.005	35
36	4.5	.0067	.0090	.0045	.0061	.0076	.004	36
37	4.25	.0064	.0085	.0040	.0054	.0068		37
38	4	.0060	.0080	.0035	.0048	.0060		38
39			.0075	.0031	.0043	.0052		39
40			.0070		.0039	.0048		40

* U.S. Standard Gage is officially a weight gage, in oz per sq ft as tabulated. The Approx. Thickness shown is the "Manufacturers' Standard" of the American Iron and Steel Institute, based on steel as weighing 501.81 lb per cu ft (489.6 true weight plus 2.5 percent for average over-run in area and thickness).

9.10.0 Volume of Vertical Cylindrical Tanks (in U.S. Gallons per Foot of Depth)

Diameter in		U. S.	Diameter in		U. S.	Diameter in		U. S.
Feet	Inches	Gallons	Feet	Inches	Gallons	Feet	Inches	Gallons
1	0	5.875	3	6	71.97	6	0	211.5
1	1	6.895	3	7	75.44	6	3	220.5
1	2	7.997	3	8	78.99	6	6	248.2
1	3	9.180	3	9	82.62	6	9	267.7
1	4	10.44	3	10	86.33	7	0	287.9
1	5	11.79	3	11	90.13	7	3	308.8
1	6	13.22	4	0	94.00	7	6	330.5
1	7	14.73	4	1	97.96	7	9	352.9
1	8	16.32	4	2	102.0	8	0	376.0
1	9	17.99	4	3	106.1	8	3	399.9
1	10	19.75	4	4	110.3	8	6	424.5
1	11	21.58	4	5	114.6	8	9	449.8
2	0	23.50	4	6	119.0	9	0	475.9
2	1	25.50	4	7	123.4	9	3	502.7
2	2	27.58	4	8	127.9	9	6	530.2
2	3	29.74	4	9	132.6	9	9	558.5
2	4	31.99	4	10	137.3	10	0	587.5
2	5	34.31	4	11	142.0	10	3	617.3
2	6	36.72	5	0	146.9	10	6	647.7
2	7	39.21	5	1	151.8	10	9	679.0
2	8	41.78	5	2	156.8	11	0	710.9
2	9	44.43	5	3	161.9	11	3	743.6
2	10	47.16	5	4	167.1	11	6	777.0
2	11	49.98	5	5	172.4	11	9	811.1
3	0	52.88	5	6	177.7	12	0	846.0
3	1	55.86	5	7	183.2	12	3	881.6
3	2	58.92	5	8	188.7	12	6	918.0
3	3	62.06	5	9	194.2	12	9	955.1
3	4	65.28	5	10	199.9			
3	5	68.58	5	11	205.7			

Figure 9.10.0 (*By permission of Cast Iron Soil Pipe Institute.*)

9.11.0 Volume of Rectangular Tank Capacities (in U.S. Gallons per Foot of Depth)

Width Feet	LENGTH OF TANK — IN FEET						
	2	2 1/2	3	3 1/2	4	4 1/2	5
2	29.92	37.40	44.88	52.36	59.84	67.32	74.81
2 1/2	—	46.75	56.10	65.45	74.81	84.16	93.51
3	—	—	67.32	78.55	89.77	101.0	112.2
3 1/2	—	—	—	91.64	104.7	117.8	130.9
4	—	—	—	—	119.7	134.6	149.6
4 1/2	—	—	—	—	—	151.5	168.3
5	—	—	—	—	—	—	187.0
	5 1/2	6	6 1/2	7	7 1/2	8	8 1/2
2	82.29	89.77	97.25	104.7	112.2	119.7	127.2
2 1/2	102.9	112.2	121.6	130.9	140.3	149.6	159.0
3	123.4	134.6	145.9	157.1	168.3	179.5	190.8
3 1/2	144.0	157.1	170.2	183.3	196.4	209.5	222.5
4	164.6	179.5	194.5	209.5	224.4	239.4	254.3
4 1/2	185.1	202.0	218.8	235.6	252.5	269.3	286.1
5	205.7	224.4	243.1	261.8	280.5	299.2	317.9
5 1/2	226.3	246.9	267.4	288.0	308.6	329.1	349.7
6	—	269.3	291.7	314.2	336.6	359.1	381.5
6 1/2	—	—	316.1	340.4	364.7	389.0	413.3
7	—	—	—	366.5	392.7	418.9	445.1
7 1/2	—	—	—	—	420.8	448.8	476.9
8	—	—	—	—	—	478.8	508.7
8 1/2	—	—	—	—	—	—	540.5
	9	9 1/2	10	10 1/2	11	11 1/2	12
2	134.6	142.1	149.6	157.1	164.6	172.1	179.5
2 1/2	168.3	177.7	187.0	196.4	205.7	215.1	224.4
3	202.0	213.2	224.4	235.6	246.9	258.1	269.3
3 1/2	235.6	248.7	261.8	274.9	288.0	301.1	314.2
4	269.3	284.3	299.2	314.2	329.1	344.1	359.1
4 1/2	303.0	319.8	336.6	353.5	370.3	387.1	403.9
5	336.6	355.3	374.0	392.7	411.4	430.1	448.8
5 1/2	370.3	390.9	411.4	432.0	452.6	473.1	493.7
6	403.9	426.4	448.8	471.3	493.7	516.2	538.6
6 1/2	437.6	461.9	486.2	510.5	534.9	559.2	583.5
7	471.3	497.5	523.6	549.8	576.0	602.2	628.4
7 1/2	504.9	533.0	561.0	589.1	617.1	645.2	673.2
8	538.6	568.5	598.4	628.4	658.3	688.2	718.1
8 1/2	572.3	604.1	635.8	667.6	699.4	731.2	763.0
9	605.9	639.6	673.2	706.9	740.6	774.2	807.9
9 1/2	—	675.1	710.6	746.2	781.7	817.2	852.8
10	—	—	748.1	785.5	822.9	860.3	897.7
10 1/2	—	—	—	824.7	864.0	903.3	942.5
11	—	—	—	—	905.1	946.3	987.4
11 1/2	—	—	—	—	—	989.3	1032.0
12	—	—	—	—	—	—	1077.0

Figure 9.11.0 *(By permission of Cast Iron Soil Pipe Institute.)*

9.12.0 Capacity of Horizontal Cylindrical Tanks

% Depth Filled	% of Capacity	% Depth Filled	% of Capacity	% Depth Filled	% of Capacity	% Depth Filled	% of Capacity
1	.20	26	20.73	51	51.27	76	81.50
2	.50	27	21.86	52	52.55	77	82.60
3	.90	28	23.00	53	53.81	78	83.68
4	1.34	29	24.07	54	55.08	79	84.74
5	1.87	30	25.31	55	56.34	80	85.77
6	2.45	31	26.48	56	57.60	81	86.77
7	3.07	32	27.66	57	58.86	82	87.76
8	3.74	33	28.84	58	60.11	83	88.73
9	4.45	34	30.03	59	61.36	84	89.68
10	5.20	35	31.19	60	62.61	85	90.60
11	5.98	36	32.44	61	63.86	86	91.50
12	6.80	37	33.66	62	65.10	87	92.36
13	7.64	38	34.90	63	66.34	88	93.20
14	8.50	39	36.14	64	67.56	89	94.02
15	9.40	40	37.36	65	68.81	90	94.80
16	10.32	41	38.64	66	69.97	91	95.50
17	11.27	42	39.89	67	71.16	92	96.26
18	12.24	43	41.14	68	72.34	93	96.93
19	13.23	44	42.40	69	73.52	94	97.55
20	14.23	45	43.66	70	74.69	95	98.13
21	15.26	46	44.92	71	75.93	96	98.66
22	16.32	47	46.19	72	77.00	97	99.10
23	17.40	48	47.45	73	78.14	98	99.50
24	18.50	49	48.73	74	79.27	99	99.80
25	19.61	50	50.00	75	80.39	100	100.00

Figure 9.12.0 *(By permission of Cast Iron Soil Pipe Institute.)*

9.13.0 Round-Tapered Tank Capacities

$$Volume = \frac{h^3}{3} \frac{[(Area_{Top} + Area_{Base}) + \sqrt{(Area_{Top} + Area_{Base}}]}{231}$$

If inches are used.

$$Volume = \frac{h}{3} [(Area_{Base} + Area_{Top}) + \sqrt{(Area_{Base} + Area_{Top}}] \times 7.48$$

If feet are used.

Sample Problem

Let d be 12" (2 ft)
D be 36" (3 ft)
h be 48" (4 ft)
Find volume in gallons.

$$Volume = \frac{48}{3} \frac{[(\pi \times 12^2) + (\pi + 18^2) + \sqrt{\pi\ 12^2 \times 18^2}]}{231}$$

Where dimensions are in inches

$$Volume = \frac{4}{3} [(\pi \times 12^2) + (\pi + 1\tfrac{1}{2}^2) + \sqrt{(\pi \times 1^2) \times \tfrac{1}{2}^2)}] \times 7.48$$

Where dimensions are in feet

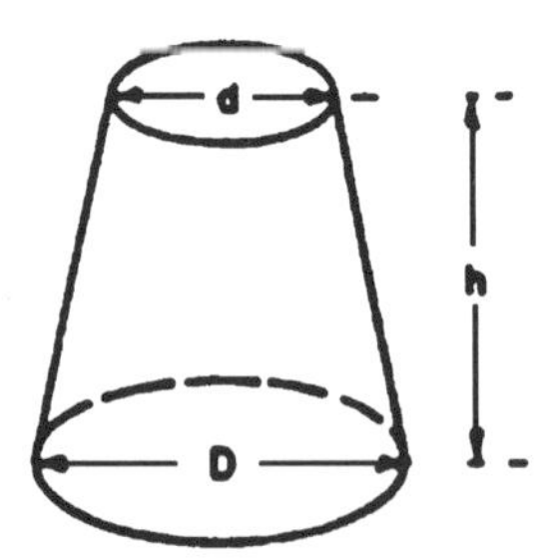

Figure 9.13.0 *(By permission of Cast Iron Soil Pipe Institute.)*

9.5.0 Useful Formulas

Circumference of a circle = π × *diameter* or 3.1416 × *diameter*

Diameter of a circle = *circumference* × 0.31831

Area of a square = *length* × *width*

Area of a rectangle = *length* × *width*

Area of a parallelogram = *base* × *perpendicular height*

Area of a triangle = ½ *base* × *perpendicular height*

Area of a circle = π *radius squared or diameter squared* × 0.7854

Area of an ellipse = *length* × *width* × 0.7854

Volume of a cube or rectangular prism = *length* × *width* × *height*

Volume of a triangular prism = *area of triangle* × *length*

Volume of a sphere = *diameter cubed* × 0.5236 (*diameter* × *diameter* × *diameter* × 0.5236)

Volume of a cone = π × *radius squared* × ⅓ *height*

Volume of a cylinder = π × *radius squared* × *height*

Length of one side of a square × 1.128 = *diameter of an equal circle*

Doubling the diameter of a pipe or cylinder increases its capacity 4 times

Pressure (in lb/sq in.) *of a column of water* = *height of the column* (in feet) × 0.434

Capacity of a pipe or tank (in U.S. gallons) = *diameter squared* (in inches) × *length* (in inches) × 0.0034

1 gal water = 8⅓ lb = 231 cu in.

1 cu ft water = 62½ lb = 7½ gal.

9.6.0 Decimal Equivalents of Inches in Feet and Yards

Inches	Feet	Yards
1	.0833	.0278
2	.1667	.0556
3	.2500	.0833
4	.333	.1111
5	.4166	.1389
6	.5000	.1667
7	.5833	.1944
8	.6667	.2222
9	.7500	.2500
10	.8333	.2778
11	.9166	.3056
12	1.000	.3333

9.9.0 Area and Other Formulas

Parallelogram	*Area = base × distance between the two parallel sides*
Pyramid	*Area = ½ perimeter of base × slant height + area of base* *Volume = area of base × ⅓ of the altitude*
Rectangle	*Area = length × width*
Rectangular prisms	*Volume = width × height × length*
Sphere	*Area of surface = diameter × diameter × 3.1416* *Side of inscribed cube = radius × 1.547* *Volume = diameter × diameter × diameter × 0.5236*
Square	*Area = length × width*
Triangle	*Area = one half of height times base*
Trapezoid	*Area = one half of the sum of the parallel sides × height*
Cone	*Area of surface = one half of circumference of base × slant height + area of base* *Volume = diameter × diameter × 0.7854 × one third of the altitude*
Cube	*Volume = width × height × length*
Ellipse	*Area = short diameter × long diameter × 0.7854*
Cylinder	*Area of surface = diameter × 3.1416 × length + area of the two bases* *Area of base = diameter × diameter × 0.7854* *Area of base = volume + length* *Length = volume + area of base* *Volume = length × area of base* *Capacity in gallons = volume in inches + 231* *Capacity of gallons = diameter × diameter × length × 0.0034* *Capacity in gallons = volume in feet × 7.48*
Circle	*Circumference = diameter × 3.1416* *Circumference = radius × 6.2832* *Diameter = radius × 2* *Diameter = square root of = (area + 0.7854)* *Diameter = square root of area × 1.1283*

9.15.0 Tap Drill Sizes for Fractional Size Threads

Approximately 65% Depth Thread / AMERICAN NATIONAL THREAD FORM

Tap Size	Threads per Inch	Hole Diameter	Drill
1/16	72	.049	3/64
1/16	64	.047	3/64
1/16	60	.046	56
5/64	72	.065	52
5/64	64	.063	1/16
5/64	60	.062	1/16
5/64	56	.061	53
3/32	60	.077	5/64
3/32	56	.076	48
3/32	50	.074	49
3/32	48	.073	49
7/64	56	.092	42
7/64	50	.090	43
7/64	48	.089	43
1/8	48	.105	36
1/8	40	.101	38
1/8	36	.098	40
1/8	32	.095	3/32
9/64	40	.116	32
9/64	36	.114	33
9/64	32	.110	35
5/32	40	.132	30
5/32	36	.129	30
5/32	32	.126	1/8
11/64	36	.145	27
11/64	32	.141	9/64
3/16	36	.161	20
3/16	32	.157	22
3/16	30	.155	23
3/16	24	.147	26
13/64	32	.173	17
13/64	30	.171	11/64
13/64	24	.163	20
7/32	32	.188	12
7/32	28	.184	13
7/32	24	.178	16
15/64	32	.204	6
15/64	28	.200	8
15/64	24	.194	10
1/4	32	.220	7/32
1/4	28	.215	3
1/4	27	.214	3
1/4	24	.209	4
1/4	20	.201	7
5/16	32	.282	9/32
5/16	27	.276	J
5/16	24	.272	I
5/16	20	.264	17/64
5/16	18	.258	F
3/8	27	.339	R
3/8	24	.334	Q
3/8	20	.326	21/64
3/8	16	.314	5/16
7/16	27	.401	Y
7/16	24	.397	X
7/16	20	.389	25/64
7/16	14	.368	U
1/2	27	.464	15/32
1/2	24	.460	29/64

Tap Size	Threads per Inch	Hole Diameter	Drill
1/2	20	.451	29/64
1/2	13	.425	27/64
1/2	12	.419	27/64
9/16	27	.526	17/32
9/16	18	.508	33/64
9/16	12	.481	31/64
5/8	27	.589	19/32
5/8	18	.571	37/64
5/8	12	.544	35/64
5/8	11	.536	17/32
11/16	16	.627	5/8
11/16	11	.599	19/32
3/4	27	.714	23/32
3/4	16	.689	11/16
3/4	12	.669	43/64
3/4	10	.653	21/32
13/16	12	.731	47/64
13/16	10	.715	23/32
7/8	27	.839	27/32
7/8	18	.821	53/64
7/8	14	.805	13/16
7/8	12	.794	51/64
7/8	9	.767	49/64
15/16	12	.856	55/64
15/16	9	.829	53/64
1	27	.964	31/32
1	14	.930	15/16
1	12	.919	59/64
1	8	.878	7/8
1 1/16	8	.941	15/16
1 1/8	12	1.044	1 3/64
1 1/8	7	.986	63/64
1 3/16	7	1.048	1 3/64
1 1/4	12	1.169	1 11/64
1 1/4	7	1.111	1 7/64
1 5/16	7	1.173	1 11/64
1 3/8	12	1.294	1 19/64
1 3/8	6	1.213	1 7/32
1 1/2	12	1.419	1 27/64
1 1/2	6	1.338	1 11/32
1 5/8	5 1/2	1.448	1 29/64
1 3/4	5	1.555	1 9/16
1 7/8	5	1.680	1 11/16
2	4 1/2	1.783	1 25/32
2 1/8	4 1/2	1.909	1 29/32
2 1/4	4 1/2	2.034	2 1/32
2 3/8	4	2.131	2 1/8
2 1/2	4	2.256	2 1/4
2 5/8	4	2.381	2 3/8
2 3/4	4	2.506	2 1/2
2 7/8	3 1/2	2.597	2 19/32
3	3 1/2	2.722	2 23/32
3 1/8	3 1/2	2.847	2 27/32
3 1/4	3 1/2	2.972	2 31/32
3 3/8	3 1/4	3.075	3 1/16
3 1/2	3 1/4	3.200	3 3/16
3 5/8	3 1/4	3.325	3 5/16
3 3/4	3	3.425	3 7/16
4	3	3.675	3 11/16

4.8.6 Common Types of Welded Wire Fabric

Style designation (W = Plain, D = Deformed)	Steel area (in ²/ft)		Approximate weight (lb per 100 sq ft)
	Longitudinal	Transverse	
4 x 4-W1.4 x W1.4	0.042	0.042	31
4 x 4-W2.0 x W2.0	0.060	0.060	43
4 x 4-W2.9 x W2.9	0.087	0.087	62
4 x 4-W/D4 x W/D4	0.120	0.120	86
6 x 6-W1.4 x W1.4	0.028	0.028	21
6 x 6-W2.0 x W2.0	0.040	0.040	29
6 x 6-W2.9 x W2.9	0.058	0.058	42
6 x 6-W/D4 x W/D4	0.080	0.080	58
6 x 6-W/D4.7 x W/D4.7	0.094	0.094	68
6 x 6-W/D7.4 x W/D7.4	0.148	0.148	107
6 x 6-W/D7.5 x W/D7.5	0.150	0.150	109
6 x 6-W/D7.8 x W/D7.8	0.156	0.156	113
6 x 6-W/D8 x W/D8	0.160	0.160	116
6 x 6-W/D8.1 x W/D8.1	0.162	0.162	118
6 x 6-W/D8.3 x W/D8.3	0.166	0.166	120
12 x 12-W/D8.3 x W/D8.3	0.083	0.083	63
12 x 12-W/D8.8 x W/D8.8	0.088	0.088	67
12 x 12-W/D9.1 x W/D9.1	0.091	0.091	69
12 x 12-W/D9.4 x W/D9.4	0.094	0.094	71
12 x 12-W/D16 x W/D16	0.160	0.160	121
12 x 12-W/D16.6 x W/D16.6	0.166	0.166	126

*Many styles may be obtained in rolls.

BAR SIZE DESIGNATION	WEIGHT	NOMINAL DIMENSIONS–ROUND SECTIONS		
	POUNDS PER FOOT	DIAMETER INCHES	CROSS-SECTIONAL AREA-SQ INCHES	PERIMETER INCHES
#3	.376	.375	.11	1.178
#4	.668	.500	.20	1.571
#5	1.043	.625	.31	1.963
#6	1.502	.750	.44	2.356
#7	2.044	.875	.60	2.749
#8	2.670	1.000	.79	3.142
#9	3.400	1.128	1.00	3.544
#10	4.303	1.270	1.27	3.990
#11	5.313	1.410	1.56	4.430
#14	7.650	1.693	2.25	5.320
#18	13.600	2.257	4.00	7.090

Figure 4.8.0.1 Concrete reinforcing bar size/weight chart.

4.8.1 ASTM Standards, Including Soft Metric

Soft metric size	Nom diam mm	Area mm^2	Weight factors		Imperial size	Nom diam inches	Area in^2	Weight factors	
			kg/m	kg/ft				lb/ft	lb/m
10	9.5	71	.560	.171	3	.375	.11	.376	1.234
13	12.7	129	.994	.303	4	.500	.20	.668	2.192
16	15.9	199	1.552	.473	5	.625	.31	1.043	3.422
19	19.1	284	2.235	.681	6	.750	.44	1.502	4.928
22	22.2	387	3.042	.927	7	.875	.60	2.044	6.706
25	25.4	510	3.973	1.211	8	1.000	.79	2.670	8.760
29	28.7	645	5.060	1.542	9	1.128	1.00	3.400	11.155
32	32.3	819	6.404	1.952	10	1.270	1.27	4.303	14.117
36	35.8	1006	7.907	2.410	11	1.410	1.56	5.313	17.431
43	43.0	1452	11.384	3.470	14	1.693	2.25	7.650	25.098
57	57.3	2581	20.239	6.169	18	2.257	4.00	13.600	44.619

Comparison of Steel Grades

Soft metric			Imperial		
Grade	mPa	psi	Grade	mPa	psi
300	300	43,511	40	257.79	40,000
420	420	60,716	60	413.69	60,000
520	520	75,420	75	517.11	75,000

Figure 4.8.1

Index

ABOUT THE AUTHOR

Sidney M. Levy is an independent construction consultant with more than 30 years in the industry. A resident of Baltimore, Maryland, he is the author of many books on construction methods and operations.

CD-ROM WARRANTY

This software is protected by both United States copyright law and international copyright treaty provision. You must treat this software just like a book. By saying "just like a book," McGraw-Hill means, for example, that this software may be used by any number of people and may be freely moved from one computer location to another, so long as there is no possibility of its being used at one location or on one computer while it also is being used at another. Just as a book cannot be read by two different people in two different places at the same time, neither can the software be used by two different people in two different places at the same time (unless, of course, McGraw-Hill's copyright is being violated).

LIMITED WARRANTY

Customers who have problems installing or running a McGraw-Hill CD should consult our online technical support site at http://books.mcgraw-hill.com/techsupport. McGraw-Hill takes great care to provide you with top-quality software, thoroughly checked to prevent virus infections. McGraw-Hill warrants the physical CD-ROM contained herein to be free of defects in materials and workmanship for a period of sixty days from the purchase date. If McGraw-Hill receives written notification within the warranty period of defects in materials or workmanship, and such notification is determined by McGraw-Hill to be correct, McGraw-Hill will replace the defective CD-ROM. Send requests to:

McGraw-Hill
Customer Services
P.O. Box 545
Blacklick, OH 43004-0545

The entire and exclusive liability and remedy for breach of this Limited Warranty shall be limited to replacement of a defective CD-ROM and shall not include or extend to any claim for or right to cover any other damages, including, but not limited to, loss of profit, data, or use of the software, or special, incidental, or consequential damages or other similar claims, even if McGraw-Hill has been specifically advised of the possibility of such damages. In no event will McGraw-Hill's liability for any damages to you or any other person ever exceed the lower of suggested list price or actual price paid for the license to use the software, regardless of any form of the claim.

McGRAW-HILL SPECIFICALLY DISCLAIMS ALL OTHER WARRANTIES, EXPRESS OR IMPLIED, INCLUDING, BUT NOT LIMITED TO, ANY IMPLIED WARRANTY OF MERCHANTABILITY OR FITNESS FOR A PARTICULAR PURPOSE.

Specifically, McGraw-Hill makes no representation or warranty that the software is fit for any particular purpose and any implied warranty of merchantability is limited to the sixty-day duration of the Limited Warranty covering the physical CD-ROM only (and not the software) and is otherwise expressly and specifically disclaimed.

This limited warranty gives you specific legal rights; you may have others which may vary from state to state. Some states do not allow the exclusion of incidental or consequential damages, or the limitation on how long an implied warranty lasts, so some of the above may not apply to you.